Springer Series in Measurement Science and Technology

For further volumes:
http://www.springer.com/series/13337

Giovanni Battista Rossi

Measurement and Probability

A Probabilistic Theory of Measurement with Applications

Springer

Giovanni Battista Rossi
DIME Laboratorio di Misure
Universita degli Studi di Genova
Genova
Italy

ISSN 2198-7807 ISSN 2198-7815 (electronic)
ISBN 978-94-017-8824-3 ISBN 978-94-017-8825-0 (eBook)
DOI 10.1007/978-94-017-8825-0
Springer Dordrecht Heidelberg New York London

Library of Congress Control Number: 2014937275

Printed on acid-free paper

Springer is part of Springer Science+Business Media (www.springer.com)

If you want to build a ship,
don't drum up the men to gather wood,
divide the work and give orders.
Instead, teach them to yearn for the vast
and endless sea.

Antoine de Saint-Exupéry, (1900–1944)
"The Wisdom of the Sands"

Preface

Measurement plays a fundamental role in both physical and behavioural sciences, as well as in engineering and technology, in that it acts as a link between abstract models and empirical reality and is a privileged method of gathering information from the real world.

This poses a challenge, therefore: is it possible to develop a single theory of measurement for the various domains of science and technology in which measurement is involved? One such theory is presented in this book, which addresses three main issues:

- what is the meaning of measurement?
- how do we measure? and
- what can be measured?

Since uncertainty plays an essential role, the theory is expressed in probabilistic terms. Some key applications are also addressed.

In writing this book, three major goals were pursued.

First of all, I set out to develop a theoretical framework that could truly be shared by scientists in different fields, ranging from physics and engineering to psychology. In the past, measurement was a controversial issue between these communities and a division took place during the last century, the consequences of which can still be felt today. Now it is time for a change: in the future it will be necessary to find new and creative methods of ensuring that human activities respect the environment and manage energy sources in a different way. This will require greater collaboration between science and technology and between different sciences. Measurement, which played a key role in the birth of modern science, can act as an essential interdisciplinary tool and language for this new scenario.

The second goal was to provide a sound theoretical basis for addressing key problems in measurement. These include perceptual measurements, the evaluation of uncertainty, the evaluation of inter-comparisons, the analysis of risks in decision-making, and the characterisation of dynamical measurements. Currently, increasing attention is paid to these issues due to their scientific, technical, economical and social impact. The book proposes a unified probabilistic approach to them which may allow more rational and effective solutions to be reached.

Lastly, I would like to reach as broad a readership as possible, including, as mentioned, people from different fields who may have different mathematical backgrounds and different understandings of measurement. For this purpose, great care was taken over language usage by giving preference to as interdisciplinary a terminology as possible and carefully defining and discussing all key terms. Concerning mathematics, all the main results are preceded by intuitive discussions and illustrated by simple examples. Moreover, precise proofs are always included in order to enable the more demanding readers to make conscious and creative use of these ideas, and also to develop new ones.

Attempts at developing a general theory of measurement have been made in the past, especially the so-called representational theory. These include some classics works, as the excellent books by Krantz, Luce, Suppes and Tversky (three volumes, 1971–1990) and by Roberts (1979), to mention just a few, that have all been recently reprinted. With respect to these, this book features two main new developments.

Firstly, it includes a general model of the measurement process which, in physics and engineering, provides a sound basis for instrument science, and, in psychology, makes it possible to consider people as 'measuring instruments'. Secondly, it uses a fully probabilistic approach which allows uncertainty to be treated as an inherent feature of measurement and helps in the systematic development of effective data processing procedures.

Finally, the book demonstrates that measurement, which is commonly understood to be a merely experimental matter, poses theoretical questions which are no less challenging than those arising in other, apparently more theoretical, disciplines.

Acknowledgments

I would like to dedicate this book to the memory of the late Prof. Ludwik Finkelstein, who encouraged and supported me greatly in this and in other projects.

I also would like to thank my colleagues at the Measurement Laboratory, Francesco Crenna, Vittorio Belotti and Luca Bovio, for the privilege of having worked with them over the years.

Lastly, I would like to thank my family, Maria, Daniele and Ester, for their continuous affectionate unconditioned support.

Genova, Italy, September, 2013 Giovanni Battista Rossi

Contents

Part I
General Problems

Chapter 1
Measurability

1.1 What Can Be Measured?

This is a key question in measurement science and it is closely related to another fundamental one: *what really is measurement* or, in other words, *what is a good definition of measurement?* Considering such a question may thus be a good starting point for an overall reflection on measurement [1, 2].

The problem of measurability was clearly stated by Norman Robert Campbell (1880–1949) in 1920 [3]. "Why can and do we measure some properties of bodies—he writes—while we do not measure others? I have before me on the table a tray containing several similar crystals. These crystals possess many properties among which may be included the following: Number, weight, density, hardness, colour, beauty. The first three of these qualities are undoubtedly capable of measurement—unless it be judged that number is to be excluded as being more fundamental than any measurement; concerning hardness it is difficult to say whether or not it can be measured, for though various systems of measuring hardness are in common use, it is generally felt that none of them are wholly satisfactorily. Colour cannot be measured as the others can, that is to say it is impossible to denote the colour of an object by a single number which can be determined with the same freedom from arbitrariness which characterises the assigning of a number to represent weight or density. The last property, beauty, can certainly not be measured, unless we accept the view which is so widely current that beauty is determined by the market value. What is the difference between the properties which determine the possibility or impossibility of measuring them?" Campbell's solution to this problem was and still is very influential. Before examining it, we should consider the earlier contribution to this same issue of a giant of the nineteenth-century science, Hermann von Helmholtz (1821–94).

G. B. Rossi, *Measurement and Probability*, Springer Series in Measurement Science and Technology, DOI: 10.1007/978-94-017-8825-0_1,

Before proceeding, however, let us establish some language rules.[1] *What we measure* is called an attribute or characteristic or property of an object (or event or body or system; in some cases it may also be a person). We prefer the terms *property* (or *characteristic*) and *object* respectively. *A measurable property* is called *quantity* or magnitude—we prefer the former. The term quantity can be understood either in a general sense, e.g., length, or in a specific sense, e.g., the length of this table. In the latter case, the term "measurand" [6] can be used. This is a quite technical term, which denotes what we want to measure in a specific situation.[2] *What we use to express the result of a measurement* is considered to be a number, a numeral or a symbol.[3] We consider it a *number* and call it *measure, measure value*, as synonyms, or *measurement value*. The distinction between this latter term and the former two will be made at a later stage.

1.2 Counting and Measuring

In 1887 Helmholtz wrote—as a part of a book published in honour of Eduard Gottlog Zeller, the famous historian of the ancient Greek thought—a *Memoire* entitled "Counting and measuring considered from the standpoint of the theory of knowledge" and now considered by many to be the first work on measurement theory [9]. Prior than Campbell, he expressly posed the problem of measurability, by investigating "the objective meaning of the fact that we express as quantities, through concrete numbers,[4] situations of real objects" and discussing "*under what circumstances we are allowed to do so*". He found a brilliant solution by establishing an *analogy between measurement and counting*. The key idea is that, in many cases, the characteristic we wish to measure is a "quantity", in the sense that it is the amount of something, and thus it may be thought of as being the sum of a number of elementary parts, or units, of that something. In these cases, *measurement is equivalent to the counting of such units*. Although measurement is not necessarily performed by

---

[1] The language issue has deserved great attention in measurement in the last thirty years. In 1984 the International Vocabulary of basic and general terms in metrology (VIM) was published as the result of a joint effort of authoritative international scientific and technical organisations [4], which has now come to the third edition, with substantial revisions [5, 6]. Yet the proposed terminology may not be general enough for accomplish the vision of measurement in behavioural sciences [7]. For this reasons, we will sometimes depart to the VIM terminology in this book [8]. For the reader's convenience a short collection of key terms is presented in the appendix at the end of the book.

[2] When modelling the measurement process, as we will do especially in Chap. 5, the terms "quantity" and measurand are often quite interchangeable. The difference concerns whether we think at the model as applied in a specific situation or intended to represent a large class of possible situations. Often both interpretation are possible and feasible.

[3] The distinction among these terms is not purely linguistic, yet we prefer not to dwell into this at this stage.

[4] In Helmholtz's language, concrete numbers are those arising from the counting of real objects.

actually counting units, Helmoltz's point is that it is *as if* it were performed in such a way.

This analogy can be used to derive the conditions for measurability, that is the conditions that must be met in order for measurement to make sense. Consider what makes counting possible: it is possible thanks to the properties of natural numbers, which are *ordered* and *can be added to each other*. Similarly, measurement is possible and well-founded whenever the objects carrying the property[5] of interest exhibit empirical counterparts of the order relation and of the addition operation. Consider, for example, mass measurement, for objects that can be put in the pans of an equal-arm balance [10]. If a and b are two such objects, we can compare and order them, by the balance, according to the empirical relation of "heavier or equally heavy". In a similar way we can also define an addition operation: if a and b, placed on the same pan, balance another object, c, put on the opposite pan, we can say that the *empirical sum* of their masses equals the mass of c. Other examples of empirical additions include the stacking of blocks in order to sum their heights or the connection of resistors in a series in order to sum their resistance.

These examples show how can we empirically establish an order relation and define an addition operation. But one thing is dealing with numbers, another is working with real-world objects. How can we be confident that such empirically defined operations are really adequate for our purposes? How can we check them? Helmholtz's idea is again simple and convincing. What characterises the relation $\geq$ and the operation $+$ among numbers? They are characterised by their formal (or logical) properties that we summarise below. Let $\mathbb{N}^*$ be the set of natural numbers, i.e., positive integers, not including zero. Then for each $i, j, k \in \mathbb{N}^*$,

(α) concerning the order relation $\geq$,

($\alpha1$) either $i \geq j$ or $j \geq i$ (completeness),

($\alpha2$) $i \geq j$ and $j \geq k$ implies $i \geq k$ (transitivity),

(β) concerning the addition operation $+$,

($\beta1$) $i + j = j + i$ (commutative property),

($\beta2$) $(i + j) + k = i + (j + k)$ (associative property),

(γ) and, linking the two,

---

5 Note that in this discussion we use the term "property" in three meanings: either to denote the property we want to measure or the empirical relations and operations that characterise it or even the formal properties of such empirical relations and operations. So, for example, length is a measurable property, characterised by the empirical properties of order and addition, and order, e.g., is characterised by the transitivity (formal) property. This is a drawback of using the term "property" for denoting what we want to measure, as in fact is recommendend by current international guidelines [6], that perhaps have not considered enough such theoretical aspects. This is why in other papers we have used the term "characteristic" instead [1]. Here we have preferred to follow the international recommendation, hoping that the discourse is still clear enough, although with some effort, to the reader.

(γ1) $i \geq j$ implies $i + k \geq j + k$ (monotonicity).

Therefore, the measurability condition that naturally arises is that the *empirical relations and operations have to satisfy similar properties*. To see how this may be applied, let us consider again mass-measurement and let us introduce some additional notation.[6] If the mass of a is greater than the mass of b, we write $a \succ b$, if they balance, $a \sim b$. If they are put in the same pan, we write $a \circ b$ and if, together, they balance a third object c, we express this as $a \circ b \sim c$. Let then A be the set of all the objects that can be put in the same pan of the balance. We may thus check whether the empirical system actually satisfies properties similar to those applying to numbers. Let us first state such properties formally and then discuss them in detail.

For each $a, b, c \in A$ the following is required:

(a) concerning the order relation $\succsim$,

(a1) either $a \succsim b$ or $b \succsim a$,

(a2) $a \succsim b$ and $b \succsim c$ implies $a \succsim c$,

(b) concerning the empirical summing operation $\circ$,

(b1) $a \circ b \sim b \circ a$,

(b2) $(a \circ b) \circ c \sim a \circ (b \circ c)$,

(c) and, linking the two,

(c1) $a \succsim b$ and $c \sim d$, implies $a \circ c \succsim b \circ d$.

It is very important to understand that we are now dealing with *empirical* properties that are similar to those applying to numbers. Since this point is essential for understanding the theory of measurement, let us practice *interpreting* them in the sphere of the mass measurement example.

Property a1 means that all pairs of objects may be compared and a2 ensures that comparisons are transitive. Property b1 states that the order in which we place the objects on the pan does not affect the result and b2 implies that we can substitute any object with any equivalent combination of objects. Lastly, c1 states that adding equivalent objects to both pans of the balance does not affect the result of the comparison. Readers are invited to check whether this picture reflects or not their current understanding of mass measurement (we hope it does!) and to practice applying these ideas to other additive measurements of their concern.

It is also important to note some differences between the properties of numbers and the corresponding properties of objects. In γ1 the same number, k, is summed on both sides of the relation $i \geq j$; in the corresponding empirical property c1, we cannot do the same, since we cannot put the same object in both pans of the balance

[6] This notation is subsequent to Helmoltz and is the one we will use routinely throughout the book. The general notation principles adopted in this book and a list of some of the main symbols, grouped by subjects, are reported in the appendix at the end of the book.

at the same time. So we have to consider two equivalent objects, c and d, instead. By the way, also note the difference between an identity and an equivalence. If a and b are objects, $a = b$ means that the symbols a and b are just *two different names for the same object*, whilst $a \sim b$ means that a and b are equivalent in respect of some property of theirs, for example they have the same mass, but they are otherwise distinct objects.

Furthermore, these ideas are not just theoretical issues, we can also note some practical applications of theirs. For instance, in the mass example, they can be used for checking whether the balance works properly. For example, from property a1 and a2 together we deduce that $a \sim b$ implies $b \sim a$. This means that if the balance is in equilibrium and we exchange the objects in the pans, the equilibrium must still hold. This is a way of checking the device and it is also the basis of the so-called *substitution method* in instrumentation. Suppose that we measure two objects using two sensors of the same kind: the result should be unaffected if we exchange the sensors. If this is not the case, we may suspect systematic deviations in sensors. Generalising, we suggest that, in this perspective, virtually all the operations included in a good practice of measurement may be interpreted as suitable for either ensuring or checking that the conditions for measurability are satisfied. We encourage readers to check this principle in other cases.

Some comments are now in order. Helmholtz's theory is based on establishing an analogy between measuring and counting. Is this possible for all kinds of measurements? He admits it is not and mentions the possibility of an indirect approach. Since this idea was later developed by Campbell, we shall defer its presentation to the next section. What is very interesting in Helmholtz's approach is his idea of assessing measurability by checking that the empirical relations that characterise the quantity of our interest satisfy the logical properties we expect from them and, in this regard, he was really a forerunner. We may perhaps consider whether *the set of empirical relations he considered*, which included addition, *is the only possible one*: this is a key point that we will amply discuss at a later stage. For the moment, let us just observe that many top-level measurements are presently performed using some kinds of counting, such as primary time and length measurements. This allows us to understand how far-reaching Helmholtz's perspective was. Let us now see how Campbell developed these ideas.

1.3 Physical Measurement

Remember Campbell's question: "What is the difference between the properties which determine the possibility or impossibility of measuring them?" He answers it by considering two kinds of quantities, *fundamental* (such as mass) and *derived* (such as temperature or density). What is the structural difference between them?

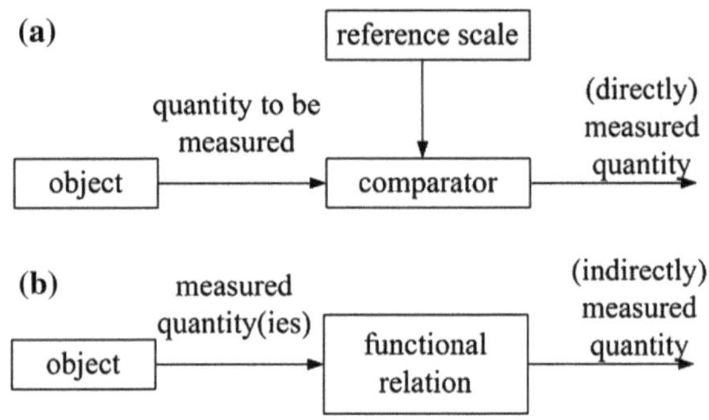

Fig. 1.1 Fundamental (**a**) versus derived (**b**) measurement

Both of them have an empirical property of order, which is an essential requirement for Campbell as well for Helmholtz for measurement.[7] But fundamental quantities also require a physical-addition operation. Why is it so important?

Because, for Campbell, it is necessary for *constructing a measurement scale*, which is a necessary prerequisite for fundamental measurement. A measurement scale basically is a series of standards with properly assigned numerical values. Once such a scale is available, measurement can be done by comparing any unknown object with the scale in order to assign a measure to it, as shown in Fig. 1.1a. Since it is based on a direct comparison with the scale, this is also called *direct measurement*.

Let us examine the role of physical addition in the construction of the measurement scale. Consider again the case of mass. We first arbitrarily select one object, u, which will serve as the unit of the scale, and we assign the number 1 to it, i.e., $m(u) = 1$, where m is the *measure* function. Then we look for another element, u', equivalent to u, i.e., such that, when put in the opposite pan of the balance, it balances it. We now sum the two elements together by putting them in the same pan of the balance and we look for a third element that can balance them, to which we assign the number 2. In this way we have constructed a multiple of the unit, and we can proceed in a similar way for the other multiples. Submultiples can also be constructed in a similar way. If we look for two equivalent elements, v and v', such that when put together on a pan, they balance u, we can assign to each of them the number 1/2. Once the scale is available, mass measurement can be performed by comparing an unknown object, r, with the elements of the scale, using the balance, until we find the equivalent element of the series, say s: we then assign $m(r) = m(s)$.[8]

Note that in the scale construction process the only arbitrary choice concerns the selection of the unitary element; then the values to be assigned to the other elements

[7] In reality, measurability under not ordered structure is also of interest, as we will see in the following of this chapter.

[8] Remember that we have assumed, in this example, that all the objects, including their combinations, can be put on each pan of the balance. Note also from this, the importance of properly defining the class of objects under consideration. In the notion of scale it is implicitly assumed that for each object there is an element in the scale which is equivalent to it. An element of a scale is called a standard.

are fully constrained by the need for conformity with the results of the summing operation. Consequently the measure may be interpreted as the *ratio* between the value of the property in object r and that of the unitary element u. In other words, $m(r) = p/q$ implies that the sum of q "copies" of r balances the sum of p unitary elements. Note that q copies of r may be realised using amplification devices, for example, by using an unequal-arm balance, with arms in a ratio $q:1$ to each other. We may thus understand Campbell's statement that only qualities "which can be determined with the same freedom from arbitrariness which characterises the assigning of a number to represent weight" fully qualify as measurable, and we may also see the rationale behind it.

What has been considered so far applies to fundamental quantities. Yet there is another way of measuring something, the way that applies to derived quantities. Consider the case of temperature, for example [11]. In this case there is not a meaningful physical addition operation for constructing a scale in a similar way as for mass or length. The other way round is to take advantage of an already existing scale, for example the length scale. This is possible provided that there is some (known) natural law that links temperature to length. This is exactly what happens in mercury-in-glass thermometers, where the law of thermal dilatation of fluids is used. Then if T is temperature, h the height of the mercury column, T_0 and h_0 their values in a reference condition, such as the liquid to solid transition of water, and α the thermal expansion coefficient for mercury, the functional relation that links T and h can be simply expressed as

$$T = T_0 + \frac{(h - h_0)}{\alpha h}. \tag{1.1}$$

Then after measuring h, a value for T can be obtained, provided that α is known.[9]

Another example, mentioned by Campbell himself, is density, ρ. For this quantity it is possible to establish an empirical order, since we can say that a is denser than b if we find a liquid in which b floats, whilst a sinks, but we are unable to perform any kind of empirical summation. The solution is to define density, ρ, as the ratio of mass, m, to volume, V, which is consistent with the overall framework of physics. Then after measuring the mass and the volume of an object, we can obtain an indirect measure of its density through the functional relation

$$\rho = m/V. \tag{1.2}$$

The general scheme for derived (or *indirect*) measurement is shown in Fig. 1.1b and compared to that of fundamental measurement. Note that in this way density "can be determined with the same freedom from arbitrariness which characterises the assigning of a number to represent weight", which is an essential feature for Campbell for a completely satisfactory measurement.

---

[9] For mercury, around room temperature $\alpha = 0.00018\,°\mathrm{C}^{-1}$ [12]. In practical application, the thermal expansion of the thermometer glass should also be accounted for, but such details are ineessential here.

To sum up, Campbell holds that measurability can be established first by proving that the characteristic under investigation involves an empirical order relation and then either by identifying an empirical addition operation, that enables the construction of a reference measurement scale, or by finding some physical law that links the quantity under investigation to other quantities, whose measurability has been already established independently.

In the case of derived measurement, the foundation of measurement is subject to acceptance of the physical, or, more generally, natural law which is invoked.

At the time Campbell presented his theory, measurements were widely used in psychology and especially in psychophysics. We must therefore also consider their foundation.

1.4 Psychophysical Measurement

According to Galileo Galilei (1564–1642), progress in science is closely related to the possibility of measuring the properties of interest [13]. "Count what is countable, measure what is measurable and what is not measurable make measurable" was one of his mottoes [14].

Two centuries later Gustav Fechner (1801–87), the father of psychophysics, established a similar research programme for this new science. "As an exact science psychophysics, like physics, must rest on experience and the mathematical connection of those empirical facts that demand a measure of what is experienced or, when such a measure is not available, a search for it", he writes in his Elements of Psychophysics, published in 1860 [15]. Yet measurement in psychophysics appeared as being more changelling than in physics. In fact, "unlike physical processes, which are external, public, objective and open to direct measurement, mental processes are internal, private, subjective and cannot be measured directly. Somehow an indirect method had to be developed" [16].

Fechner's indirect method aims at measuring the increments of mental activity by measuring the increments in the energy of the physical phenomenon that causes them. For this to be possible it is necessary to assume a law linking them. For finding such a law, Fechner started from the earlier experimental results obtained by Weber [17], who found that often the increment of the physical stimulus, $\delta\varphi$, needed to cause a just perceptible variation in the corresponding sensation, is proportional to the intensity of the stimulus, φ, that is

$$\delta\varphi = k\varphi, \tag{1.3}$$

where k is a constant that depends upon the sensation considered [18]. This result is known as Weber's law, and concerns what happens at the stimulus's side. But how to model the sensation counterpart of this? Fechner somewhat refers to the counting paradigm, and looks for a zero condition and a unitary increment for sensation intensity. He makes a natural and convincing choice, by assuming as zero the perception

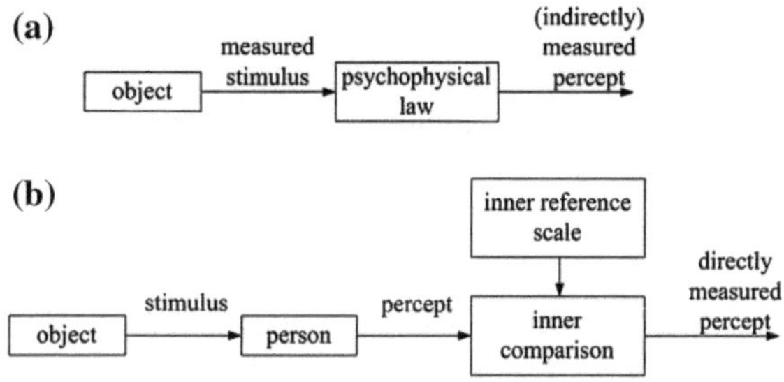

Fig. 1.2 Indirect (**a**) versus direct (**b**) psychophysical measurement

threshold, and as the unit the increment of sensation corresponding to a just perceivable variation in the stimulus. The underlying assumption is that such variation is constant, regardless of the value of the stimulus. Thus, sensation intensity results to be the sum of a number of elementary variations of sensation intensity, all equal and corresponding to just noticeable variations in the stimulus. If, in turn, just noticeable variations in the stimulus follow Weber's law, a logarithmic law ensues,

$$\psi = \alpha \ln \varphi + \beta, \tag{1.4}$$

where ψ is the sensation intensity and α and β are constant parameters characterising the response of an *average observer* to the stimulus. So it is possible to indirectly measure sensation intensity, ψ, by measuring the intensity of the stimulus, φ, and applying the psychophysical law [19]. This is illustrated in Fig. 1.2a. The similarity with the procedure for indirect physical measurement, illustrated in Fig. 1.1b, can be noted.

To sum up, in classical psychophysical measurement we measure a property of a physical object (or event), which consists in its capacity of evoking a sensation in individuals, with different possible degrees of intensity [19]. Measurement can be indirectly performed by measuring the (physical) intensity of the stimulus and by applying the psychophysical law. Fechner's law is based on Weber's law and on the hypothesis that just noticeable variations in the intensity of the stimulus evoke mutually-equivalent variations in sensation intensity. The resulting law depends upon upon two parameters that characterise the behaviour of an average observer. Individual differences are considered as a noise effect, in this approach. In fact, the assumption of Weber's law, although mathematically convenient for its simplicity, is not strictly necessary. What is needed is a relation between just perceivable variations and the intensity of the stimulus,

$$\delta\varphi = l(\varphi), \tag{1.5}$$

which can be experimentally derived. Once such a relation is available, the corresponding psychophysical law can be obtained by integrating it. On the other hand Fechner's hypothesis is crucial and it was subject to criticism, as we will see in the following.

The debate on the foundation of measurement, initiated by these key contributions in the nineteenth century, took a further turn in the first part of the twentieth century, when the scientific communities that claimed an experimental basis for their theories, compared their views.

1.5 The Debate at the British Association for the Advancement of Science

In 1932 the British Association for the Advancement of Science appointed a Committee of physicists and psychologists to consider and report on the possibility of making quantitative estimates of sensory events. After years of discussion the Committee admitted the impossibility of reaching a common view [20].

Let us try and summarise the positions of the two parties, although it is not easy. We have seen that psychophysical experiments can be performed to obtain a curve that links just perceptible variations in a stimulus to its intensity, corresponding to Eq. (1.5). Often the result is a smooth curve that seems to reveal some empirical relation between them. The point is if such a function, that links two physical quantities, the intensity variation, $\delta\varphi$, and the overall intensity, φ, *fully describes the facts or not*. The psychologists claimed the need for a quantitative description of the intensity of the evoked sensation, ψ, in our notation. As we have seen, one such description can be obtained, according to Fechner, by further assuming that each just noticeable step in the intensity of the stimulus produces a step in sensation and *that such steps are all equal*, which leads to the law expressed by formula (1.4). The physicists argued that the possibility of mathematically obtaining such a law does not "prove" the measurability of sensation intensity, since *it requires an assumption that cannot be verified, unless it is possible to independently measure the intensity of sensation*. But this is not possible, in their view, since no additive operation can be defined for sensations.

The main case against the measurability of sensation intensity therefore was ultimately the impossibility of satisfactorily defining an addition operation for it. On the other hand, the psychologists claimed that although Fechner's position could not be fully supported, the result of psychophysical experiments cannot, in general, be expressed only in terms of intensity of stimulus. In fact, the particular intensity changes which are being plotted "*are only obtained by referring to the sensation they evoke and are based on the equalityor inequality of these sensation intensities*".

The Committee's Report [21] had an enormous influence in the following years and we can say that it led to an essentially parallel development of measurement science, with physical sciences on one side and behavioural sciences on the other. Basically, two main points were raised against the measurability of sensory events: the impossibility of directly measuring sensation intensity and the need for additivity as a necessary property for measurement in general. During the twentieth century, answers were found for both points, mainly thanks to the contribution of Stanley S. Stevens [17].

1.6 A Turning Point: Stevens's Twofold Contribution

1.6.1 Direct Measurement of Percepts

Concerning the first point, the possibility of directly assessing sensation intensity, he introduced other measurement or scaling methods, in which magnitude or ratio of sensations are estimated directly by the subjects [22]. These are called magnitude estimation or production and ratio estimation or production.

Present a line of a given length—he writes—and tell the observer to call it some number, say, 10. Then present a line of some other length and say: 'If the first line was 10, what would you call the second line? Use any number that seems appropriate—fractional, decimal, whole number—but try to make the number proportional to the apparent length as you see it'. Then proceed to other lengths, in irregular order, and ask for similar estimates of apparent length [23].

This is *magnitude estimation*. Similar methods include *magnitude production*, where—referring to the above example—the observer is asked to alter the length in order to match numbers; *ratio estimation*, where the observer is asked to estimate the apparent ratio between several length and a standard length, that is maintained constant during the experiment, and *ratio production*, where the observer is asked to change a variable length in order to match given ratios with respect to the standard one. A scheme of the magnitude estimation method is shown in in Fig. 1.2b, to be compared with Fig. 1.1a. It appears how in this approach persons, individually or in a group, act as a measuring instruments.[10]

These new methods were instrumental in the development of psychophysics. In fact, Stevens checked Fechner's law and ultimately rejected it, proposing the so called "power law" as a substitute. This demonstrates that the role of measurement in the development of psychophysics, has been similar than in physics, and in agreement with the general scientific method as envisaged by Galileo. Indeed, in the perspective of modern science measurement and experimentation provide the tools for checking scientific theories.

The power law reads as

$$\psi = \eta\varphi^{\theta}, \tag{1.6}$$

---

[10] Although magnitude estimation can be considered a measurement method, the other three are rather scaling procedures. The difference is subtle, often overlooked, but substantial. In a scaling method, the aim is to assigning numbers to a set of objects, in order to constitute a scale for the quantity of interest. Instead in measurement the goal is to assign numbers to one, or more objects, in respect of a previously established scale. In the case of magnitude estimation, the previously established scale is assumed to be some inner scale in the subject(s). In ratio estimation or production it is not necessary to dispose of such a scale, but only to be able to perceive ratios. This difference will be easier to understand after reading Chaps. 3–5 of this book. We will consider ratio estimation or production at a later stage in this section, when dealing with the classification of measurement scales.

where φ is the physical stimulus, ψ is the corresponding perceived quantity, η and θ are constants that depend upon the kind of sensation considered.

After this finding, the general indirect approach to measurement, illustrated in Fig. 1.2a, can be maintained, by simply replacing the logarithmic with the power law. Yet an important step forward was made, since with Fechner the law had been assumed as an *hypothesis* about the way sensation arises, whilst here, with Stevens, it was been experimentally confirmed by directly measuring both the stimuli and the related sensations, thanks to the new direct-measurement approach.

1.6.2 Classification of Measurement Scales

Concerning the second crucial point in the debate at the British Association, that is additivity, Stevens provided a key contribution for understanding the notion of measurement scale. "In the meantime,—he notes, commenting on the British Association report, in a beautiful paper on measurement and psychophysics—unaware that the British committee was trying to settle the issue, some of us at Harvard were wrestling with similar problems... What I gained from these discussions was a conviction that a more general theory of measurement was needed and that the definition of measurement should not be limited to one restricted class of empirical operations. The best way out seemed to approach the problem from another point of view, namely, that of invariance, and to classify scales of measurement in terms of the group of transformations that leave the scale form invariant... A fourfold classification of scales based on this notion was worked out sometime around 1939 and was presented to the International Congress for the Unity of Science in 1941. World War II then came along, and publication was delayed until 1946 [23]". Stevens's classification of measurement scales is presented in Table 1.1 [24].

This is really a milestone in measurement science, of invaluable merit. It is innovative in different regards, all of clue importance. For Campbell, there was just one kind of fully satisfactory measurement scales, basically corresponding to ratio scales in the table. Stevens's instead extends the notion of scale to other three important and general cases. For doing so he holds that a measurement scale is characterised by two *complementary* properties: *the class of empirical operations it accounts for*, as appears in the second column of the table, and *the class of its admissible transformation.*[11] This second aspect is usually recognised as his main contribution: a scale transformation is admissible if, intuitively, it can be safely applied to the scale without altering the meaning of the measurement. For example, in the case of ordinal scale, monotonic increasing transformations may be applied, since they are order preserving. To be more precise, if m is a suitable measure function, m' will also be good, provided that, for each object a, they are linked by

[11] In fact this term was introduced later on—Stevens rather spoke of "mathematical group structure"—but we prefer it since it is perhaps easier to understand.

Table 1.1 Classification of measurement scales, according to Stevens

Scale	Basic empirical operations	Admissible transformations	Examples
Nominal	Determination of equality	One-to-one	Auditory thresholds. Automatic defects detection. Automatic pattern recognition
Ordinal	Determination of greater or less	Monotone increasing	Hardness of minerals. Earthquake intensity. Wind intensity
Interval	Determination of the equality of intervals or of differences	Linear positive	Celsius or Fahrenheit temperature. Position (as measured, e.g. by the Global Positioning System (GPS) [25]. Time, as date, presently worldwide provided by the Coordinated Universal Time (UTC) system [26]
Ratio	Determination of the equality of ratios	Similarity	Length, mass, density, time intervals as measured by chronometers. Thermodynamic temperature (kelvin). Loudness (sone). Brightness (brill)

In respect of the original formulation we have made some, non substantial, changes. We have used the term "admissible transformations" instead of "mathematical group structure" as he did. Furthermore we have partially changed the examples from the original, to mention modern technologies, that demonstrate the actuality of Stevens's approach

$$m'(a) = \phi(m(a), \tag{1.7}$$

where ϕ is monotonic increasing, that is

$$u > v \Leftrightarrow \phi(u) > \phi(v). \tag{1.8}$$

In reality, this part of his contribution has probably been overestimated in respect of what is, in my opinion, even more relevant. In fact he really contributed to developing "a more general theory of measurement" as he claimed, and he did this especially by noting that "the definition of measurement should not be limited to one restricted class of empirical operations". His main idea can be probably summarised in this way: we can say that we have a meaningful measurement scale whenever we can identify a class of empirical operation of scientific and practical interest and we can express them by describing objects by numbers.[12] Such scales will have a different degree of "conventionality"—what Campbell called "arbitrariness" in a deprecative way—corresponding to the class of their admissible transformations.

For better understanding and appreciating his approach let us briefly look at and comment upon the scales in Table 1.1.

[12] This paved the way to the representational theory of measurement, as we will see in a moment.

Nominal scales are related to classification operations and numbers serve only to distinguish one class of objects from another. Any one-to-one substitution is permissible since identification is still possible. Thus admissible transformations are those that satisfy

$$u \neq v \Leftrightarrow \phi(u) \neq \phi(v). \tag{1.9}$$

Ordinal scales allow a rank ordering of objects and remain invariant under monotonic increasing transformations, as just discussed.

Interval scales entail a constant unit of measurement, that is, they introduce a metric, and thus allow *empirical differences* to be meaningfully expressed. They remain invariant under positive linear transformations, such as

$$m'(a) = \alpha m(a) + \beta, \tag{1.10}$$

with $\alpha > 0$.

Ratio scales also feature a constant unit of measurement, but they also allow *empirical ratio* to be properly expressed since an absolute zero exists. They are invariant under any simply multiplicative transformation:

$$m'(a) = \alpha m(a), \tag{1.11}$$

still with $\alpha > 0$. Note that Stevens does not mention empirical addition when dealing with these scales and substitutes it with empirical ratio, that is to say the possibility of the empirical "determination of equality of ratios". This point has been somewhat overlooked when considering Stevens's contribution, probably because he did not provide a formal theory for this new way of looking at ratio scales. In fact he only provided methods for experimentally evaluating ratios, that is the procedures for ratio estimation and production that we have already mentioned.[13] In fact the theory needed was provided at a later stage, mainly by Krantz et al. [27] and Miyamoto [28]. That theory provided the basis for properly dealing with intensity quantities, that is those quantity that describe the intensity of a phenomenon or of a sensation. In my opinion they are still not completely understood and, for these reason, we will amply discuss them in Chap. 3.

This approach of Stevens's to measurement scales opened the path to a vast class of studies, usually referred to as the representational theory of measurement [27, 29–34]. This theory provided a first noteworthy systematisation of measurement.

[13] Thus, continuing the discussion in Footnote 11, whilst magnitude estimation may be regarded as a measurement method, ratio estimation and production are rather scaling procedures, which allow obtaining a ratio scale even when there is no empirical addition operation, as usually happens with perceptual quantities. We will discuss in dept this important and conceptually difficult point in Chap. 3.

1.7 The Representational Theory

The main idea, traceable to Helmoltz, is that the numbers we obtain through measurement *represent* empirical relations, hence the name of the theory. This holds true not only for fundamental physical measurement, as intended by Campbell, but also in other cases, as envisaged by Stevens. The representational viewpoint can be best illustrated by hardness measurement, as addressed by Friedrich Mohs (1773–1839) in 1812, who proposed the famous hardness scale, named after him [35]. The key idea of his method was that the hardness of a mineral may be characterised by its ability to scratch other minerals. So Mohs identified a series of 10 reference materials collectively suited to express the different degrees of hardness we may encounter in nature. They were, in increasing order of hardness, talc, gypsum, calcite, fluorite, apatite, orthoclase, quartz, topaz, corundum and diamond. He then assigned numbers from 1 to 10 to them, thus fully defining a reference scale. Call a *standard* each element in the scale and *measure* the corresponding number. Then the hardness of any piece of mineral, r, can be measured by comparing it to the scale and identifying the standard it is equivalent to, in the sense that it neither scratches nor is scratched by it. Suppose, for simplicity, that you always find one and one only such standard and denote it with s. We therefore give it the same measure of s, viz., $m(r) = m(s)$. The key point to be noted is that the number (measure) assigned in this way really represents hardness in that we expect r to have the same hardness as any other object obtaining the same measure, to be harder of any object obtaining a lesser measure and vice-versa. Remember that we did the same for mass measurement, according to Campbell, *but here the additivity property is no longer required for constructing the measurement scale.*

In general, a *representation theorem* for a given scale states how, for that scale, empirical relations are mapped into corresponding numerical ones. For example, in an ordinal scale, as in the case of Mohs hardness, it reads

$$a \succsim b \Longleftrightarrow m(a) \geq m(b), \tag{1.12}$$

that is, an empirical order between two objects, $a \succsim b$, holds if and only if the corresponding numerical order holds between their measures, $m(a) \geq m(b)$. In the case of mass measurement, instead, the representation theorem reads

$$a \sim b \circ c \Longleftrightarrow m(a) = m(b) + m(c), \tag{1.13}$$

that is object a is equivalent to the empirical sum of objects b and c, if and only if its measure equals the (numerical) sum of the measures of b and of c.

A *uniqueness theorem* instead is one that identifies the classes of admissible transformations, that is those transformations that may be safely applied to the scale, without altering the meaning of the measurement, exactly as Stevens did, and we have illustrated in formulae (1.7–1.11).

The representational theory has been developed mainly in the field of behavioural sciences but was brought to the attention of physicists and engineers in the 1970s, mainly by Finkelstein, who supported its feasibility for all kinds of measurements [30]. He famously defines measurement as "a process of empirical, objective assignment of symbols to attribute of objects and events of the real world, in such a way as to represent them or to describe them" [35]. This theory later also received contributions from that community and it actually constitutes an excellent starting point for a unified theory of measurement [36, 37].

We will discuss the representational approach, in its classical, essentially deterministic, formulation, in Chaps. 3, and 4, where a probabilistic reformulation will be provided.

1.8 The Role of the Measuring System

"Seeing, in the science of our time, means, almost exclusively, to interpret signs produced by instruments"—Paolo Rossi (1923–2012) writes, highlighting the importance of scientific and measuring instrument in modern science [13]. In spite of this, quite surprisingly the notion of a measuring system is almost completely absent in the representational theory of measurement. This can probably be explained by considering that such a theory has been mainly developed in the area of behavioural sciences, where the concept of a measuring system has not traditionally been considered as particularly important.[14] Instead, it seems necessary to explicitly account for its role in a theory of measurement [37, 39, 40].

Indeed, once we have constructed a reference scale, we have to consider how to measure an object, a, that is not included in the scale. This can be done by comparing it with scale, which in turn, can be done either directly or indirectly using a calibrated measuring system, as illustrated in Fig. 1.3a.

In both cases, we will call *measurement process* the process of performing a measurement once the reference scale has been given, and *measuring system* the device we use for that.

When we focus on the measurement process, we need considering a specific object and the way it manifests the quantity we want to measure, hence the term "measurand" appears.[15] An overall picture of the measurement process is presented in Fig. 1.3b. It can be noted that such a scheme includes, as special cases, those in Figs. 1.1 and 1.2.

Since we wish to develop an interdisciplinary theory, we must look for a sufficiently general definition of a measuring system that can also be applied to perceptual

[14] In contrast to this, the role of persons as "measuring instruments" has been recently highlighted [38]. We will amply discuss this subject in Chap. 8.

[15] We introduced the term "measurand" in Sect. 1.1 [6]. See also the glossary in the appendix at the end of the book.

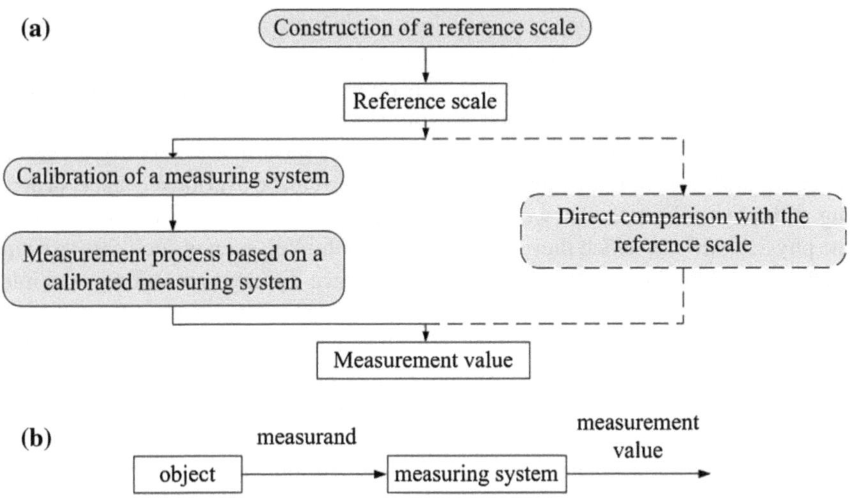

Fig. 1.3 **a** Direct versus indirect measurement process. **b** General scheme of a measurement process

measurement. We thus propose to define it as an empirical system capable of interacting with objects that manifest the characteristic under investigation and, as a result of such interaction, capable of producing signs that can be used to assign a measurement value to the measurand (the characteristic to be measured), on the basis of a previously defined reference scale.[16] We suggest that this definition can be used not only in physics and engineering, where the measuring system or instrument is usually a physical device, but also in psychophysics, where people act as measuring instruments, or even in psychometrics, where the instrument is a procedure based on test items.

The measurement process will be amply investigated in Chap. 5. Thus we not pursue it any more here.

1.9 The Proposed Approach

After discussing measurability in its historical and logical implications we now propose a measurability criterion [1, 41, 42].

On the basis of what we have discussed so far, we can see that, generally speaking, in order to measure something we must successfully complete the following steps:

1. define the class of objects manifesting the property,
2. identify the empirical properties that define the property,
3. construct a reference measurement scale, and
4. devise at least one measuring system based on that reference scale.

[16] Note how the idea of interpreting signs produced by instruments, suggest by Rossi, comes into play.

Then we say that *a property x of a class of objects A is measurable if the above four-step procedure can be successfully applied.*

Simple as it may seem, this approach is quite demanding in reality. To see that, let us now briefly comment on each step of the procedure.

When considering a candidate new quantity we have first to identify a class of objects that manifests it. The importance of this step can be overlooked since, in dealing with physical quantities, we usually assume that they are measurable whenever the physical laws in which they occur are applicable. Yet, even in such cases, from a metrological standpoint, care should be taken, since different measuring principles are used in different ranges. For example, we cannot measure length in the same way in the range of everyday objects, that are of the order of the metre, as in the atomic scale or, at the opposite, as in the astrophysical scale. In principle, we should ensure that there is some overlap between adjacent measurement regions and that results in overlapping regions, obtained using different principles, are consistent.

Even more critical are quantities related to perception. A sensible approach in this case is to consider a limited class of objects first and then, if measurability is achieved there, progressively extend the class. For example, in the case of loudness, we can start with pure tones, then move to stationary and even to non stationary sounds.[17]

After identifying the class of objects, we have to look for a set of empirical properties that characterise the quantity. We may have different ways of defining them, each leading to a different definition of the scale. In the case of temperature, for example, we can define it on the basis of the order relation of "hotter than" when touching an object, or on the variation of height in a liquid-in-glass thermometer, or on the variation of pressure in a constant-volume thermometer [11]. The choice of the empirical relations has an impact on achieving a lower uncertainty. This has been one of the main driving forces under the development of the international system of units: the different definitions of the metre, and consequently of the scale of length, have led to a progressive diminution of the related uncertainty and the same has happened for the other fundamental quantities. In dealing with a candidate new quantity we may again adopt a progressive approach. We may first consider a set of properties giving rise to an order scale, check measurability on it, and, if it is satisfied, try with an interval scale and so on. Furthermore, we may consider different sets of empirical properties—for example, in the case of hardness, we may consider indentation instead of scratching as the basic empirical property—and obtain different scales. In that case, checking for their agreement may be a powerful and highly recommendable validation strategy. In the case of loudness, for example, the results of a procedure based on adjusting sounds in such a way that they seem twice as loud as others, may be compared to the results of a procedure based on comparing monaural and binaural listening to the same sounds [22].

The scale construction issue concerns the development, maintenance and dissemination of reference scales. This point has been given great attention in physics and engineering and was the driving force behind the institution of the international system of metrology, with the Metre Convention of 1875 [43]. Much less attention has

[17] We will discuss loudness measurement in some detail in Chap. 8.

been paid to it in psychology and behavioural sciences, and this is still a point of great difference in the sensitivity of the two communities. Progress is required in this area by both parties [44].

Concerning the measuring system, we have already discussed its importance as well as the need of having an open but rigorous view of it.

Lastly, it may seem that the approach presented here is biased towards what Campbell called "fundamental" quantities. Yet we will discuss derived quantities in dept in Chap. 3 and we will see that the main difference between them and fundamental ones consists in the way empirical properties are defined. But once this difference is clarified, it will be apparent that the above procedure essentially applies to derived quantities as well.

To sum up, the steps of the procedure may be more synthetically grouped in two main classes: (a) construction of the reference measurement scale (Steps 1–3 above) and (b) measurement (given a reference measurement scale), Step 4.

The measurement scale will be discussed in a deterministic framework in Chap. 3 and in probabilistic terms in Chap. 4. Multidimensional extensions will be briefly addressed in Chap. 7. The measurement process will be described in Chaps. 5 and 6. But before moving on to present the theory, we have to consider another main problem in measurement: how to deal with uncertainty? This is the subject of the next chapter.

References

1. Rossi, G.B.: Measurability. Measurement **40**, 545–562 (2007)
2. Mari, L.: Measurability. In: Boumans, M. (ed.) Measurement in Economics, pp. 41–77. Elsevier, Amsterdam (2007)
3. Campbell, N.R.: Physics—The Elements. Reprinted as: Foundations of Science. (1957). Dover, New York (1920)
4. BIPM, IEC, OIML, ISO: International Vocabulary of Basic and General Terms in Metrology. ISO, Genève (1984)
5. BIPM, IEC, IFCC, ISO, IUPAC, IUPAP, OIML: International Vocabulary of Basic and General Terms in Metrology, 2nd edn (1993)
6. ISO: ISO/IEC Guide 99:2007 International Vocabulary of Metrology—Basic and General Terms (VIM). ISO, Geneva (2007)
7. Galanter, E., et al.: Measuring the Impossible—Report of the MINET High-Level Expert Group. EU NEST, Bruxelles (2010)
8. Rossi, G.B.: Cross-disciplinary concepts and terms in measurement. Measurement **42**, 1288–1296 (2009)
9. von Helmholtz, H.: Zählen und Messen Erkenntnis—theoretisch betrachtet. Philosophische Aufsätze Eduard Zeller gewidmet, Fuess, Leipzig (1887)
10. Kisch, B.: Scales and Weights. Yale University Press, New Haven (1965)
11. Ellis, B.: Basic Concepts of Measurement. Cambridge University Press, Cambridge (1968)
12. Nicholas, J.V., White, D.R.: Traceable Temperatures. Wiley, Chichester (1994)
13. Rossi, P.: La nascita della scienza moderna in Europa. Laterza, Roma (1997)
14. Aumala, O.: Fundamentals and trends of digital measurement. Measurement **26**, 45–54 (1999)
15. Fechner, G.: Elements of Psychophysics, Leipzig. English edition: (1966) (trans: Adler, H.E.). Holt, New York (1860)
16. Wozniak, R.H.: Classics in Psychology, 1855–1914: Historical Essays. Thoemmes Press, Bristol (1999)

17. Jones, F.N.: History of psychophysics and judgement. In: Carterette, E.C., Friedman, M.P. (eds.) Handbook of Perception, vol. 2. Academic Press, New York (1974)
18. Baird, J.C., Noma, E.: Fundamentals of Scaling and Psychophysics. Wiley, New York (1978)
19. Berglund, B.: Measurement in psychology. In: Berglund, B., Rossi, G.B., Townsend, J., Pendrill, L. (eds) Measurement with Persons. Taylor and Francis, London, pp. 27–50 (2012)
20. Ferguson, A., Myers, C.S., Bartlett, R.J.: Quantitative estimates of sensory events. Report Br Assoc Adv Sci **108** (1938)
21. Ferguson, A., Myers, C.S., Bartlett, R.J.: Quantitative estimates of sensory events. Report Br Assoc Adv Sci **2**, 331–349 (1940)
22. Stevens, S.S.: The direct estimation of sensory magnitudes: loudness. Am J Psychol **69**, 1–25 (1956)
23. Stevens, S.S.: Measurement, psychophysics and utility. In: Churchman, C.W., Ratoosh, P. (eds.) Basic Concepts of Measurements. Cambridge University Press, Cambridge, pp 1–49 (1959)
24. Stevens, S.S.: On the theory of scales and measurement. Science **103**, 667–680 (1946)
25. Misra, P., Enge, P.: Global positioning system, 2nd edn. Ganga-Jamuna Press, Lincoln, MA, (2011)
26. Essen, L.: Time scales. Metrologia **4**, 161–165 (1968)
27. Krantz, D.H., Luce, R.D., Suppes, P., Tversky, A.: Foundations of Measurement, vol. 1. Academic Press, New York (1971)
28. Miyamoto, J.M.: An axiomatization of the ratio difference representation. J Math Psychol **27**, 439–455 (1983)
29. Roberts, F.S.: Measurement theory, with applications to decision-making, utility and the social sciences. Addison-Wesley, Reading. Digital Reprinting (2009). Cambridge University Press, Cambridge (1979)
30. Finkelstein, L., Leaning, M.S.: A review of the fundamental concepts of measurement. Measurement **2**, 25–34 (1984)
31. Narens, L.: Abstract measurement theory. MIT Press, Cambridge (1985)
32. Suppes, P., Krantz, D.H., Luce, R.D., Tversky, A.: Foundations of Measurement, vol. 2. Academic Press, New York (1989)
33. Luce, R.D., Krantz, D.H., Suppes, P., Tversky, A.: Foundations of Measurement, vol. 3. Academic Press, New York (1990)
34. Luce, R.D., Suppes, P.: Representational measurement theory. In: Stevens' Handbook of Experimental Psychophysics, vol. 4. Wiley, New York (2002)
35. Finkelstein, L.: Theory and philosophy of measurement. In: Sydenham, P.H. (ed.) Handbook of Measurement Science, vol. 1, pp. 1–30. Wiley, Chichester (1982)
36. Muravyov, S., Savolainen, V.: Special interpretation of formal measurement scales for the case of multiple heterogeneous properties. Measurement **29**, 209–224 (2001)
37. Mari, L.: Beyond the representational viewpoint: a new formalization of measurement. Measurement **27**, 71–84 (2000)
38. Berglund, B., Rossi, G.B., Townsend, J., Pendrill, L. (eds.): Measurement with Persons. Taylor and Francis, London (2012)
39. Gonella, L.: Measuring instruments and theory of measurement. In: Proceedings of XI IMEKO World Congress, Houston (1988)
40. Rossi, G.B.: A probabilistic model for measurement processes. Measurement **34**, 85–99 (2003)
41. Finkelstein, L.: Widely, strongly and weakly defined measurement. Measurement **34**, 39–48 (2003)
42. Finkelstein, L.: Problems of measurement in soft systems. Measurement **38**, 267–274 (2005)
43. BIPM: The International System of Units, 8th edn. STEDI, Paris (2006)
44. Rossi, G.B., Berglund, B.: Measurement of quantities involving human perception and interpretation. Measurement **44**, 815–822 (2011)

Chapter 2
Uncertainty

2.1 Why are Measurement Results not Certain?

Suppose that you go to your physician and that he measures your blood pressure. He will probably repeat the measurement a few times and will usually not obtain exactly the same result through such repetitions. This is an example of measurement uncertainty. There are many other examples of this limited *repeatability* which is a good reason for being, to some extent, uncertain about the measurement value. Moreover, even in cases in which the reading is stable, for example when we read the temperature in a room from a wall thermometer, we cannot be totally certain about its value, since a thermometer typically has some "tolerance", for example $\pm 1\,^{\circ}\mathrm{C}$, so that if we read $23\,^{\circ}\mathrm{C}$, we may be quite confident that room temperature will be somewhere between 22 and $24\,^{\circ}\mathrm{C}$, but we would not "bet" on any precise value with a high expectation of winning.

Uncertainty does not only concern simple, everyday measurements: researchers at National Metrology Institutes, such as the National Institute for Standards and Testing (NIST) in USA or the Istituto Nazionale di Ricerca in Metrologia (INRiM) in Italy, spend a considerable amount of time in dealing with uncertainty, even though they work at the highest levels of precision. Measurements in psychophysics are also affected by an even more apparent variability, due to intra- and inter-individual variations.

So it is worth considering uncertainty as an inherent characteristic of measurement. As such, it is has been studied deeply since the beginning of the modern measurement science, and it is therefore useful to give a brief overview of the development of uncertainty theory.

G. B. Rossi, *Measurement and Probability*, Springer Series in Measurement Science and Technology, DOI: 10.1007/978-94-017-8825-0_2,

2.2 Historical Background

2.2.1 Gauss, Laplace and the Early Theory of Errors

The importance of accuracy in measurement has probably been recognised since ancient times. Classical and even scriptural texts warn about the incorrectness of faulty measurement. The concern for reliable measurement units is another side of the problem. Modern scientists have been aware of the need for accurate measurement in order to check scientific theories. Yet the explicit treatment of measurement "errors" was only begun at the beginning of the nineteenth century, when the development of scientific theories and of measuring instruments have required an explicit evaluation of instrumental and measurement performance.

In his *Theoria motus corporum coelestium* [1], Carl Friedrich Gauss (1777–1855) discusses how to obtain estimates of the orbital parameters of heavenly bodies on the basis of a set of observations. He distinguishes between *systematic* and *random* errors. This distinction, already mentioned in his *Theoria motus*, is more clearly expressed in the subsequent *Theoria combinationis observationum erroribus minimis obnoxiae* [2]. Due to the importance of this issue, it is worth reading the original text.

"Certain causes of error—he writes—are such that their effect on any one observation depends on varying circumstances that seem to have no essential connection with the observation itself. Errors arising in this way are called *irregular* or *random*... On the other hand, other sources of error by their nature *have a constant effect on all observations of the same class*. Or if the effect is not absolutely constant, its size varies regularly with circumstances that are essentially connected with the observations. These errors are called *constant* or *regular*". Gauss further observes that "this distinction is to some extents relative and *depends on how broadly we take the notion of observations of the same class*". He explicitly excludes the consideration of systematic (regular, in his terminology) errors in his investigation and warns that "of course, it is up to the observer to ferret out all sources of constant error and remove them". This choice of neglecting systematic errors characterises the classical *theory of errors*, and it is probably its main limitation [3]. We shall see later that the need to overcome this limitation has been the driving force behind the studies on uncertainty in the second half of the twentieth century [4].

To see how Gauss deals with random errors, let us consider the measurement of a single constant quantity x by N-repeated observations. We may model the ith observation by

$$y_i = x + v_i, \tag{2.1}$$

where y_i is the ith observed and recorded value, x is the measurand, which remains constant during the observation process, v_i is the (unknown) value assumed by the probabilistic (or random) variable v during the ith observation, and $i = 1, 2, ..., N$. The random variable v accounts for the scattering that we observe in the data.[1]

This can also be more compactly expressed in vector notation, as

$$\mathbf{y} = x + \mathbf{v}, \tag{2.2}$$

where $\mathbf{y}$ is a vector of observations, and $\mathbf{v}$ is the vector of random measurement errors.

At this point, Gauss needs an explicit expression for the probability distribution of the errors,[2] p_v, and he thus assumes some properties that correspond to the common understanding of measurement errors. He assumes that p_v is symmetric, maximum in its origin and decreasing on each side of the origin. Furthermore, he assumes that the most probable value for x, once the observations $\mathbf{y}$ have been acquired, is the arithmetic mean of the observed values, since "it has been customary certainly to regard as an axiom the hypothesis that if any quantity has been determined by several direct observations, made under the same circumstances and with equal care, the arithmetic mean of the observed values affords the most probable value, if not rigorously, yet very nearly at least, so that it is always safe to adhere to it". If we denote "the most probable value for x" by $\hat{x}$,[3] this key assumption may be explicated as follows:

$$\hat{x} = \overline{y} \triangleq N^{-1} \sum_i y_i. \tag{2.3}$$

On the basis of this assumption, Gauss could derive the famous distribution named after him. In modern notation, if we introduce the standard normal (Gaussian) distribution, with zero mean and unitary variance defined by

$$\varphi(\xi) = (2\pi)^{-1/2} \exp(-\xi^2/2), \tag{2.4}$$

[1] Henceforth, we need the notion of *probabilistic* or *random* variable (we prefer the former term, although the latter is more common). Though we assume that the reader has a basic knowledge of probability theory, for the sake of convenience, we present a brief review of the probability notions used in this book in Sect. 4.1. Note in particular the notation, since we often use a shorthand one. We do not use any special conventions (such as capital or bold characters) for probabilistic variables. So the same symbol may be used to denote a probabilistic variable or its specific value. For example the probability density function of v can be denoted either as $p_v(\cdot)$ or, in a shorthand notation, as $p(v)$. For notational conventions, see also the Appendix at the end of the book, in particular under the heading "Generic probability and statistics".

[2] A definition of probability distribution, also (more commonly) called the probability density function for continuous variables, is provided in Sect. 4.1.8.

[3] In general the "hat" symbol is used to denote an estimator or an estimated value. If applied to the measurand, it denotes the measurement value.

the result can be compactly expressed as

$$p(v) = \sigma^{-1}\varphi(\sigma^{-1}v), \tag{2.5}$$

where σ is the *standard deviation* [5].

A similar result was reached, in a different way by Pierre-Simon Marquis de Laplace (1749–1827), in his *Théorie analytique des probabilités* [6]. Let us consider the case of repeated measurement once more, and let us still assume that the errors v_i are independent and equally distributed. We now also assume that their distribution $p(v)$ is symmetric about the origin and has a finite support. Let $\hat{x} = \overline{y}$ be the selected estimate for x and

$$e = \hat{x} - x \tag{2.6}$$

the estimation error. Then Laplace showed that e is asymptotically normally distributed with a variance proportional to N^{-1}. In this sense, the normal distribution is regarded as the distribution of the estimation error, for a long series of observations.

It is also possible to consider the problem from another favourable viewpoint, traceable once again to Laplace [7]. Indeed, if we consider the measurement error as deriving from the contribution of a large sum of small independent error sources,

$$v = \sum_j w_j, \tag{2.7}$$

if none of them prevails over the others, the distribution of the resulting error tends to be normal provided that the number of the error sources increases.

In conclusion, the classical measurement error theory, developed mainly thanks to the contributions of Gauss and Laplace, concerns random errors only and results in a probabilistic model, the normal distribution, whose validity can be supported by different arguments.

We will reconsider the measurement error theory at a later stage and will discuss its merits and limitations, and how to overcome them. But we shall now go back to consider the problem of uncertainty from a totally different perspective.

2.2.2 Fechner and Thurstone: The Uncertainty of Observed Relations

The problem of measurement uncertainty was also considered, in around the middle of the nineteenth century, by Fechner in a even more fundamental way [8]. For him, *the only reliable judgements that an observer may express with respect to his sensations are either equality or ordered inequalities* (greater than). His law, in fact, was formulated on the basis of such results. In general, people's responses, which we will call "indications", must be regarded as an expression of *non-deterministic* phenomena, since, for the same pair of stimuli, we may obtain different responses from different subjects (inter-subjective variability) or even from the same subject, by repeating the test (intra-subjective variability).

A typical experiment in early psychophysics consists in the determination of the just noticeable difference between two stimuli. We already know from Chap. 1 that Fechner's law was developed from such differences. Let us discuss this in further detail. Let ϕ_0 denote the physical intensity of a reference, fixed, stimulus, for example a sound at 1 kHz, with a sound intensity level of 60 dB, and let ϕ be a the variable stimulus of the same kind, having a slightly higher intensity than ϕ_0.[4] Let ψ_0 and ψ be the perceived intensities associated with ϕ_0 and ϕ. Suppose now that we make an experiment with different subjects over repeated trials, in which we wish to determine the minimum value for ϕ that gives rise to a perceivable (positive) variation. In practice, we keep ϕ_0 fixed and, we vary ϕ until the subject listening to both stimuli notices a difference between the two, that is he/she perceives the sensation ψ, associated with ϕ, as being more intense than the sensation ψ_0, associated with ϕ_0.[5] This will not always occur at the same value of ϕ, due to differences in the responses of different people or even to differences in the responses of the same person, when the trial is repeated. The result of one such experiment can therefore be expressed and summarised by the conditional probability[6]

$$P(\psi \succcurlyeq \psi_0|\phi), \tag{2.8}$$

that is the probability that the sensation ψ is "greater" ($\succcurlyeq$) than the sensation ψ_0. This probability is a function of ϕ, which is varied during the experiment (whilst ϕ_0 is kept fixed), and may qualitatively look as shown in Fig. 2.1.

On the basis of this experimental result, the *differential threshold* can be estimated, conventionally but reasonably, by the value $\delta\phi$ at which $P(\psi \succcurlyeq \psi_0|\phi_0 + \delta\phi) = 0.75$ [9].

More generally, if we consider two objects a and b and the property ψ associated with them, we can consider the probability $P(\psi_b \succcurlyeq \psi_a)$, or, in a shorthand notation, $P(b \succcurlyeq a)$. The important point here is that *the empirical relation holding between two sensations is recognised as being probabilistic*. This is a somewhat more fundamental perspective than that of the early theory of errors, since *uncertainty is here ascribed to empirical relations rather than to measurement values*. Since empirical relations play a fundamental role in measurement, uncertainty is understood here as affecting the very roots of measurement.

---

[4] We will discuss loudness measurement in some detail in Chap. 8. Readers who are unfamiliar with acoustic quantities may consult the initial section of that chapter for some basic ideas.

[5] In the practical implementation of the experiment, there are different ways of varying the stimulus, either through series of ascending or descending values, or as a random sequence. The variation can be controlled by the person leading the experiment or by the test subject [9, 10] . In any case, such technicalities do not lie within the sphere of this discussion.

[6] For the notion of conditional probability, see Sects. 4.1.1–4.1.3 of Chap. 4, in this book, as well as any good textbook on probability theory [11].

Fig. 2.1 Probability that $\psi \succcurlyeq \psi_0$, as a function of the stimulus ϕ

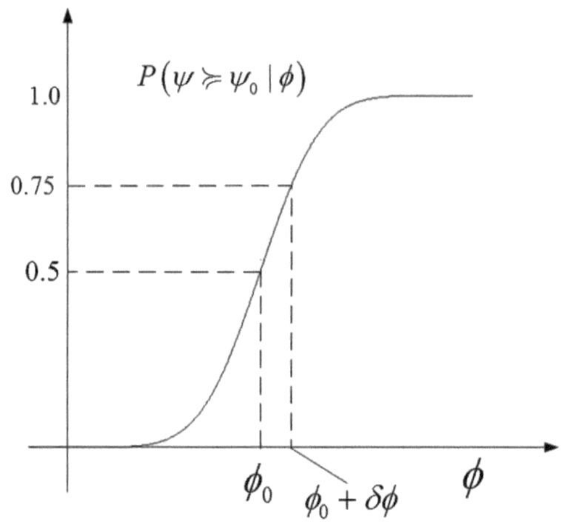

Once that we have recognised that empirical relations have a probabilistic nature, the challenge is how to represent that in a numerical domain. The solution to this problem will be shown and fully discussed in Chap. 4. For the moment, let us just mention an approach related to the *law of comparative judgement* developed by Luis Leon Thurstone (1887–1955) [12].

Let us then look for a numerical representation of sensations ψ_0 and ψ_1, evoked by objects a and b, respectively, that complies with the empirical evidence that $P(\psi_1 \succcurlyeq \psi_0) = p$, or, equivalently, $P(b \succcurlyeq a) = p$, where p is a probability value, $p \in [0, 1]$.

If we describe ψ_0 and ψ_1 with two independent probabilistic variables, x_a and x_b, whose probability distributions, $p_{x_a}(\psi)$ and $p_{x_a}(\psi)$, are Gaussian, with expected values $\hat{\psi}_0$ and $\hat{\psi}_1$, respectively, and equal variance, σ^2, our condition can be satisfied, provided that

$$\hat{\psi}_1 - \hat{\psi}_0 = z_{10}\sqrt{2}\sigma, \tag{2.9}$$

where z_{10} is such that

$$\int_0^{z_{10}} \varphi(\xi)\mathrm{d}\xi = p - \frac{1}{2}, \tag{2.10}$$

where

$$\varphi(\xi) = (2\pi)^{-1/2}\exp(-\xi^2/2) \tag{2.11}$$

is the standard normal distribution (with zero mean and unitary variance).

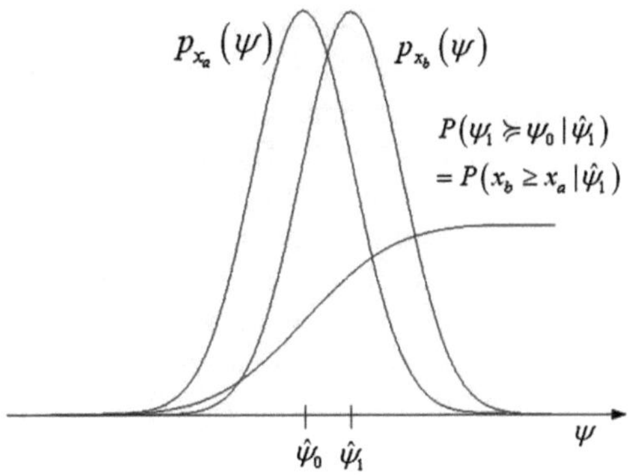

Fig. 2.2 Representation of sensations ψ_0 and ψ_1 by probability distributions on the ψ axis, corresponding to $P(\psi_1 \succcurlyeq \psi_0) = 0.75 = P(x_b \geq x_b)$

Let us briefly show how this result can be obtained. Let us introduce the probabilistic variable $u = x_b - x_a$, which will have mean value $\hat{u} = \hat{\psi}_1 - \hat{\psi}_0$ and standard deviation $\sigma_u = \sqrt{2}\sigma$.[7] Then

$$p = P(x_b \geq x_a) = P(u \geq 0) = \int_0^\infty p(u)\mathrm{d}u = \int_0^\infty (\sigma_u)^{-1}\varphi\left((u - \hat{u})/\sigma_u\right)\mathrm{d}u. \tag{2.12}$$

Making the substitution $v = u - \hat{u}$, we obtain

$$p = \int_{-\infty}^{\hat{u}} (\sigma_u)^{-1}\varphi(v/\sigma_u)\mathrm{d}v = \frac{1}{2} + \int_0^{\frac{\hat{u}}{\sigma_u}} \varphi(\xi)\mathrm{d}\xi. \tag{2.13}$$

Then defining z_{10} as in (2.10), we obtain $z_{10} = \hat{u}/\sigma_u$, from which (2.9) follows.[8] This is illustrated in Fig. 2.2, for $p = 0.75$.[9]

Note that, in this case, $z_{10} = 0.6745$; thus if we denote as $\delta\psi$ the increment in the sensation scale corresponding to a just noticeable increment, $\delta\phi$, in the stimulus, we obtain, approximately,

$$\delta\psi \cong \sigma, \tag{2.14}$$

[7] In fact the variance of the sum (or of the difference) of two independent probabilistic variables equals the sum of their individual variances. Thus, in our case, $\sigma_u^2 = \sigma_{x_b}^2 + \sigma_{x_a}^2 = 2\sigma^2$.

[8] The device of using the abscissae of the standard normal distribution, usually called z-points, is widely used in probability and statistics and, consequently, in psychophysics too.

[9] Interestingly enough, Link notes that Fechner proposed a similar (but not identical) approach, which is very close the signal-detection method of the 1950s. Applying this approach, one would obtain $\hat{\psi}_1 - \hat{\psi}_0 = 2z_{10}\sigma$, instead of the result in (2.9) [13].

which indicates an interesting link between the dispersion of sensorial responses, expressed by σ, and the resolution of the sensation scale, expressed by $\delta\psi$.[10]

As a numerical example, consider again the case of a sound at 1 kHz, with a sound intensity level of 60 dB. The differential threshold value in this case is, roughly, $\delta\phi$ =1 dB [14]. The corresponding loudness values can be obtained by defining the loudness scale for pure tones. We will briefly present such a scale in Chap. 8, Sect. 8.2.2, formulaes 8.17 and 8.11. The measurement unit for that scale is the *sone*. We obtain $\psi_0 = 4$ sone, $\psi_1 \cong 4.3$ sone, that is $\delta\psi$ =0.3 sone, and $\sigma \cong 0.3$ sone.

Thus the two sensations can be represented on a sone scale as probabilistic variables, with mean values $\psi_0 = 4.0$ sone, $\psi_1 \cong 4.3$ sone, and standard deviation $\sigma \cong 0.3$ sone.

2.2.3 Campbell: Errors of Consistency and Errors of Methods

Let us now go back to physical measurement and take a look at Campbell's position. We have already encountered Campbell in the first chapter, as the first (and one of the few) proposers of a comprehensive theory of measurement, at least for physical measurement [15]. In his theory, he also considers measurement errors and distinguishes, as Gauss does, between two kinds of them, which he calls *errors of consistency* and *errors of method*. The former are those that occur when the same measurement is repeated several times under the same conditions and correspond to Gauss's random errors, the latter correspond to systematic errors. It is interesting to see the way he introduces methodical errors: they appear as violations of empirical relations, in particular as violation of "equality", or equivalence, in a more modern language. Equivalence should be transitive, yet in "real" measurement it is often possible to find three objects, a, b and c, such that $a \sim b$ and $b \sim c$, but not $a \sim c$. *How is it possible to reconcile this evidence with the possibility of making fundamental measurements of the quantity for which this happens?*

One way to do this is to consider probabilistic rather than deterministic relations. So the solution to this (fundamental) problem raised but not solved by Campbell comes from an idea that is ultimately traceable to Fechner and Thurstone, that is from the other side of the barricade, in the perspective of the Report of the British Association for the Advancement of Science!

This is further evidence in favour of the need for a unified, interdisciplinary theory of measurement. Prior to discussing probabilistic relations in greater depth, we shall review some subsequent important contributions to the treatment of measurement data.

---

[10] The resolution of a measurement scale is the minimum variation that can be expressed with that scale (see also the glossary, in the Appendix, at the end of the book).

2.2.4 The Contribution of Orthodox Statistics

Orthodox statistics—the term was coined by E. T. Jaynes (1922–1998) [3]—is a school whose principal exponent was Ronald Aylmer Fisher (1890–1962). During the first part of the twentieth century, he made an important contribution to the development of probabilistic-statistical models by providing a store of methods for their use in conjunction with experimentation [16]. Interestingly enough, this approach makes it possible, in some cases, *to model systematic effects*. To understand how this can be achieved, suppose that we have a measuring instrument that may be affected by a residual (additive) calibration error. If we have just one instrument, the calibration error will give rise to a systematic effect since it will remain constant, at least for some time. But if we have a set of independently calibrated instruments, the calibration error will vary randomly amongst the instruments. Consider now an experiment in which we measure the same fixed quantity x with a set of m independently calibrated measuring instruments of the same type, repeating the measurement n times for each instrument and collecting a total of $N = nm$ observations. The experiment can thus be modelled as follows:

$$y_{ij} = x + \theta_i + v_{ij}, \tag{2.15}$$

where

- $i = 1, \ldots, m$ is the index denoting the instruments,
- $j = 1, \ldots, n$ is the index denoting the repetitions,
- θ_i is a probabilistic variable representing the residual calibration error of each instrument and
- v_{ij} is an array of probabilistic variables, representing random samples from a probabilistic variable, v, that models the random error.

In this framework, the residual calibration error θ gives rise to a systematic error, if we consider the indications of a single instrument as "observations of the same class", whilst it varies randomly if we sample instruments from the class of all the instrument of the same type. From the mathematical point of view, to select a single instrument we fix index i to a constant value, i_0, whilst to sort different instruments, we let it vary within a range from 1 to m. Consider now the following averages: the overall

$$\bar{y} = \frac{1}{N} \sum_{ij} y_{ij}, \tag{2.16}$$

and the average per instrument

$$\bar{y}_i = \frac{1}{n} \sum_{j} y_{ij}. \tag{2.17}$$

Then the measurand x can be estimated as

$$\hat{x} = \overline{y}, \tag{2.18}$$

whilst the systematic deviation of the ith instrument by

$$\hat{\theta}_i = \overline{y}_i - \overline{y}. \tag{2.19}$$

Interestingly enough, in this experiment *it is possible to quantify both the effect of random variations and of systematic deviations.* In fact, the variance of the random variations can be estimated as

$$\hat{\sigma}_v^2 = \frac{1}{N - m} \sum_{ij} (y_{ij} - \overline{y}_i)^2, \tag{2.20}$$

whilst the variance of the calibration error by

$$\hat{\sigma}_\theta^2 = \frac{1}{m - 1} \sum_i (\overline{y}_i - \overline{y})^2. \tag{2.21}$$

So if it is possible to develop experiments in which the quantities that normally give rise to systematic effects are allowed to vary at random, it is possible to quantitatively evaluate their effect. Unfortunately, this is not the general case in measurement and when this approach is not applicable we have to look for another solution which we will describe at a later stage.

2.2.5 Uncertainty Relations in Quantum Mechanics

A decisive contribution to a deeper understanding of measurement uncertainty came, in the twentieth century, from quantum mechanics [17, 18]. As an example, we briefly mention the basic idea behind the celebrated Heisenberg uncertainty relation. Consider the single-split experiment schematically illustrated in Fig. 2.3.

Suppose we have a beam of electrons impinging on a screen with a thin split. The electrons passing through the split will reach a second, photo-sensitive, screen and form an image on it. If the split is very thin, diffraction will occur, and the image will be wider than the split. We can consider the motion of the electrons passing through the split, as characterised by their position and velocity along the y axis. Interestingly enough, this apparatus will reduce the experimenter's uncertainty as regards position but will increase that concerning velocity. Indeed, before reaching the first screen, the position of the electrons is somewhere in the interval D, whilst their velocity in the y direction is equal to zero. The positions of the electrons passing through the screen lies within the interval d, much smaller than D, but, their velocity v_y is no

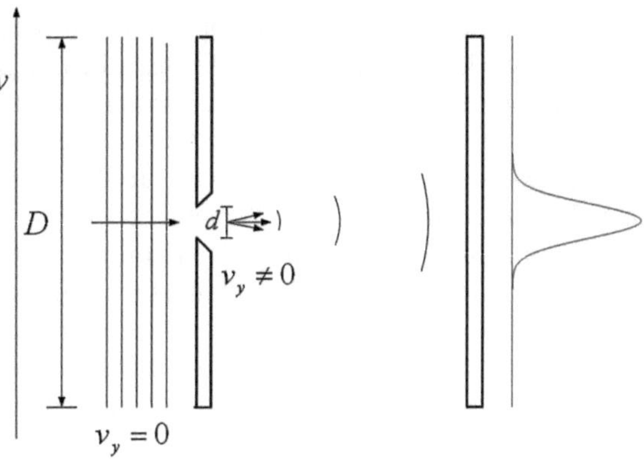

Fig. 2.3 The single-split experiment [18]

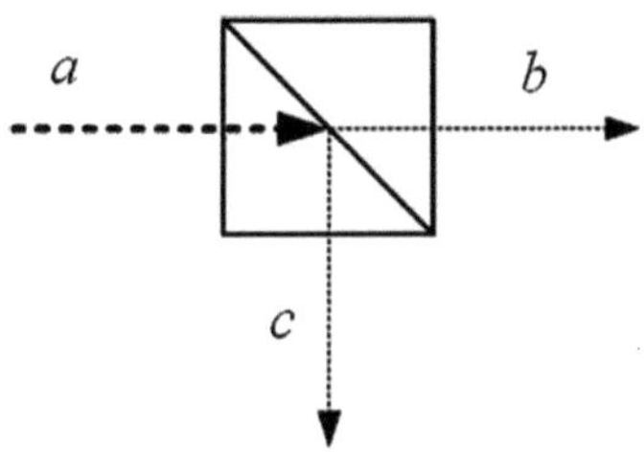

Fig. 2.4 Beam splitter: *a* incident light; *b* transmitted light; *c* reflected light

longer null. If we (informally) denote the uncertainties concerning y and v_y with Δy and Δv_y, respectively, Heisenberg's principle states that

$$\Delta x \Delta v_x \geq \frac{h}{m}, \tag{2.22}$$

where h is Planck's constant and m is the mass of the electron.[11] This is an example of interaction of the measuring system with the object under observation—a system of particles—, which gives rise to a kind of fundamental uncertainty. This suggests that numerical representations in this field must be regarded as inherently probabilistic. In contrast with the classical theory of errors, where, in the absence of systematic effects, measurement precision can be, at least in principle, indefinitely increased, here this is no longer possible, and probabilistic representations are definitely needed.

In fact there is another way, even more important for our purpose, in which quantum mechanics departs from the "classical" perspective of the early theory of errors. To see this in the simplest way, consider a beam splitter, that is an optical device that splits a beam of light in two, as shown in Fig. 2.4.

It is quite easy to describe the macroscopic behaviour of such a device: one half of the incident light is transmitted, whilst the other half is reflected. But now consider a single photon of light: what happens in this case? It cannot be split any further,

[11] This formulation is somewhat qualitative but sufficient for the purpose of this informal discussion.

since it is an elementary, indivisible, entity; it may therefore either pass through or be reflected. This situation can be described in probabilistic terms by assigning a probability of 0.5 to each of these two possibilities. But note now an important difference with respect to the early theory of errors, which was developed within the framework of classical mechanics. In that theory, probability was essentially intended as accounting for our "ignorance" (partial knowledge). Ideally, should we dispose of all the information needed to fully describing the system, we would have no "error". But here, in the new quantum mechanics framework, we cannot describe the behaviour of an elementary item, such as a photon or a particle, better than in probabilistic terms. This makes a big difference. It is often said that in this case probability has a non-epistemic nature ("epistemic" means: related to our state of knowledge).

An important lesson can be learned from quantum mechanics. The approach to uncertainty made in this book will consider measurement as an inherently uncertain process, and we will develop the theory in this perspective. We will also briefly discuss the "nature" of probability in Sect. 4.1.1 and will take a position in that regard.

2.2.6 The Debate on Uncertainty at the End of the Twentieth Century

In the late 1970s, the metrological community recognised the need to reach an internationally agreed way of expressing measurement uncertainty. It also recognised the need to accompany the reporting of measurement results by some quantitative indications of its quality, not only in primary metrology, but also in everyday measurements. In 1978, therefore, the Bureau International des Poids et Mesures (BIPM) carried out an investigation on a large number of laboratories and prepared a recommendation, INC-1 (1980), which was also adopted by the Comité International des Poids et Mesures (CIPM).[12] An international working group was then established, under the guidance of the International Organization for Standardization (ISO), for the purpose of developing a detailed technical Guide. One of the major scientific problems to be faced was the composition of random and systematic effects causing uncertainty. The work of the group was paralleled by intensive scientific debate on these issues. In 1993, the "Guide to the expression of uncertainty in measurement" (GUM) [19, 20] was published. The document had a great impact on both technical and scientific aspects and further stimulated international debate on measurement uncertainty and related topics.

[12] The BIPM and the CIPM are two of the main bodies in the international system of metrology and were established when the Metre Convention was signed (1875). A concise introduction to the organisation of the system is made in Sect. 3.7.4. and additional details on how it works are given in Sect. 10.1.

Let us briefly review some of its main points. Firstly, the GUM recognises, several possible sources of uncertainty, including the following:

1. incomplete definition of the measurand;
2. imperfect realisation of the definition of the measurand;
3. non-representative sampling—the sample measured may not represent the defined measurand;
4. inadequate knowledge of the effects of environmental conditions on the measurement or imperfect measurement of environmental conditions;
5. personal bias in reading analogue instruments;
6. finite instrument resolution or discrimination threshold;
7. inexact values of measurement standards and reference materials;
8. inexact values of constant or other parameters obtained from external sources and used in the data-reduction algorithm;
9. approximations and assumptions incorporated in the measurement method and procedure;
10. variations in repeated observations of the measurand under apparently identical conditions.

Then, addressing uncertainty evaluation, the GUM adopts the paradigm of indirect measurement which has already been mentioned in Chap. 1. In this kind of measurement, the value of the measurand is not obtained directly from the measuring instrument, but by first measuring other quantities that are functionally related to the measurand, and then processing data according to this functional relation. This may be expressed as

$$x = g(\mathbf{z}), \tag{2.23}$$

where x is the measurand, $\mathbf{z}$ a vector of quantities functionally related to the measurand and g a function.[13] We shall call this expression the *(GUM) evaluation model* or *formula*. The quantities appearing in it are treated as probabilistic (or random) variables and their standard deviation, here known as standard uncertainty and denoted with u, is of special interest. Basically the formula allows the uncertainties on the quantities $\mathbf{z}$ to be "propagated" to the measurand x, as we will see in a moment. In turn, these uncertainties may be evaluated on the basis of different pieces of information, which the GUM classifies into two main categories: those coming from a series of observations (type A) and those coming from other sources, such as information provided by the instrument manufacturers, by calibration, by experience, and so on (type B). Note that, in this approach, the focus moves from the type of the uncertainty sources (systematic vs. random) to the type of information concerning them (type A vs. type B). Consequently, it is possible to pragmatically support a common treatment for both of them.

[13] We do not use the GUM's notation here, since we wish to be consistent with the notation used in this book. See the Appendix for further details.

Let us now see how can we apply this approach to the basic case in which we obtain the measurement result directly from a measuring system. We can interpret one of the z_i, for example the first one, as the indication, y, of the measuring system, that is $z_1 = y$, and the remaining z_i as "corrections" that should ideally be applied to correct the effect of the various error sources. The (possible) spread of the indications is accounted for by considering the variability of the probabilistic variable y. The evaluation procedure for the standard uncertainty then proceeds as follows. Since the variables appearing in the evaluation formula are regarded as probabilistic, if $\hat{\mathbf{z}}$ is the expected value[14] of $\mathbf{z}$, that is $\hat{\mathbf{z}} = \mathrm{E}(\mathbf{z})$, $\Sigma_{\mathbf{z}}$ the covariance of $\mathbf{z}$ and $\mathbf{b}$ the vector of the "sensitivities" of x with respect to $\mathbf{z}$, calculated for $\mathbf{z} = \hat{\mathbf{z}}$, that is

$$b_i = \left. \frac{\partial g}{\partial z_i} \right|_{\mathbf{z}=\hat{\mathbf{z}}}, \tag{2.24}$$

then an "estimate" of x may be obtained as

$$\hat{x} = g(\hat{\mathbf{z}}), \tag{2.25}$$

and the standard uncertainty, u, to be associated with $\hat{x}$ is

$$u = \sqrt{\boldsymbol{b}^T \sum\nolimits_{\mathbf{z}} \boldsymbol{b}}. \tag{2.26}$$

At present, the GUM is an important international reference for the evaluation of measurement uncertainty. Yet, as we have seen, the proposed solution is based on a pragmatic agreement reached within the working group that developed it, rather than on a coherent measurement theory. In this book, we will attempt to do the opposite, that is, derive the rules for evaluating and expressing measurement uncertainty from an overall probabilistic theory of measurement. Due to its importance, uncertainty evaluation will be specifically addressed in Chap. 9. We will also consider the extension of these ideas to all the domains of science, including experimental psychology in particular. Some indications in this sense will be provided in Chap. 8.

2.3 The Proposed Approach

In the above brief historical review, we have learned to distinguish between random variations in observations and systematic effects in the measurement process. We have seen how the former may be modelled according to the classical theory of errors, whilst the latter requires a different approach. Orthodox statistics has provided

---

[14] In the GUM, the expected value of a quantity is regarded as a "best estimate" of that quantity.

a model for randomising these effects, where practically possible, in order to gain some control on the variables affecting the experiment. Although this is not the general case in measurement, this method is certainly useful, when it is applicable; otherwise a different approach is needed. We have also seen that in psychophysics empirical relations are understood to have a probabilistic character and that in quantum mechanics quantities are regarded as inherently probabilistic. Lastly, we have seen how internationally recognised guidelines are devoted to the evaluation and expression of measurement uncertainty.

In this book, we will develop a general probabilistic approach to measurement that enables uncertainty to be considered and treated in all its forms, in rigorous probabilistic terms.

In Chap. 1, we have seen that in order to measure something a reference scale must first be established and then at least one measuring system based on that scale devised. In dealing with uncertainty, we will follow the same pattern, distinguishing between uncertainty mainly related to the scale and uncertainty mainly related to the measurement process.

2.3.1 Uncertainty Related to the Measurement Scale and to Empirical Relations

A measurement scale is characterised by the empirical relations that can be mapped into corresponding numerical ones. For example, in the case of an ordinal scale, the representation reads

$$a \succsim b \Leftrightarrow m(a) \geq m(b). \tag{2.27}$$

But this is a deterministic description, since it really implies that, whenever we observe the pair of objects a and b, we always observe either $a \succ b$ or $a \sim b$ or $b \succ a$, and the measurement reflects this state of affairs. Is this what really *always* happens? I suggest readers think of any measurable property of their interest and check if this is the case. I think that making such a statement—a definite relation holds for a, b—is only possible if, intuitively, a and b are "far apart". Instead, if they are "close" to each other it may be, in general, impossible to establish a definite relation between them. To be more precise let us introduce the notion of "comparator", here intended as a device that is capable of establishing an order relation for pairs of elements, with respect to the characteristic under investigation. Let us also consider the notion of "repeatability", this being the ability of a device to produce the same result when operated in the same conditions. Operatively, "same conditions" means that they are undistinguishable for the operator. Repeatability is usually characterised by a standard deviation that quantifies the dispersion of the observations and that can be assessed by a proper calibration test. For example, if we say that a length measuring device has a repeatability of, say, 10 μm, we mean that when repeating the measurement of an object in undistinguishable conditions, we observe a spread in the instrument indication, with a standard deviation of 10 μm.

So, going back to the issue of comparison, I suggest that if we compare two objects, a and b, whose difference is comparable with the repeatability of the comparator, and we repeat the comparison several times, we may sometimes observe $a \succ b$, sometimes $a \sim b$ and sometimes even $b \succ a$. If this happens, we can say that empirical relations are uncertain and we can describe this situation by means of probability. Very simply, we can assign a probability to each of the possible observed relations, that is

$$P(a \succ b), P(a \sim b), P(a \prec b), \tag{2.28}$$

satisfying the condition

$$P(a \succ b) + P(a \sim b) + P(a \prec b) = 1. \tag{2.29}$$

We will see later on in this book how to treat the notion, here just presented intuitively, of the probability of a relation in rigorous terms.

To complete this quick look at uncertain relations, we mention that there is another way in which empirical relations may be uncertain. Suppose that we have two equally reliable comparators, C and D, and suppose that, when comparing a and b,

- with C we obtain $a \succ_C b$, whilst
- with C we obtain $a \sim_D b$.

We can interpret this evidence in different ways. We may think that either $a \succ b$ or $a \sim b$ is true and one of the two comparators is wrong, but we do not know which one. Or we may think that the two objects interact with the comparators, in such a way that there are state changes in them, but we are unable to define their states outside these comparisons. Although this uncertainty condition is completely different from the one concerning the issue of repeatability, yet both of them can be described in probabilistic terms. Indeed, in both cases, we can consider $a \succ b$ and $a \sim b$ as uncertain statements characterised by a probability figure.

In Chap. 4, we will see that this yields a probabilistic representation, such as

$$P(a \succsim b) = P\left(m(a) \geq m(b)\right), \tag{2.30}$$

that replaces formula (2.27). In Chap. 4, we will systematically derive these relations for the scales that are of the greatest interest.

To sum up, I have suggested that the first, in a logical order, sources of uncertainty occurring in measurement may be found in the scale construction phase and that they are related to empirical relations. We may be uncertain about them both due to the lack of perfect repeatability of observations and as a consequence of systematic deviations in what we observe. In both cases, uncertainty can be expressed by probabilistic statements.

2.3.2 *Uncertainty Related to the Measurement Process and the Measuring System*

The second major part of a theory of measurement concerns the measurement process. We have seen in Chap. 1 that in order to measure something we must first devise a reference scale and then we need a device for comparing unknown objects with the reference scale. We have called such a device a measuring system, and we have also provided a definition for it, as an empirical system capable of interacting with objects incorporating the property under investigation and of producing, as the result of such interaction, signs, on the basis of which it is possible to assign a value to the object to be measured. We can thus model measurement as a process that maps (the properties of) objects in measurement values, that is

$$\hat{x} = \gamma(a), \tag{2.31}$$

where a is an object (considered in respect of a quantity x of its) and $\hat{x}$ is the measurement value that we obtain as the result of the measurement process. Consider now the question of whether such a description is satisfactory or not. Consider what it really implies. It requires that, given any object a, it is always possible to assign to it a measurement value, $\hat{x}$, that exactly describes it.

I do not think that this is generally possible, for reasons similar to those just considered. Again, if we repeat the measurement of the same object several times in equivalent conditions for the experimenter, we may obtain different values,

$$\hat{x}_1, \ldots, \hat{x}_N, \tag{2.32}$$

or, in another scenario, if we measure the same object with two equally reliable devices, R and S, we may repeatedly obtain two different values, one for each system,

$$\hat{x}_R = \gamma_R(a), \quad \hat{x}_S = \gamma_S(a). \tag{2.33}$$

Again it is possible to express such evidence in probabilistic terms. Basically we interpret the *statement* $\hat{x} = \gamma(a)$, where $\hat{x}$ is a number in a set X, as an *uncertain* one, to which a probability can be assigned,

$$P\left(\hat{x} = \gamma(a)\right). \tag{2.34}$$

This probabilistic representation may be interpreted as the probability of obtaining the measurement value $\hat{x}$, for $\hat{x}$ varying in a set of possible values, when object a is measured. This can also be expressed by the conditional probability distribution

$$P\left(\hat{x}|a\right), \tag{2.35}$$

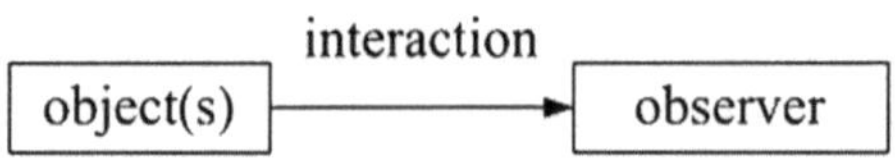

Fig. 2.5 Ideal communication between the object(s) and the observer

where the conditioning event may be modelled as the "random extraction" of object a from the set A. In Chap. 5, we will derive a general expression for such a distribution, based on a general characterisation of the measuring system.

2.3.3 Information Flux Between the Objects(s) and the Observer

Let us now consider a different standpoint that is transversal to the above consideration.

In the case of an observed empirical order relation, $a \succcurlyeq b$, we have a pair of objects, a and b, and a relation between them, established by a comparator. We can thus consider an objects/observer scheme in which the comparator "observes" the empirical relation that holds between the objects. Similarly, in the case of a ternary relation, $a \succcurlyeq b \circ c$, we have three objects, a, b and c, and a device that can establish whether the relation holds or not. So, by extension, we will call *comparator* the device that allows us to establish an empirical relation, and we regard it (including the operator that handles it, in the case of manually operated systems) as an *observer*.

In the case of a measurement process, we have the object to be measured and the measuring system (plus the operator where applicable), in the role of the observer. In both cases, therefore, we can synthetically depict the ideal communication situation as shown in Fig. 2.5.

With respect to this ideal situation, uncertainty sources may affect either the object(s) or the observer or their interaction. The latter hypothesis includes the case, very important in many measurements, in which the measuring system modifies the state of the object.[15] This means that the interaction between object and observer is no longer uni-directional, as in the ideal case,—the object modifies the state of the observer and information is transmitted thanks to this modification—but the observer also modifies the object, and so the state that we actually observe is no longer the original one. All these possibilities are illustrated in Fig. 2.6.

To sum up, I have proposed a taxonomy of uncertainty sources, based on three conceptual coordinates, considering uncertainty either as

- related to empirical relations or to the measurement process,
- referring to random variations or to systematic deviations,

[15] This is usually called "loading effect" in the technical literature.

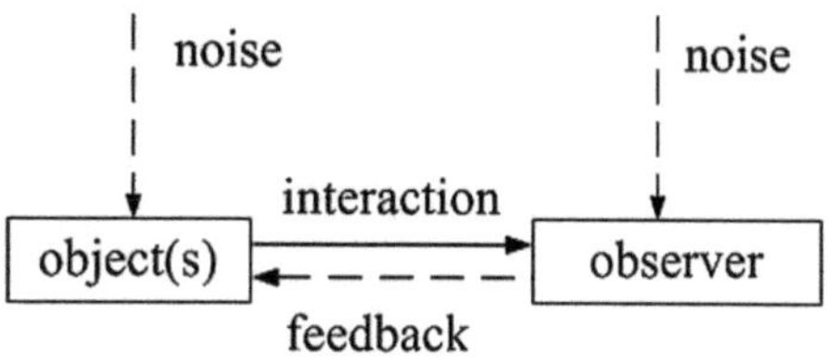

Fig. 2.6 Real communication between the object(s) and the observer, affected by uncertainty sources

- related to the information flux in different ways, that is to say either affecting the object(s) or the observer or their interaction.

I hope that this taxonomy can help in the identification of uncertainty sources as this is the first, and often the most critical, step in uncertainty evaluation. In the second part of the book, we will develop a probabilistic theory, for dealing with uncertainty in general terms, whilst, in the third part, we will discuss some important application issues.

References

1. Gauss, C.F.: Theoria motus corporum coelestium in sectionibus conicis solem ambientium. Hamburg. English edition: (2004) (trans: Davis C.H.). Dover (1809)
2. Gauss, C.F.: Theoria combinationis observationum erroribus minimis obnoxiae. Gottingen. English edition: (1995) (trans: Stewart G.W.). SIAM, Philadelphia (1823)
3. Costantini, D.: I fondamenti storico-filosofici delle discipline statistico probabilistiche. Bollati Boringhieri, Torino (2004)
4. Rossi, G.B.: Probability in metrology. In: Pavese, F., Forbes, A. (eds.) Data Modeling for Metrology and Testing in Measurement Science. Birkhauser-Springer, Boston (2009)
5. Sheynin, O.B.: C. F. Gauss and the theory of errors. Arch. Hist. Exact Sci. **20**, 21–72 (1979)
6. Laplace, P.S.: Theorie analytique des probabilités. Courcier, Paris. In: Oeuvres Complètes de Laplace, vol. 7. Gauthier-Villars, Paris (1812)
7. Sheynin, O.B.: Laplace's theory of errors. Arch. Hist. Exact Sci. **17**, 1–61 (1977)
8. Nowell Jones, F.: History of psychophysics and judgement. In Carterette, E.C., Friedman, M.P. (eds.) Handbook of Perception, vol. 2. Academic Press, New York (1974)
9. Baird, J.C., Noma, E.: Fundamentals of Scaling and Psychophysics. Wiley, New York (1978)
10. Gescheider, G.A.: Psychophysics: The Fundamentals. 3rd edn., Erlbaum, New York (1997)
11. Monti, M., Pierobon, G.: Teoria della probabilità. Zanichelli, Bologna (2000)
12. Thurstone, L.L.: A law of comparative judgements. Psychol. Rev. **34**, 273–286 (1927)
13. Link, S.W.: Rediscovering the past: Gustav Fechner and signal detection theory. Psychol. Sci. **5**, 335–340 (1994)
14. Zwicker, E., Fastl, H.: Psycho-acoustics. Springer, Berlin (1999)
15. Campbell, N.R.: Physics— the Elements. Reprinted as Foundations of Science (1957). Dover, New York (1920)
16. Fisher, R.A.: Statistical Methods and Scientific Inference. Oliver and Boyd, Edinburgh (1956)
17. Heisenberg, W.: Physics and Philosophy. George Allen and Unwin Edition, London (1959)
18. Ghirardi, G.C.: Un'occhiata alle carte di Dio. Il Saggiatore, Milano (2003)
19. BIPM, IEC, IFCC, ISO, IUPAC, IUPAP, OIML: Guide to the Expression of Uncertainty in Measurement. ISO, Geneva, Switzweland. Corrected and reprinted 1995, ISBN 92-67-10188-9 (1993)
20. BIPM: Mutual Recognition. STEDI, Paris (2008)

Part II
The Theory

Chapter 3
The Measurement Scale: Deterministic Framework

3.1 What is the Meaning of Measurement?

Suppose that we want to compare the working environments of two offices. We would probably measure the width of the working space allocated to each person, and the temperature, humidity and noise level, in typical working conditions. We could also measure more sophisticated characteristics, such as the visual appearance (a combination of colour, gloss, texture) of desk surfaces. With such measures, we can compare the two offices to each other, or the characteristics of each of them with some standard values, recommended by ergonomists. *Such comparisons are possible, provided that reference scales exist for the properties under consideration.* Thanks to them, measurement serves its scope, that is to enable comparisons and support decision-making. Furthermore, it depends upon the type of such scales whether statements based on measurement results are meaningful or not. We can say that the working space for each person in one office is 20 % greater than in the other, but we cannot say that the Celsius temperature on some Summer day is 20 % higher, whilst we can say that it is 4° higher (in the sense that there is a difference of 4° between the temperatures of the two offices).[1]

The notion of measurement scale, as we already know, has been deeply investigated in the representational theory. It is now time, for us also, to undertake such a study.

The term scale has two meanings, specific and general. In the specific sense, a scale is a series of standard objects, to which a value has been properly assigned: we will call this a reference scale. This is the current way the term scale is understood. For example, in the International Vocabulary of Metrology, we find the following definition: "ordered set of values of quantities of a given kind, continuous or discrete, used in arranging quantities of the same kind by magnitude" [1]. Then, there is another, more technical, meaning, typical of the representational theory. In this sense, when we say that a scale exists for some quantity, we mean that the set of objects

[1] The meaningfulness of statement concerning measurement on a given scale depends upon the *uniqueness* conditions for that scale, as briefly mentioned in Sect. 1.7 and as will be discussed in detail, in the following of this chapter.

G. B. Rossi, *Measurement and Probability*, Springer Series in Measurement Science and Technology, DOI: 10.1007/978-94-017-8825-0_3,

manifesting the quantity is defined, as well as are some empirical relations, and that it is possible to assign numbers to the objects in such a way as to reproduce, amongst the numbers, the same relations that hold amongst the objects [2]. So, the scale, in this general meaning, includes everything that is necessary for measurement to be possible, apart from the measuring instrument.

For example, in the case of Mohs hardness, the scale, in the general sense, is defined once that we specify that hardness concerns minerals in their natural state, we define it as the ability of a mineral to scratch another, and we establish a rule for assigning numbers representing such a property. The scale in the specific sense, instead, is the series of the standard materials selected by Mohs, each with one number assigned.

In this chapter, we study in detail both aspects for three most important kinds of scales, ordinal, interval and ratio.

We will consider *finite structures* only. At first glance, this may seem a limitation. In fact, we suggest it is not, since, in reality, in the realisation of a reference scale, we cannot experimentally attain an infinite *resolution*,[2] since this would imply the possibility of detecting infinitely small variations. On the other hand, for any *fixed* class of measurement problems, it will also be possible to assume a maximum value for the quantity under consideration. For example, if we consider the important example of length measurement for the dimensional control of workpieces, a typical resolution could be $\Delta x_r = 1\,\mu\text{m}$ and the maximum length involved may be, e.g., $x_{\text{max}} = 10\,\text{m}$. So we have, in this case, we need to have $n = x_{\text{max}}/\Delta x_r = 10^7$ elements in a reference scale, appropriate for this class of problems.[3] This is a big number but still a finite one: being finite does not imply being small![4] Note that we can properly imagine a scale in this case as a ruler and the "standard" objects as traits on that ruler, and that Δx_r is the distance between any two successive traits. As other examples, for the measurement of the mass of persons, we may assume $\Delta x_r = 0.05$ kg and $x_{\text{max}} = 220\,\text{kg}$, which yields $n = 4{,}400$; for the measurement of temperature in a living or working environment, we may assume $\Delta x_r = 0.1\,°\text{C}$, $x_{\text{min}} = 0\,°\text{C}$ and $x_{\text{max}} = 100\,°\text{C}$, yielding $n = 1{,}000$.

---

[2] The resolution, for a reference scale, is the difference between two adjacent standards. For a measuring system, it is the capability of detecting and measuring small differences between different objects in respect of the quantity under consideration. In both cases, such differences can be very small but they must be *finite*. The infinitely small, as well the infinitely large, are not attainable in actual experimentation.

[3] Then, for any measurement problem, we can think at it as belonging to a class of similar problems, for which an appropriate finite reference scale can be established.

[4] The attentive reader could object that when we consider a finite structure, we mean that the total number of measurable objects is finite. So, for example, if we have, for each element of the reference scale, m objects that have the same length of it, the total number of object is $N = nm$. Thus, our assumption actually implies that N is finite. But, after some reflection, it is easy to get persuaded that *what really matters* is the number n of elements of the reference scale, which is also the number of distinct ways (or states) in which the quantity manifests itself. We will see in the following of the chapter that this is the number of the *equivalence classes* for the quantity. It is indeed immaterial how many objects, equivalent to each elements of the scale, are there. So, we can safely assume that they are in a finite number, and even that that number is the same, m, in all cases.

Furthermore, by now, we assume that all the relations involved have a well-defined true/false state, that is we follow an essentially *deterministic* approach. This will provide a theory for ideal, uncertainty-free measurement. In the next chapter, we will remove this assumption and will account for uncertainty that constitutively affects measurement. This will require an appropriate logic for dealing with uncertainty, and so we will adopt a probabilistic approach, apt to properly accounting for any possible source of uncertainty.

3.2 The General Framework

3.2.1 Overview

We have seen, informally, that the definition of a measurement scale requires the specification of

1. the set of objects manifesting the characteristic we are interested in and of the empirical relations that hold amongst them,
2. a set of numbers and of relations amongst them, and
3. the mapping that allows us to assign numbers to objects in such a way as to reproduce, amongst the numbers, the same relations that hold amongst the objects.

In the case of Mohs hardness, for example, we may denote by A the set of objects, here the minerals in their natural state, by the symbol $\succcurlyeq$ the empirical relation of weak order, that here means "not scratched by".[5] Then, the couple $\mathcal{A} = (A, \succcurlyeq)$ is called empirical relational system. We will then consider a corresponding numerical relational system $\mathcal{B} = (I, \geq)$, where I is the set of the first ten integer numbers. Then, we can identify the reference scale and define the measure function m that associate to each object the appropriate measurement value (by comparing the object with the reference scale). The measure function thus establishes a correspondence between the empirical and the numerical relational system. All this also provides a *meaning* to measurement, since in consequence of this, measurement may be understood as *a number assignment to objects, in such a way as to reproduce, in the numerical domain, the relations holding in reality amongst the objects*. The whole apparatus considered up to now, the empirical relational system, the numerical relational system and the measure function, constitutes a measurement scale, in the general sense. The main characteristic of a scale is expressed by a representation theorem, that in the case of an ordinal scale reads as follows:

for each $a, b \in A, a \succcurlyeq b \Leftrightarrow m(a) \geq m(b)$.

For each scale, we will also consider a *uniqueness theorem* that identifies the class of admissible transformations, as we have discussed in Chap. 1, Sect. 7. Uniqueness is linked to meaningfulness, through this criterion: on a given scales, only statements

[5] In fact, $a \succ b$ means "a scratches b", $a \sim b$ means "a neither scratches nor is scratched by b", then $a \succcurlyeq b$ means "a is not scratched by b".

that are unaffected by admissible transformations on that scale are meaningful [3]. So for example, if the hardness of a is 2 and the hardness of b is 4, we can say that a is harder than b, since any admissible transformation will conserve this inequality, but we cannot say that a is twice as hard as b, since ratio will not be, in general, maintained by admissible transformations of the Mohs scale.

Let us now express these ideas in a formal way.

3.2.2 Some Formal Statements

Let us consider a set A of objects that manifest the characteristic x. We will consider the following concepts as primitive: *object, property (or characteristic) of an object, empirical relation, empirical system, observable output.*

Let us now introduce the notion of empirical relational system.

Definition 3.1 An empirical relational system is a structure $\mathcal{A} = (A, R_1, R_2 \ldots R_p)$, where A is a set of objects and each R_i is an m_i-ary empirical relation on A.

Note that m_i here denotes the "arity" of the ith relation considered and should not be confused with the measure function.[6] Similarly, we introduce the notion of numerical relational system.

Definition 3.2 A numerical relational system is a structure $\mathcal{B} = (\mathbb{R}, S_1, S_2 \ldots S_p)$, where $\mathbb{R}$ is the set of real numbers and each S_i is an m_i-ary relation on $\mathbb{R}$.

We have seen that measurement constitutes a correspondence between an empirical relational system and a numerical one, in such a way that empirical relations are mapped into correspondent numerical ones. The notion of scale summarises this idea.

Definition 3.3 (*Measurement scale*) A (measurement) scale is a triple $\mathcal{S} = (\mathcal{A}, \mathcal{B}, m)$, where $\mathcal{A}$ is an empirical relational system, $\mathcal{B}$ is a numerical relational system and m is a function, $m \colon A \to \mathbb{R}$, that we call a measure (function) that satisfies the following property: for every $i = 1, \ldots p$ and for every a_j, $R_i(a_1, a_2, \ldots, a_{m_i}) \Leftrightarrow S_i[m(a_1), m(a_2), \ldots, m(a_{m_i})]$. Such a property is called a representation theorem.

A measurable property or quantity is a property for which a measurement scale exists.

Definition 3.4 (*Quantity*) A measurable property or quantity, x, is a property of objects in a class A to which a scale $\mathcal{S}_x$ is associated.

If no ambiguity arises, we will usually omit the index x.

Lastly, we also define a uniqueness condition.

Definition 3.5 (*Uniqueness*) A uniqueness condition for a scale $\mathcal{S}$ is the specification of the class of measure functions that satisfy it.

---

[6] In practice, we will consider *binary*, *ternary* or *quaternary* relations.

Table 3.1 Summary of the main properties of the empirical structures and related measurement scales

Empirical structure	Empirical relations	Scale type	Admissible transformations
Order	Order amongst objects	Ordinal	Monotone increasing
Difference	Order and differences amongst objects	Interval	Linear positive
Intensive	Order, difference and ratio amongst objects	Ratio	Similarity
Extenive	Order amongst objects and an addition operation	Ratio	Similarity

3.2.3 Overview of the Main Types of Scales

We will consider the following types of empirical relational systems and related measurement scales:

1. *order structures*, giving rise to *ordinal scales*;
2. *difference structures*, giving rise to *interval scales*;
3. *intensive structures*, giving rise to *ratio scales*;
4. *extensive structures*, giving rise to *ratio scales* as well.

Their main characteristics are summarised in Table 3.1.

We do not treat nominal structures and related scales here, since they are particularly important in the case of multidimensional measurement, and thus, they will be treated in Chap. 7.

Order structures are characterised by an empirical relation of weak order which plays a fundamental role in measurement, as we will show in the next section, and is present in all the structures we are considering now. The corresponding ordinal scales are unaffected by monotone increasing transformations.

With difference structures, we are mainly concerned with differences amongst objects. Related interval scales may safely undergo linear positive transformations.

Intensive structures include an empirical relation of ratio between objects, whilst extensive structures include an empirical operation of summation. Both of them can be represented on ratio scales that are invariant under multiplication by a positive constant. Such a transformation, called similarity, occurs when we change the measurement unit. Intensive structures, although less known then extensive, are not less important, since, for example, they are essential in psychophysics.

3.3 Ordinal Scales

3.3.1 Motivations for Dealing with Ordinal Scales

Ordinal scales are important for two reasons. Order is perhaps the most basic property that characterises numbers and so, if measurement is intended to map things into numbers, it is natural to assume order as the essential condition for measuring.

On this, both Helmoltz and Campbell agreed, although they did not draw the natural conclusion, that is that order measurement makes sense, even without additional properties. With the representational approach and its emphasis on representation, it has become more natural to accept that. In the International Vocabulary of Metrology (VIM), an ordinal quantity is defined as a "quantity, defined by a conventional measurement procedure, for which a total ordering relation can be established, according to magnitude, with other quantities of the same kind, but for which no algebraic operations among those quantities are defined" [1].[7]

The second reason of interest is that we have real scales of that kind, such as the intensity of natural phenomena as wind or earthquakes. The VIM provides, as example of ordinal scales, Rockwell C hardness, octave number for petroleum fuel and earth quake strength on the Richter scale. In general, we can say that order scales occur when we want to quantify the intensity of a phenomenon, where a limited number of states are of interest, and each of them is defined by a plurality of features. The Beaufort wind-intensity scale, summarised in Table 3.2, is a good example [4]. Note that wind intensity, as measured on a Beaufort scale, is different from wind speed, as measured by an anemometer (on a velocity scale). They convey a different kind of information, intended to address different applications. For example, if we want to estimate how much energy can we obtain from a wind turbine, we will be mainly interested in the former, whilst if we want to properly drive a sailboat, the latter would probably provide a more useful information.

3.3.2 Serialising and Numbering Objects

The characterising relation of an order system is, obviously, order. Yet, we have different kinds of order: for example the order $\geq$, that holds amongst natural numbers and in other cases of our interest, is called *simple* and is such that only one item may be put in each position in the ordering: there is only one number between, say, 3 and 5. Conversely, in ordering objects, we may find different items fitting in the same place, for example different minerals having equal Mohs' hardness, say, 4. We call this a *weak order*.

The reference paradigm for an order system is provided by objects that can be put in a series according to some property. Objects that manifest the property in the same way must be grouped together and share the same place in the series.

Let us illustrate this in a simple way. Consider a set $A = \{a, b, c, d, e\}$. The relation may be represented by a matrix, R, whose generic element r_{ij} denotes the relation holding between elements a_i (row) and element a_j (column) of A. Suppose, e.g., that the matrix looks as in Table 3.3.

---

[7] I do not want to comment here this, somewhat questionable, definition, but just note that ordinal quantities have been defined and consequently accepted, in this environment.

Table 3.2 The Beaufort wind scale

Beaufort number	Wind speed (km/h)	Beaufort wind scale	
		Description	Sea effects
0	<1	Calm	Water is mirror-like
1	1–5	Light air	Small ripples appear on water surface
2	6–11	Light breeze	Small wavelets develop, crests are glassy
3	12–19	Gentle breeze	Large wavelets, crests start to break, some whitecaps
4	20–28	Moderate breeze	Small waves develop, becoming longer, whitecaps
5	29–38	Fresh breeze	White crested wavelets (whitecaps) form, some spray
6	39–49	Strong breeze	Larger waves form, whitecaps prevalent, spray
7	50–61	Moderate or near gale	Larger waves develop, white foam from breaking waves begins to be blown
8	62–74	Gale or fresh gale	Moderately large waves with blown foam
9	75–88	Strong gale	High waves (6 m), rolling seas, dense foam, blowing spray reduces visibility
10	89–102	Whole gale or storm	Large waves (6–9 m), overhanging crests, sea becomes white with foam, heavy rolling, reduced visibility
11	103–117	Violent storm	Large waves (9–14 m), white foam, visibility further reduced
12	118	Hurricane	Large waves over 14 m, air filled with foam, sea white with foam and driving spray, little visibility

Sea effects only are included; land effects can also be considered, but are here omitted for the sake of brevity

Table 3.3 A simple example of an order system

	a	b	c	d	e
a		$\prec$	$\prec$	$\prec$	$\prec$
b	$\succ$		$\sim$	$\prec$	$\prec$
c	$\succ$	$\sim$		$\prec$	$\prec$
d	$\succ$	$\succ$	$\succ$		$\prec$
e	$\succ$	$\succ$	$\succ$	$\succ$	

For example, looking at the crossing between row 3 and column 4, we see that $c \prec d$. We can serialise the elements of A, as shown in Fig. 3.1, where the following rules have been followed:

- elements in the same column are equivalent, whilst
- an element at the right of another is greater,

Elements in the same column belong to the same *equivalence class*. We may now number objects, assigning the same number to elements in the same column and increasing numbers to the different columns, from left to right. Such numbers provide an order measure. In this example, we have four equivalence classes and we can

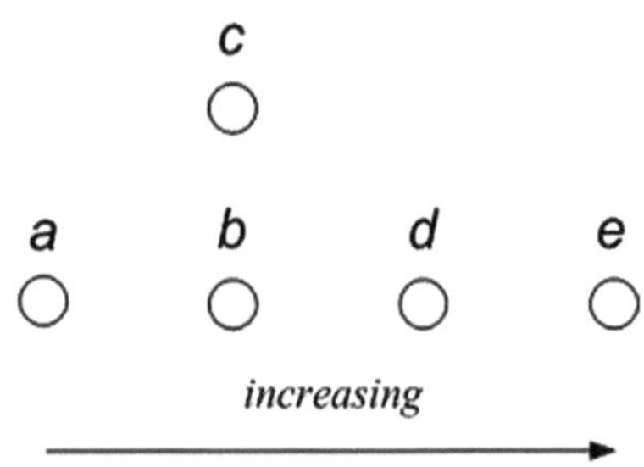

Fig. 3.1 Serialising objects

assign $m(a) = 1$, $m(b) = m(c) = 2$, $m(d) = 3$, $m(e) = 4$. It is intuitive that such a procedure always works, when A is a finite set.

3.3.3 Representation for Order Structures

We are now ready to present the above ideas in a more formal fashion. Let us first define an order structure.

Definition 3.6 (*Order structure*) Let A be a finite (not empty) set of objects carrying the property x. Then, $(A, \succeq)$ is an (empirical) order structure if $\succeq$ is an (empirical) binary relation on A that satisfies, for each $a, b, c \in A$,

6.1 $a \succeq b$ or $b \succeq a$,
6.2 $a \succeq b$ and $b \succeq c \Rightarrow a \succeq c$.

The relation $\succeq$ is a weak order.

The first property, usually called *completeness* or *connectivity*, ensures that each object can be compared with all the others; the second, *transitivity*, is what mainly characterises order. The relation of weak order can be split into two, that we will denote by $\sim$ and $\succ$, respectively.

Definition 3.7 Let $(A, \succeq)$ be an order structure. Then, for each $a, b \in A$,

7.1 $a \sim b \Leftrightarrow a \succeq b$ and $b \succeq a$,
7.2 $a \succ b \Leftrightarrow a \succeq b$ and not $b \succeq a$.

It is easy to check that $\sim$ is an *equivalence relation*, i.e., it is *reflexive*, *symmetric* and *transitive*, and $\succ$ is a *strict order*, that is it is *complete*, *transitive* and *asymmetric* [3]. Note that the *simple order*, $\geq$, can also be split in two, the *strict simple order*, $>$, and the *equality* relation, $=$, in a similar way.

As we have seen in the previous section (Fig. 3.1), the basic idea in constructing the scale is thus the following. We partition A by grouping together elements that are mutually equivalent through $\sim$, each such group constituting an equivalence class. Then, we order equivalence classes in an ascending order and assign number 1 to all the elements in the first (lowest) class, 2 to those in the second and so forth. It is easy to check that the numbers (measures) so assigned do satisfy the representation condition. The representation theorem for a finite order structure thus runs as follows.

Theorem 3.8 (Representation for order structures) *Let A be a finite (not empty) set of objects carrying the property x and let $(A, \succcurlyeq)$ be an order structure. Then, there exists a measure function $m: A \to \mathbb{R}$ such that, for each $a, b \in A$,*

$$a \succcurlyeq b \Leftrightarrow m(a) \geq m(b).$$

Proof Let $(A, \succcurlyeq)$ be a finite order structure and let $\sim$ be the corresponding equivalence relation, according to Definition 3.7. For each $a \in A$, $a^* = \{b \in A | a \sim b\}$ is the equivalence class containing a. The class A^* of all the equivalence classes in A according to $\sim$ constitutes a partition of A. Since A is finite, A^* is also finite; let its cardinality[8] be n. Define also, for notation convenience, the following sets: $I = \{1, 2, \ldots, n\}$ and $I' = \{1, 2, \ldots, n - 1\}$. Then, pick one (whatever) element from each equivalence class in A^* and form with them the set S. Note that the relation $\succcurlyeq$, if applied to the elements of S only, becomes a *simple order*, since for each $r, s \in S$, not $r \sim s$.[9] Let us then label the element of S according to the (simple strict) ordering $\succ$, that is $S = \{s_i \in A | i \in I \text{ and, for } i \in I', s_i \prec s_{i+1}\}$. We call S a *series of standards* for A. Define now the measure function $m: A \to I$ as follows:

- for each $s_i \in S$, $m(s_i) = i$,
- for each $a \in (A - S)$, there will be one and only one $s_i \in S$ such that $a \sim s_i$; then let $m(a) = m(s_i) \Leftrightarrow a \sim s_i$.

Consider now any $a, b \in A$. Then, there will be $s_i, s_j \in S$ such that $a \sim s_i$, $b \sim s_j$. If $a \succcurlyeq b$, then $s_i \geq s_j$ and $i \geq j$. Then, $a \succcurlyeq b \Rightarrow m(a) = m(s_i) = i \geq j = m(s_j) = m(b)$. Conversely, $m(a) \geq m(b) \Rightarrow m(s_i) \geq m(s_j) \Rightarrow s_i \succcurlyeq s_j \Rightarrow a \succcurlyeq b$, which completes the proof.[10] □

Let us now consider uniqueness.

Theorem 3.9 (Uniqueness for order structures) *Let A be a finite (not empty) set of objects carrying the property x, let $(A, \succcurlyeq)$ be an order structure and $m: A \to \mathbb{R}$ a measure function for it. Then, any other measure function m' is such that, for each $a \in A$,*

$$m'(a) = \phi(m(a)),$$

[8] The cardinality of a (finite) set is the number of its elements.

[9] We could introduce a special symbol for this relation in S, but we avoid to do that, since we think that a proliferation of symbols is rather confusing than clarifying. In fact, an alternative way of formulating the representation theorem is to consider an *isomorphism* between A^* and I, as Narens, e.g., does, [5], rather than an *homomorphism* between A and I, as we, and many others [3, 6], do.

[10] The symbol □ denotes the end of a proof.

where ϕ is a monotone increasing function, that is

$$\phi(u) > \phi(v) \Leftrightarrow u > v.$$

Proof We have to first prove that if m is a valid measure function, $m' = \phi(m)$, with ϕ monotonic increasing, also is. Indeed, for each $a, b \in A$,

$$a \succsim b \Leftrightarrow m(a) \geq m(b) \Leftrightarrow \phi(m(a)) \geq \phi(m(b)) \Leftrightarrow m'(a) \geq m'(b).$$

Then, we have to show that if both m and m' are valid measure functions, then their exists a monotonic increasing function ϕ, such that $m' = \phi(m)$. In fact, if both m and m' are valid measure functions,

$$m(a) \geq m(b) \Leftrightarrow a \succsim b \Leftrightarrow m'(a) \geq m'(b).$$

Then, the function ϕ, defined, for each $a \in A$, by $m'(a) = \phi(m(a))$, is monotonic increasing. □

The notions of *series of standards* and of *reference scale* just introduced are of general interest and apply, in a similar way, to other types of scales. So, we pick them up in two formal statements.

Definition 3.10 (*Series of standards*) Let $\mathcal{A} = (A, \succsim)$ be a finite order structure and A^* the class of all the equivalence classes in A according to the equivalence relation $\sim$. Let n be the cardinality of A^*, $I = \{1, 2, \ldots, n\}$ and $I' = \{1, 2, \ldots, n-1\}$. Then, there exist a (in general not unique) subset S of A, such that $S = \{s_i \in A | i \in I$ and, for $i \in I', s_i \prec s_{i+1}\}$ and, for each $a \in A$, there exists one and only one element $s \in S$ such that $a \sim s$. We call S a series of standards for $\mathcal{A}$.

Definition 3.11 (*Reference scale*) Let $\mathcal{A} = (A, \succsim)$ be a finite order structure, S a series of standards and m a measure function for $\mathcal{A}$. Then, the set $R = \{(s_i, m(s_i)), i \in I\}$ is a reference scale for $\mathcal{A}$.

As an example, consider again the system illustrated in Table 3.3 and Fig. 3.1. Here, we can take, as reference scale:

$$R = \{(s_1 = a, 1), (s_2 = b, 2), (s_3 = d, 3), (s_4 = e, 4)\}.$$

If we measure c by comparing it with the elements of the scale, we observe $c \sim b$ and consequently we assign $m(c) = 2$.

3.4 Interval Scales

3.4.1 Dealing with Intervals

So far we have considered individual objects, $a, b, c, \ldots$, we have now to turn our attention to intervals.[11] For doing so, it is necessary first to understand well this notion, which may cause some difficulty, because of our familiarity with intervals of reals, that are a different thing. Here, an *interval* is basically *an ordered pair of objects* and, as such, we can denote it by, say, ab: note that ab is different from ba. The objects a and b are the extremes of the interval; whilst in an interval of reals, there are always points between the extremes—actually there are infinitely many—here, there are can be no object at all or a finite number of them. In this regard, they are similar to intervals of integers. As with intervals of integers, we will outline elementary, unitary, intervals and we will "measure" the width of intervals by counting the number of unitary elements they include.

We will consider different properties of intervals. The simplest is the order between the extremes, $a \succsim b$, and we are already able to do so, since we have studied in depth order amongst objects in the previous section. Another important property is the *distance* between the extremes, which conveys the idea of *how far is a from b*, or, equivalently, b from a. When we want to highlight some specific property of an interval, we use a different notation; in the case of distance, we write δ_{ab}. Note anyway that δ_{ab} refers to the *empirical distance* which is a property of the interval. As such it has still to be understood as *a pair of objects*: in fact δ_{ab} still denotes the interval ab but considered in regard of the distance between its extremes. Do not confuse empirical distance with its numerical representation: to avoid confusion, we denote this last by d_{ab} and d_{ab} is *a positive number*. Anyway, we will not treat distance in this chapter; we will briefly touch on them in Chap. 7, that deals with multidimensional measurement.

Here, we will consider *difference* instead, which is somewhat a "signed" distance, and will be denoted by Δ_{ab}.

For characterising differences, we have to compare them and we need a notation for that. Formally, we denote a weak order for differences by $\succsim_d$, and we write, e.g., $\Delta_{ab} \succsim_d \Delta_{cd}$. This notation is rigorous but somewhat redundant, since we convey the idea that we are comparing intervals in respect of differences in two ways: through the symbols Δ_{ab} and Δ_{cd}, that imply we are focussing on differences, and through the symbol $\succsim_d$, that specifies that the weak order concerns intervals. Alternatively, the shorthand notations $ab \succsim_d cd$ or $\Delta_{ab} \succsim \Delta_{cd}$, can be used, which are not ambiguous, since they also make clear that we are comparing intervals in respect of differences.

Another important property of intervals is the empirical *ratio* between their extremes. We denote it by a/b; to avoid ambiguity, we use the solidus, /, to denote

[11] The notion of interval is one of the most fundamentals in measurement. This is why we dwell somewhat on it. The concept of interval, together with that—at another level of discourse—of probability, constitutes the pillars on which most of the theory presented in this book is constructed.

empirical ratio, whilst we use the horizontal line for the ratio between numbers, as in the expression $\frac{i}{j}$, where i, j are integers. Similarly to how we have done for differences, we denote a weak order for ratio by $a/b \succsim_r c/d$, or by $ab \succsim_r cd$ or $a/b \succsim c/d$, as shorthand notations. We will treat ratios in a later section on intensive structures.

Furthermore, note that since intervals, as well as the related empirical distances, differences or ratios are, mathematically, ordered pairs of objects, that is elements of the Cartesian product $A \times A$ or A^2, an order amongst intervals, or amongst distances, differences or ratios, can be viewed, equivalently, as a *binary* relation on $A \times A$ or as a *quaternary* relation on A.

Lastly, let us show how we can convey the idea that a difference is a signed distance and how we can fix the sign. We say that an interval is *null* if its extremes coincide, e.g. *aa* or *bb*, or if they are equivalent objects. So, for example, if $a \sim b$ then *ab* is a null interval. All null intervals are mutually equivalent. Similarly, we define null distances and null differences. How can we then express the idea that a distance is a "positive" and symmetric characteristic, without using its numerical representation, but only its formal properties? We simply say that $\Delta_{ab} \succsim \Delta_{aa}$ (positiveness) and $\Delta_{ab} \sim \Delta_{ba}$ (symmetry). How can we state that differences are "signed" characteristics instead? We first say that Δ_{ab} is positive if $\Delta_{ab} \succ \Delta_{aa}$; then, we require that $\Delta_{ab} \succ \Delta_{aa}$ implies $\Delta_{ba} \prec \Delta_{aa}$. Lastly, we have to specify whether Δ_{ab} is to be interpreted as the difference between a and b or vice versa: we choose the former, that is $\Delta_{ab} \succ \Delta_{aa}$ if $a \succ b$.

Now, we are ready to discuss the main ideas underlying difference measurement and interval scales.

3.4.2 Measuring Differences

For properly representing difference, we can imagine each object, in respect of the property under consideration, as a point placed in an appropriate position along a straight line, as in Fig. 3.2. The position of the points on the line expresses both the order and the distances amongst them.

How can we assign numbers (measures) to them in such a way that they properly express both the order and the distances? For doing so, we have to establish an equally spaced graduation along the line, as in Fig. 3.3; then, assuming that the position of each object correspond to one trait of the graduation, after fixing an *arbitrary*[12] "origin", we can number the traits in respect of it and associate to each object the number (measure) the corresponds to its position. Such numbers will properly represent distances and differences amongst the objects. In the case in the figure, if we fix the origin, e.g., in correspondence to c, we obtain $m(c) = 0, m(b) = 3$

---

[12] Note the origin is an arbitrary point since the line extends to infinity both to the left and to the right, and there is no special point on it that merits to be origin. Different will be the case for ratio scales, to be treated at a later stage.

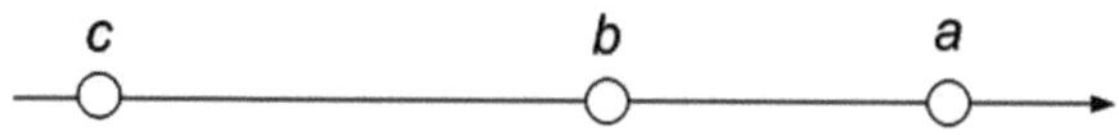

Fig. 3.2 Objects as points on a *straight line*

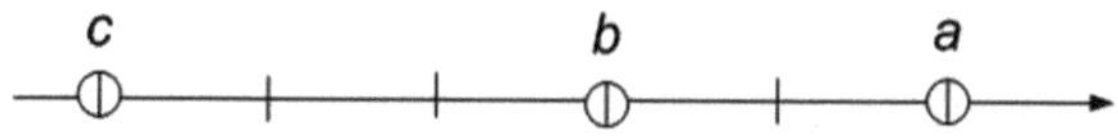

Fig. 3.3 Graduating the reference *straight line*

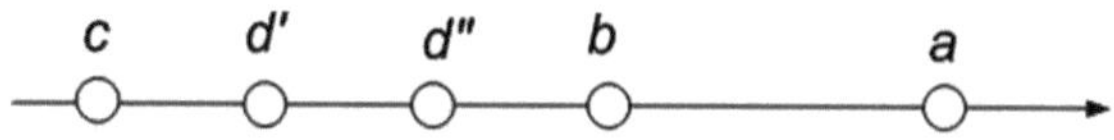

Fig. 3.4 Illustration of the solvability condition

and $m(a) = 5$. Such measures correctly express the order amongst the objects and also the fact that, say, interval bc is greater than ab.[13]

It is now important to understand, in an intuitive fashion first, then in precise formal terms, *what empirical properties must be satisfied for this representation to be possible.*

We have to first assume a weak order amongst objects and also a weak order amongst intervals, yet this is not enough. We have seen that we need to establish a graduation along the reference line: in empirical terms, this implies finding elements, in the set A of all the elements that carry the quantity under consideration, that correspond to each treat of the graduation. So, we have to assume that A *includes all the needed elements*. The empirical property that ensures that, is called a *solvability condition*, and can be illustrated as follows. Look at Fig. 3.2: since bc is greater than ab, we proceed towards establishing a graduation, by looking for two elements—call them d' and d''—such that both bd' and $d''c$ match ab, as shown in Fig. 3.4. *The solvability condition simply requires that whenever we have two not equivalent intervals, this is always possible.*

To complete the graduation, note that now, in Fig. 3.4, ab is no longer the smallest interval, since, e.g., bd''—or $d''d'$ or $d'c$, which are all equivalent to each other—is smaller. Then, we look now for two more elements—call them e' and e''—such that both ae' and $e''b$ match bd'', that must exist thanks to the solvability condition. This is shown in Fig. 3.5: this time e' coincides with e'' and the graduation is now complete.

Generally speaking, the solvability condition requires that if $\Delta_{bc} \succ \Delta_{ab}$ there always exist d' and d'' such that both $\Delta_{bd''} \sim \Delta_{ab}$ and $\Delta_{d'c} \sim \Delta_{ab}$ hold true.[14] This condition ensures that the afore procedure always works, for finite sets.

Once a graduation has been established, we have noted that we can assign a measure to the elements in A that correctly represent the order amongst the objects

---

[13] Note that if we fix the origin in any other point, the procedures still works.

[14] The term "solvability" suggests that, for $\Delta_{bc} \succ \Delta_{ab}$, the equations

- $\Delta_{bd''} \sim \Delta_{ab}$ and
- $\Delta_{d'c} \sim \Delta_{ab}$,

where d' and d'' are the unknowns, always have a solution.

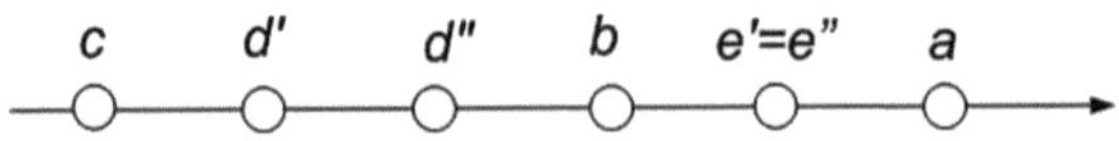

Fig. 3.5 Obtaining a equally space graduation of the reference axis, thanks to the solvability condition

Fig. 3.6 The monotonicity condition

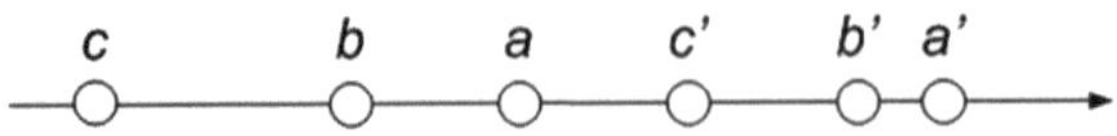

and distances and differences amongst them. But why does this construction work? Why, for instance, an interval, that includes three equally wide intervals should be considered greater than one that includes only two? The underlying assumption is that each interval can be obtained by assembling adjacent intervals, and that

- if we add to an interval another non-null interval, we obtain a greater interval and
- adding together equivalent intervals yields equivalent intervals, irrespective of where they are placed along the reference scale.

This property is called *weak monotonicity*[15] and is illustrated in Fig. 3.6. In terms of differences, it can be stated as follows: if $\Delta_{ab} \succsim \Delta_{a'b'}$ and $\Delta_{bc} \succsim \Delta_{b'c'}$, then also $\Delta_{ac} \succsim \Delta_{a'c'}$ holds true.

So, summarising, the main properties that we have used, are the following:

(i) a weak order for objects in A,
(ii) a weak order for intervals of elements of A,
(iii) a rule for composing intervals (weak monotonicity), and
(iv) a solvability condition.

For taking a further step towards the axiomatisation of difference systems, we can note that the weak order amongst objects can be deduced from the order amongst intervals. In fact, consider again Fig. 3.2: it is apparent that $a \succ b \succ c$, but how can we obtain that from the order amongst intervals? We see that c is the smallest element, since both $\Delta_{ac} \succ \Delta_{ca}$, and $\Delta_{bc} \succ \Delta_{cb}$, which means that c is the "switching point" between positive and negative intervals.[16] Thus, both $a \succ c$ and $b \succ c$ hold true. Then, starting from c, we obtain $\Delta_{ac} \succ \Delta_{bc}$ that naturally leads to $a \succ b$, which completes the ordering. Since the order amongst the objects can be deduced from the order on intervals, axiom (i) is not necessary and can be dropped.

---

[15] In fact, the term "monotonicity" suggests that adding equivalent intervals to two intervals does not change the order that exists between them. A monotonic transformation, in general, is one that does not alter order.

[16] Remember the discussion about the sign of empirical differences, at the end of the previous section.

3.4.3 Representation for Difference Structures

Let us now formalise the notion of difference structure.

Definition 3.12 (*Difference structure*) Let A be a finite (not empty) set of objects carrying the property x. Let $\succsim_d$ be a binary relation on $A \times A$ that satisfies, for each $a, b, c, d, a', b', c', d' \in A$, the following axioms:

12.1 $\succsim_d$ is a weak order;
12.2 if $\Delta_{ab} \succsim_d \Delta_{cd}$, then $\Delta_{dc} \succsim_d \Delta_{ba}$;
12.3 if $\Delta_{ab} \succsim_d \Delta_{a'b'}$ and $\Delta_{bc} \succsim_d \Delta_{b'c'}$, then $\Delta_{ac} \succsim_d \Delta_{a'c'}$;
12.4 if $\Delta_{ab} \succsim_d \Delta_{cd} \succsim_d \Delta_{aa}$, then there exists $d', d'' \in A$ so that $\Delta_{ad'} \sim_d \Delta_{cd} \sim_d \Delta_{d''b}$.

Then, $(A, \succsim_d)$ is a (finite empirical) difference structure.

From the discussion in the previous section, we are now able to understand the rationale of the axioms. Axiom 12.1 assumes an order on intervals: we have seen that the order amongst objects can be deduced from it. Axiom 12.2 defines a sign for differences. Axiom 12.3 concerns weak monotonicity, which constitutes the basic concatenation rule for intervals. Axiom 12.4 expresses the solvability condition.

Let us now formalise the deduction of the order amongst objects from the order amongst intervals, through the following Lemma.[17]

Lemma 3.13 *Let $(A, \succsim_d)$ be a difference structure and let $\succsim$ be a binary relation on A, such that for each $a, b, c \in A$, $a \succsim b \Leftrightarrow \Delta_{ac} \succsim_d \Delta_{bc}$. Then $\succsim$ is a weak order.*

Proof We have to show that:

1. $\forall a, b \in A$, either $a \succsim b$ or $b \succsim a$;
2. $\forall a, b, c \in A$, if $a \succsim b$ and $b \succsim c$, then $a \succsim c$.

Indeed, $\forall a, b \in A$, let $c \in A$. Then, since $\succsim_d$ is a weak order, either $\Delta_{ac} \succsim_d \Delta_{bc}$ or $\Delta_{bc} \succsim_d \Delta_{ac}$. Consequently, either $a \succsim b$ or $b \succsim a$, as needed.

Furthermore, $\forall a, b, c \in A$, if $a \succsim b$ and $b \succsim c$, for any $d \in A$, $\Delta_{ad} \succsim_d \Delta_{bd}$ and $\Delta_{bd} \succsim_d \Delta_{cd}$. But since $\succsim_d$ is a weak order, also $\Delta_{ad} \succsim_d \Delta_{cd}$ holds true and, consequently, $a \succsim c$, which completes the proof. □

We will occasionally denote concatenation (or, informally, addition) of adjacent intervals or of differences of adjacent intervals, with the symbol "$\circ$", that is we will write, e.g., $ac = ab \circ bc$, or $\Delta_{ac} = \Delta_{ab} \circ \Delta_{bc}$.

Let us now outline the basic properties of this concatenation.

Informally, we will see that

- all null intervals are mutually equivalent,

[17] Proofs in this section, as well as in following of this chapter, are somewhat technical and can be omitted in a first reading, without substantial loss of continuity.

- adding equivalent intervals to equivalent intervals still results into equivalent intervals,
- adding a null interval does not change the "width" of an interval, whilst
- adding a positive interval increases it.

These properties can be formally stated as follows.

Lemma 3.14 *Let* $(A, \succsim_d)$ *be a difference structure. Then,*

- $\Delta_{aa} \sim_d \Delta_{bb}$;
- *if* $\Delta_{ab} \sim_d \Delta_{a'b'}$ *and* $\Delta_{bc} \sim_d \Delta_{b'c'}$, *then* $\Delta_{ac} \sim_d \Delta_{a'c'}$;
- *if* $\Delta_{bc} \sim_d \Delta_{bb}$, *then* $\Delta_{ac} = \Delta_{ab} \circ \Delta_{bc} \sim_d \Delta_{ab}$;
- *if* $\Delta_{bc} \succ_d \Delta_{bb}$, *then* $\Delta_{ac} = \Delta_{ab} \circ \Delta_{bc} \succ_d \Delta_{ab}$.

Proof For proving the first statement, note that either $\Delta_{aa} \succsim_d \Delta_{bb}$ or $\Delta_{bb} \succsim_d \Delta_{aa}$. If $\Delta_{aa} \succsim_d \Delta_{bb}$, then, for Axiom 12.2, $\Delta_{bb} \succsim_d \Delta_{aa}$, which implies $\Delta_{aa} \sim_d \Delta_{bb}$. Conversely, if $\Delta_{bb} \succsim_d \Delta_{aa}$, then $\Delta_{aa} \succsim_d \Delta_{bb}$, which again implies $\Delta_{aa} \sim_d \Delta_{bb}$.

Concerning the second statement, note that

$$\Delta_{ab} \sim_d \Delta_{a'b'} \Leftrightarrow \Delta_{ab} \succsim_d \Delta_{a'b'} \text{ and } \Delta_{a'b'} \succsim_d \Delta_{ab},$$
$$\Delta_{bc} \sim_d \Delta_{b'c'} \Leftrightarrow \Delta_{bc} \succsim_d \Delta_{b'c'} \text{ and } \Delta_{b'c'} \succsim_d \Delta_{bc}.$$

Then, for Axiom 12.3,

$$\Delta_{ac} \succsim_d \Delta_{a'c'} \text{ and } \Delta_{a'c'} \succsim_d \Delta_{a'c'},$$

which implies that $\Delta_{ac} \sim_d \Delta_{a'c'}$.

For the third statement, note that, trivially, $\Delta_{ab} \sim_d \Delta_{ab}$, so, if also $\Delta_{bc} \sim_d \Delta_{bb}$, for the statement just proved, $\Delta_{ac} \sim_d \Delta_{ab}$.

Finally, since $\Delta_{ab} \sim_d \Delta_{ab}$, if $\Delta_{bc} \succ_d \Delta_{bb}$, then for Axiom 12.3, $\Delta_{ac} \succsim_d \Delta_{ab}$. But $\Delta_{ac} \sim_d \Delta_{ab}$ cannot hold, since if it held, we would have $\Delta_{ab} \circ_d \Delta_{bc} = \Delta_{ac} \sim_d \Delta_{ab} = \Delta_{ab} \circ \Delta_{bb}$, which implies $\Delta_{bc} \sim_d \Delta_{bb}$, which contradicts the hypothesis. Thus, we conclude that $\Delta_{ac} \succ_d \Delta_{ab}$. □

Now, we are ready to approach the representation theorem. We will follow the same pattern as for ordinal scales and we will construct a series of standards S. This is possible since a difference structure includes a weak order amongst the objects. The point here is that in such a scale the differences between any two pairs of consecutive elements are equivalent: we then say that S is an *equally spaced* (for differences) series of standards. We thus introduce the following.

Definition 3.15 Let $(A, \succsim_d)$ be a difference structure and let $\succsim$ be the weak order induced on A by $\succsim_d$. Define $I_0 = \{0, 1, \ldots, n\}$ and $I'_0 = \{0, 1, \ldots, n-1\}$, and let $S = \{s_0, s_1, \ldots s_n\}$ be a series of standards for $(A, \succsim)$, as in Definition 3.10. Then, we say that S is equally spaced for differences if for each $i \in I'_0$, $\Delta_{s_{i+1}s_i} \sim_d \Delta_{s_1 s_0}$.

An intuitive property of an equally spaced series of standards is that any set of consecutive elements along the series defines an interval that is equivalent to the interval formed by any other set including the same number of consecutive elements, placed everywhere along the series. This occurs thanks to the weak monotonicity axiom, which we have informally discussed in the previous section. We thus now state this formally and prove it.

Lemma 3.16 *Let S be a series of standards equally spaced for differences. Then for* $i, j, k, l \in I_0$, $i \geq j$ *and* $k \geq l$,

$$\Delta_{s_i s_j} \succsim_d \Delta_{s_k s_l} \Leftrightarrow i - j \geq k - l.$$

Proof Let $i - j = p \geq 0$ and $k - l = q \geq 0$.
Then, the thesis can be rewritten as

$$\Delta_{s_{j+p} s_j} \succsim_d \Delta_{s_{l+q} s_l} \Leftrightarrow p \geq q.$$

We prove this statement by bi-argumental induction: we first fix $q = 0$ and prove it for every p, then we prove that it also holds true for every q.[18]

1. Let $q = 0$. If also $p = 0$, we obtain $\Delta_{s_{j+p} s_j} = \Delta_{s_j s_j}$ and $\Delta_{s_{l+q} s_l} = \Delta_{s_l s_l}$, which satisfies the right-to-left implication; the left-to-right implication is also, trivially, satisfied. We proceed by induction on p, by showing that, if the statement is true for p, it is also true for $p + 1$. Indeed, if $\Delta_{s_{j+p} s_j} \succsim_d \Delta_{s_l s_l} \Leftrightarrow p \geq q$, we also obtain $\Delta_{s_{j+p+1} s_j} \succ_d \Delta_{s_{j+p} s_j} \succsim_d \Delta_{s_l s_l}$ and $(p + 1) > p \geq q$.
2. Since we have just proved that the statement holds for $q = 0$, we have just to prove that if it holds for q, then it also holds for $q + 1$. So, assume that $\Delta_{s_{j+p} s_j} \succsim_d \Delta_{s_{l+q} s_l} \Leftrightarrow p \geq q$. We have three possible cases: $p > q$, $p = q$ and $p < q$.
   (a) If $p > q$ we have $\Delta_{s_{j+p} s_j} \succ_d \Delta_{s_{l+q} s_l} \Leftrightarrow p > q$. Then, $\Delta_{s_{l+q+1} s_l} \succ_d \Delta_{s_{l+q} s_l} \precsim_d \Delta_{s_{j+p} s_j}$ and $(q + 1) > q \leq p$, which is in agreement with our statement.
   (b) If $p = q$ we have $\Delta_{s_{j+p} s_j} \sim_d \Delta_{s_{l+q} s_l} \Leftrightarrow p = q$. Then, $\Delta_{s_{l+q+1} s_l} \succ_d \Delta_{s_{l+q} s_l} \sim_d \Delta_{s_{j+p} s_j}$ and $(q + 1) > q = p$, which is still in agreement with our statement, which implies $\Delta_{s_{j+p} s_j} \prec_d \Delta_{s_{l+q} s_l} \Leftrightarrow p < q$.
   (c) If $p < q$ we have $\Delta_{s_{j+p} s_j} \prec_d \Delta_{s_{l+q} s_l} \Leftrightarrow p < q$. Then, $\Delta_{s_{l+q+1} s_l} \succ_d \Delta_{s_{l+q} s_l} \succ_d \Delta_{s_{j+p} s_j}$ and $(q + 1) > q > p$, which is still in agreement with our statement, which implies $\Delta_{s_{j+p} s_j} \prec_d \Delta_{s_{l+q} s_l} \Leftrightarrow p < q$.

[18] The induction principle is a well-established argument in mathematics [7]. It states that if ϕ is some proposition predicated of i where i is a natural number, and if:

- $\phi(0)$ holds true and
- whenever $\phi(j)$ is true, then $\phi(j + 1)$ is also true

then $\phi(i)$ hold true for all i. Bi-argumental induction is the same as simple induction, but concerns statements predicated for a couple of argument, $\phi(i, k)$.

This completes the induction with respect to q and the proof. □

We are now finally ready to formulate and proof the representation theorem.

Theorem 3.17 (Representation for difference structures) *Let A be a finite (not empty) set of objects carrying the property x and let $(A, \succsim_d)$ be a difference structure. Then, there exists a function $m: A \rightarrow \mathbb{R}$, called a measure function, such that, for each $a, b, c, d \in A$,*

$$\Delta_{ab} \succsim_d \Delta_{cd} \Leftrightarrow m(a) - m(b) \geq m(c) - m(d).$$

Proof The difference structure $(A, \succsim_d)$ also includes a weak order $\succsim$ amongst the elements of A, as previously proved. So let S be the series of standards defined according to the weak order amongst the elements. We prove now that such a series is equally spaced with respect to differences. Consider, for $i \in I_0'$, the interval $\Delta_{s_{i+1}s_i}$ and let us compare it with $\Delta_{s_1 s_0}$. We find $\Delta_{s_{i+1}s_i} \sim_d \Delta_{s_1 s_0}$. In fact, suppose we had $\Delta_{s_{i+1}s_i} \succ_d \Delta_{s_1 s_0}$. Then, for Axiom 12.4, there should exist s_i' such that $\Delta_{s_i' s_i} \sim_d \Delta_{s_1 s_0}$. But in this case, s_i' would be intermediate between s_i and s_{i+1}, which is impossible for the way we have constructed the series of standards S. Analogously, $\Delta_{s_{i+1}s_i} \prec_d \Delta_{s_1 s_0}$ would imply the existence of an element s_1' intermediate between s_0 and s_1, which is also impossible. So we conclude that for $i \in l'$, $\Delta_{s_{i+1}s_i} \sim_d \Delta_{s_1 s_0}$, i.e., the series of standards is *equally spaced* according to Definition 3.15.

Then, Lemma 3.16 applies and so, for $i, j, k, l \in I_0$, $i \geq j$, $k \geq l$, $\Delta_{s_i s_j} \succsim_d \Delta_{s_k s_l} \Leftrightarrow i - j \geq l - m$. On the other hand, for each $a, b, c, d \in A$, there are $s_i, s_j, s_k, s_l \in S$ such that $a \sim s_i$, $b \sim s_j$, $c \sim s_k$, $d \sim s_l$.

If we now assign $m(a) = m(s_i), m(b) = m(s_j), m(c) = m(s_k)$ and $m(d) = m(s_l)$, we finally obtain

$$\begin{aligned} &\Delta_{ab} \succsim_d \Delta_{cd} \Leftrightarrow \Delta_{s_i s_j} \succsim_d \Delta_{s_k s_l} \\ &\Leftrightarrow i - j \geq k - l \Leftrightarrow m(a) - m(b) \geq m(c) - m(d), \end{aligned}$$

which completes the proof. □

Note that the measure function also allows a representation of the order-amongst-objects relation, as per Theorem 3.8, since a difference structure also includes an order structure, as shown in Lemma 3.13.

Lastly, let us consider uniqueness.

Theorem 3.18 (Uniqueness for difference structures) *Let A be a finite (not empty) set of objects carrying the property x, let $(A, \succsim_d)$ be a difference structure and $m: A \rightarrow \mathbb{R}$ a measure function for it. Then, any other measure function m' is such that, for each $a \in A$,*

$$m'(a) = \alpha m(a) + \beta,$$

with $\alpha > 0$.

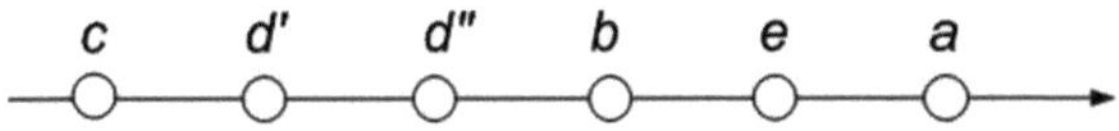

Fig. 3.7 An example of difference structure

Proof We have first to prove that, if m is a proper measure function, also $m'(a) = \alpha m(a) + \beta$, with $\alpha > 0$, is appropriate. In fact, if $m(a) - m(b) \geq m(c) - m(d)$, then also $[\alpha m(a) + \beta] - [\alpha m(b) + \beta] \geq [\alpha m(c) + \beta] - [\alpha m(d) + \beta]$ holds true.

Then, we have to prove that if both m and m' satisfy the representation, then they must be related by $m' = \alpha m + \beta$. Let $S = \{s_0, \ldots, s_n\}$ be a series of standards for the structure under consideration and m any proper measure function for it. Since S is equally spaced for differences, the difference $m(s_i) - m(s_{i-1})$ must be a positive constant for each i, $1 \leq i < n$, call it $d > 0$. Then, for each i, $m(s_i) = m(s_0) + id$. Similarly, if m' is another valid measure, we will also have $m'(s_i) = m'(s_0) + id'$, where $d' = m'(s_i) - m'(s_{i-1})$. Now for each $a \in A$, there will be a standard $s_i \in S$ such that $a \sim s_i$. Then, $m(a) = m(s_i) = m(s_0) + id$ and also $m'(a) = m'(s_i) = m'(s_0) + id'$. Thus,

$$m'(a) = \frac{d'}{d}m(a) + m'(s_0) - \frac{d'}{d}m(s_0) = \alpha m(a) + \beta,$$

where $\alpha = \frac{d'}{d} > 0$. This completes the proof. □

Lastly, let us practice applying the afore theory to the structure represented in Fig. 3.5, with just one minor modification: we consider a single element e instead of the two equivalent objects, e' and e''. This is simply for slightly reducing the number of elements in A, which simplifies things; we obtain the situation in Fig. 3.7.

The set of objects is thus $A = \{a, b, c, d', d'', e\}$, which includes $N = 6$ objects. The total number of intervals is thus $N \times N = 36$; yet, we can omit from our considerations null intervals, such as $aa, bb, \ldots$; there are 6 of them. Furthermore, once we have considered, say, ab, we know that ba is its reversed-sign version, and its properties are easily deduced from those of ab. Excluding such reverse intervals, we half the number of the remaining ones, and thus, we have to consider just $(N \times N - N)/2 = 15$ intervals in total. For listing them systematically, we can use the alphabetical order. In this way, we obtain the following list:

$$ab, ac, ad', ad'', ae, bc, bd', bd'', be, cd', cd'', ce, d'd'', d'e, d''e.$$

Yet, here, there are "negative" intervals also, such as, be or cd'. We prefer to have positive intervals only, so we reverse the negative ones; we obtain

$$ab, ac, ad', ad'', ae, bc, bd', bd'', eb, d'c, d''c, ec, d''d', ed', ed''.$$

Now, we have the (most significant) intervals. For completing the description of this structure, we need to specify the weak order amongst them. This can be done by a matrix, as in Table 3.3 before, and is shown in Table 3.4.

Table 3.4 An example of difference structure, corresponding to Fig. 3.7

	ab	*ac*	*ad′*	*ad″*	*ae*	*bc*	*bd′*	*bd″*	*eb*	*d′c*	*d″c*	*ec*	*d″d′*	*ed′*	*ed″*
ab		$\prec$	$\prec$	$\prec$	$\succ$	$\prec$	$\sim$	$\succ$	$\succ$	$\succ$	$\sim$	$\prec$	$\succ$	$\prec$	$\sim$
ac	$\succ$		$\succ$	$\succ$	$\succ$	$\succ$	$\succ$	$\succ$	$\succ$	$\succ$	$\succ$	$\succ$	$\succ$	$\succ$	$\succ$
ad′	$\succ$	$\prec$		$\succ$	$\succ$	$\succ$	$\succ$	$\succ$	$\succ$	$\succ$	$\succ$	$\sim$	$\succ$	$\succ$	$\succ$
ad″	$\succ$	$\prec$	$\prec$		$\succ$	$\sim$	$\succ$	$\succ$	$\succ$	$\succ$	$\succ$	$\prec$	$\succ$	$\sim$	$\succ$
ae	$\prec$	$\prec$	$\prec$	$\prec$		$\prec$	$\prec$	$\sim$	$\sim$	$\sim$	$\prec$	$\prec$	$\sim$	$\prec$	$\prec$
bc	$\succ$	$\prec$	$\prec$	$\sim$	$\succ$		$\succ$	$\succ$	$\succ$	$\succ$	$\succ$	$\prec$	$\succ$	$\sim$	$\succ$
bd′	$\sim$	$\prec$	$\prec$	$\prec$	$\succ$	$\prec$		$\succ$	$\succ$	$\succ$	$\sim$	$\prec$	$\succ$	$\prec$	$\sim$
bd″	$\prec$	$\prec$	$\prec$	$\prec$	$\sim$	$\prec$	$\prec$		$\sim$	$\sim$	$\prec$	$\prec$	$\sim$	$\prec$	$\prec$
eb	$\prec$	$\prec$	$\prec$	$\prec$	$\sim$	$\prec$	$\prec$	$\sim$		$\sim$	$\prec$	$\prec$	$\sim$	$\prec$	$\prec$
d′c	$\prec$	$\prec$	$\prec$	$\prec$	$\sim$	$\prec$	$\prec$	$\sim$	$\sim$		$\prec$	$\prec$	$\sim$	$\prec$	$\prec$
d″c	$\sim$	$\prec$	$\prec$	$\prec$	$\succ$	$\prec$	$\sim$	$\succ$	$\succ$	$\succ$		$\prec$	$\succ$	$\prec$	$\sim$
ec	$\succ$	$\prec$	$\sim$	$\succ$	$\succ$	$\succ$	$\succ$	$\succ$	$\succ$	$\succ$	$\succ$		$\succ$	$\succ$	$\succ$
d″d′	$\prec$	$\prec$	$\prec$	$\prec$	$\sim$	$\prec$	$\prec$	$\sim$	$\sim$	$\sim$	$\prec$	$\prec$		$\prec$	$\prec$
ed′	$\succ$	$\prec$	$\prec$	$\sim$	$\succ$	$\sim$	$\succ$	$\succ$	$\succ$	$\succ$	$\succ$	$\prec$	$\succ$		$\succ$
ed″	$\sim$	$\prec$	$\prec$	$\prec$	$\succ$	$\prec$	$\sim$	$\succ$	$\succ$	$\succ$	$\sim$	$\prec$	$\succ$	$\prec$	

The specification of the weak order amongst the intervals completely specifies the structure. The reader is invited to practice checking that the axioms for the representation are satisfied. From the order amongst the intervals, we can deduce the order amongst the objects. We note that all the intervals that have c as second extreme are greater that all other intervals having the same first element, e.g., $ac \succ ab$, $bc \succ bd'$ and so forth. So, c is the minimum element in A. If we now consider all the intervals having c as the second extreme, we observe that $ac \succ ec \succ bc \succ d''c \succ d'c$, and thus, it is easy to conclude that the ordering of the elements of A is simply: $a \succ e \succ b \succ d'' \succ d' \succ c$. Furthermore, we note that this series is equally spaced, since $ae \sim eb \sim bd'' \sim d''d' \sim d'c$ and thus it forms a series of standard, in this very simple example including all the elements of A. If we consider instead the original example in Fig. 3.5, we have to choose whether including e' or e'' in the series, since both choices are correct.

Then, a proper measure function is, e.g.,[19]

$$m = \{(c, 0), (d', 1), (d'', 2), (b, 3), (e, 4), (a, 5)\}.$$

Another appropriate measure function would be, e.g.,

$$m' = \{(c, 1), (d', 2), (d'', 3), (b, 4), (e, 5), (a, 6)\}.$$

3.5 Ratio Scales for Intensive Structures

3.5.1 Is Empirical Addition Necessary for Establishing a Ratio Scale?

After interval scales, ratio scales come. They allow meaningful statements on ratios to be made. We can say that a building is twice higher than another or that the mass of the pollutants produced by a plant this year is 20 % lower than last year. Both these quantities, length and mass, are empirically additive: we can increase the height of a building by adding one or more flats, and the amount of pollutants in a year is the sum of those produced in each month. But what about temperature or loudness or pain intensity? Thermodynamics ensures that saying that an object at 60 K is twice hotter than one at 30 K makes sense, since there is a unique zero-kelvin condition, that can be properly defined as a limiting state, although not really reachable. Yet, there is in general no way of assembling two objects at 30 K, in order to obtain one at 60 K.

---

[19] Remember that a generic function $f: X \to Y$, where X and Y are finite sets, can be defined by listing, in a set, all the pairs (x, y) that satisfy it, that is, $f = \{(x, y) | x \in X, y \in Y, y = f(x)\}$.

Fig. 3.8 Geometrical representation of an extensive and an intensive quantity by a segment, *S*, and a point, *P*, respectively

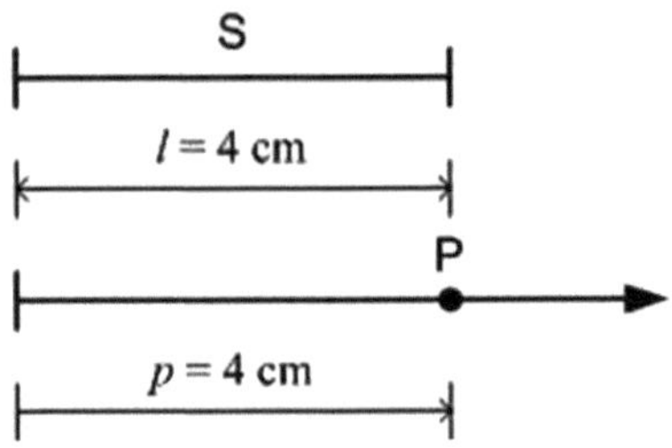

Psychophysics also ensures that it makes sense to say that loudness in an airport is 50 % higher than in an office, although it is unlikely that two such offices can be assembled in order to match the loudness of the airport.

Lack of additivity was the main argument against the measurability of perceptual properties raised in the Committee of the British Association, as we have seen in Chap. 1. "I submit that any law purporting to express a quantitative relation between sensation intensity and stimulus intensity is not merely false but it is in fact meaningless unless and until a meaning can be given to the concept of addition as applied to sensation"—Guild, one of the members of the Committee writes in the final Report [8]. Is he right? Is addition really needed, if not to measure at all, at least to attain a ratio scale?

Stevens in his classification of measurement scales mentioned the possibility of empirically assessing *both equality of differences and equality of ratio* as a way for attaining a ratio scale, even in the absence of an empirical addition property.

This idea, whose importance was somewhat underestimated in my opinion, is instead very fruitful, and we will pursue it in the following. In this way, we will be able to properly define and measure intensive quantities.

3.5.2 Extensive and Intensive Quantities

In his "Critique of pure reason", Immanuel Kant (1724–1804) writes: "I call extensive that quantity in which the representation of parts enables the representation of the whole", whilst "A quantity which is understood as a unity and in which the multiplicity may be represented only as proximity to its negation =0, I call an intensive quantity" [9]. In other words, an extensive quantity may be thought as a sum of parts, whilst an intensive as the grade of a sensation. Both can be represented by numbers, but the meaning of such numbers is different. In the former case, the number expresses the multiplicity of the parts, in the latter the multiplicity of the grades that separate the present state to the zero, that is to the absence of sensation.

A geometrical illustration of these ideas is presented in Fig. 3.8. (A single realisation of) an extensive quantity can be geometrically represented as a segment, *S*; an intensive one by a point, *P*, on an oriented semi-straight line.

The associated geometrical features are length, l, and position, p: they are both expressed in metres, but their meaning is noteworthy different. Even in physics, it is possible to distinguish between extensive and intensive quantities. The former are closely related to the space–time extension of bodies, the latter are not. For example, the mass of an homogeneous body is proportional to its space extension (its volume), whilst its density is independent of it.

Let us now come back to the problem of how to measure an intensive quantity on a ratio scale. The classical answer, provided by Campbell [10], is based on the distinction between *fundamental* and *derived* quantities: non-additive intensive quantities, such as density, can be measured only indirectly, as derived quantities.

Stevens provided two key contributions to this subject. Firstly, he introduced, as we know, the magnitude estimation method, for directly measuring the intensity of a sensation [11]. Secondly, he indicated "equality of differences" together with "equality of ratios" as distinctive empirical properties that allow measurement on a ratio scale [12]. Each of these two contributions gave rise to a distinctive line of research.

Magnitude estimation has been studied thenceforth both experimentally and theoretically. Axiomatisations have been attempted [3]; Narens provided a conspicuous contribution [13], developing an approach that has been checked experimentally [14]. This research line also includes the investigation of conditions for having persons to reliably perform as "measuring instruments" [15].

Even more relevant for our purposes is the second line of research, out-springing from Stevens's claim that the empirical assessment of *both* ratios and differences can yield a ratio scale, even when there is no empirical addition. Such studies have led to the axiomatisation of ratio/difference representations [6, 16, 17], which have also been studied experimentally, in psychophysics, to some extent [18, 19].

We also follow this line of though in developing a representation for intensive structures [20].

3.5.3 Scaling Intensities

We want now to understand how a ratio scale can be attained when intervals can be ordered both in terms of differences and of ratios between their extremes. We should first note that both differences and ratios can be represented on an interval scale and the corresponding representation theorems are, for differences:

$$\Delta_{ab} \succsim \Delta_{cd} \Leftrightarrow m(a) - m(b) \geq m(c) - m(d),$$

and for ratios:

$$a/b \succsim c/d \Leftrightarrow \frac{m(a)}{m(b)} \geq \frac{m(c)}{m(d)}.$$

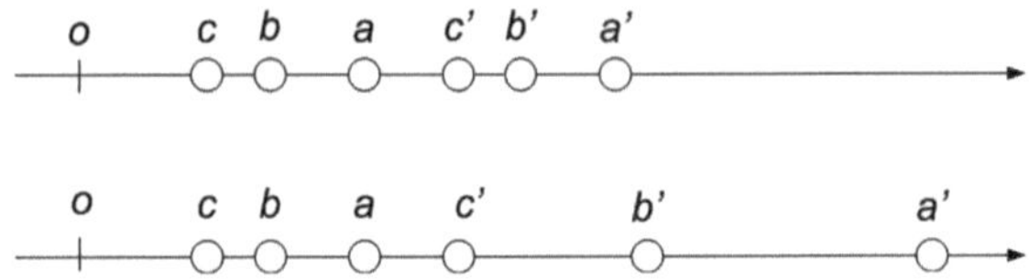

Fig. 3.9 Concatenation of intervals in respect of differences (*upper part*) and of ratios (*lower part*)

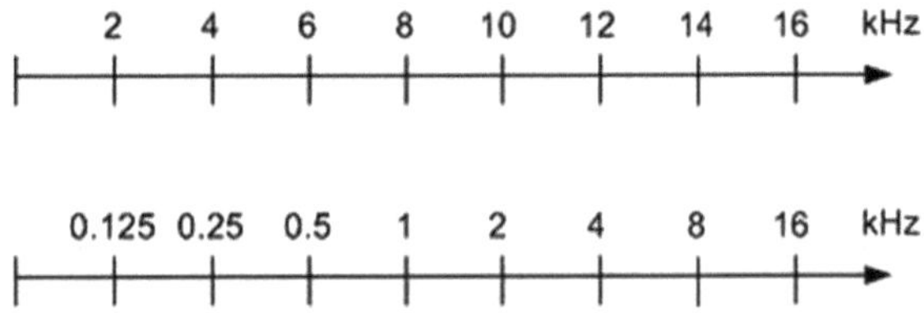

Fig. 3.10 Linear and logarithmic frequency scales

This is possible since the formal properties for both representations are the same [6]. Remember that for differences, they include (1) order amongst intervals, (2) weak monotonicity, and (3) a solvability condition. The key property is monotonicity, which basically implies that concatenation of adjacent intervals does not depend on where they are placed along a reference axis. Interestingly enough, this property can be stated in a formally equivalent way and do make sense *both for differences and for ratios*. For better understanding it, let us formulate it in terms of equivalences, rather than in terms of weak order as we have done in the previous section. For differences, it reads:

if $\Delta_{ab} \sim \Delta_{a'b'}$ and $\Delta_{bc} \sim \Delta_{b'c'}$, then also $\Delta_{ac} \sim \Delta_{a'c'}$,

whilst for ratios it becomes:

if $a/b \sim a'/b'$ and $b/c \sim b'/c'$, then also $a/c \sim a'/c'$.

These two properties are illustrated in Fig. 3.9.

In the upper part of the figure, the difference between the extremes of ab, Δ_{ab}, is equivalent to the difference between those of $a'b'$, $\Delta_{a'b'}$, and the same happens for Δ_{bc} and $\Delta_{b'c'}$. This ensures that the differences Δ_{ac} and $\Delta_{c'c'}$ are equivalent as well.

In a similar way, in the lower part of the figure, an alternative interpretation of the same monotonicity principle is offered in terms of ratios. Here, the ratio of the extremes of ab, a/b, is now equivalent to that of $a'b'$, a'/b', and the same holds true for bc and $b'c'$. Hence, the concatenation ac of ab and bc is equivalent, in respect of ratio, to the concatenation $a'c'$ of $a'b'$ and $b'c'$. A good example of these two approaches is frequency scales for spectrum measurement (Fig. 3.10).

In such a measurement, it is possible to consider either a

- Constant-resolution spectrum, where spectral lines represent power associated to equal frequency intervals and the width of such intervals constitutes a measure of the resolution of the analyser, or

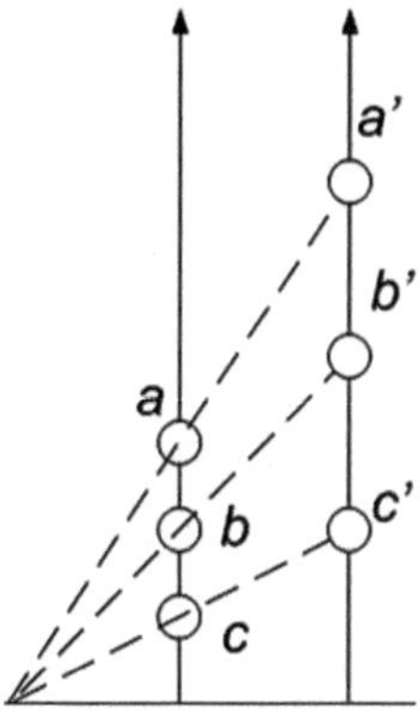

Fig. 3.11 The compatibility condition implies that the scaling of intervals does not affect the ordering of differences

- Constant-relative-resolution spectrum, where spectral lines represent power associated to frequency intervals, where the ratio of the extremes of each interval equals (a fraction of) an octave.[20]

In the figure, it appears how the concatenation of adjacent intervals can be done either according to differences or ratios.

Yet, for attaining a ratio scale, it is not sufficient that differences and ratio can be assessed independently. They also have to satisfy two compatibility conditions. The first is quite obvious: the ordering of elements induced by the ordering of differences is consistent with that induced by ratios. This means that if, e.g., the difference between a and b is positive,[21] their ratio is also "greater than one".

The second is, instead, substantial and it is *what really characterises an intensive structure*. It requires that "scaling" of intervals does not affect their difference ordering. In numerical terms, this property corresponds to the following. If x, y, z are numbers, and if $x - y \geq y - z$ then, e.g., $2x - 2y \geq 2y - 2z$. If now we set $x' = 2x$, $y' = 2y$ and $z' = 2z$, we obtain that if $\frac{x'}{x} = \frac{y'}{y} = \frac{z'}{z}$, then $x' - y' \geq y' - z'$. Stated in terms of empirical relations, this becomes that if $a'/a \sim b'/b \sim c'/c$, then $\Delta_{ab} \succsim \Delta_{bc}$ implies $\Delta_{a'b'} \succsim \Delta_{b'c'}$. This is illustrated in Fig. 3.11. Note that if b is mid-point between a and c, that is $\Delta_{ab} \sim \Delta_{bc}$, then also b' will be mid-point between a' and c', i.e., $\Delta_{a'b'} \sim \Delta_{b'c'}$.

With this additional property, it is possible to obtain a representation over a continuous support [17]. For finite structures, which are particularly interesting for us,[22] things are somewhat different. Let us consider them in some detail.

We proceed as follows. As far as differences are involved, we regard the minimum element in the set as the "zero" or the "origin" of the scale. So, we denote the standard

---

[20] The term "octave" (Latin: *octavus*, eighth) comes from musical acoustics and denotes the interval between two sounds whose pitch is one the double of the other. This happens when the second sound is the eighth in a musical (heptatonic) scale starting from the first.

[21] Remember the way we have defined positive differences in Sect. 3.4.1.

[22] On finiteness, remember discussion in Sect. 3.1.

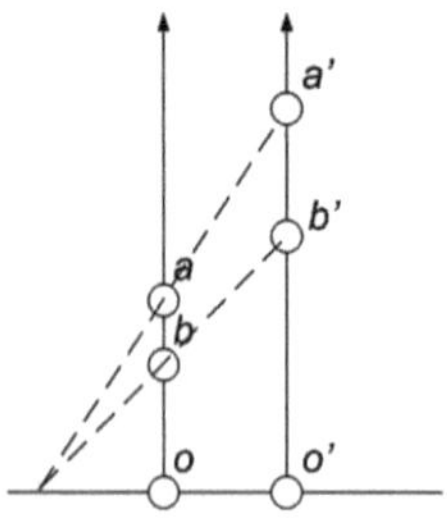

Fig. 3.12 The compatibility condition for finite structures

series as $S = \{s_0, s_1 \ldots s_n\}$. On the other hand, we exclude this initial element from that standard series for ratios. Then, we require that the ordering of elements induced by the ordering of differences is the same as the one induced by ratios (first compatibility condition). With these two positions, *the series of standards* for ratio will be simply $S' = \{s_1 \ldots s_n\}$, that is it *will coincide with the one for differences, apart from the exclusion of the origin.*

The second compatibility condition for finite structures can be stated as follows:

if $a \sim s_i$, $b \sim s_j$, $a' \sim s_{k \cdot i}$, $b' \sim s_{k \cdot j}$, then $a/b \sim_r a'/b'$.

This idea is illustrated in Fig. 3.12.

Here, we have $a \sim s_3$, $b \sim s_2$, $a' \sim s_6$ and $b' \sim s_4$. Then, the ratio a/b is the same as the ratio a'/b'.

The way for obtaining the representation can also be explained by geometrical considerations. Suppose that we want to compare the ratio a/b with the ratio c/d. For doing so, we map b and d in the same point, $b' \sim d'$, and then, we map a into a' and c into c', in such a way that the original ratios are maintained, that is $a/b \sim a'/b'$ and $c/d \sim c'/d'$. Then, we compare the differences $\Delta_{a'b'}$ and $\Delta_{c'd'}$. If, for example, $\Delta_{a'b'} \succ \Delta_{c'd'}$, we conclude that also $a/b \succ c/d$. If this happens, it can be proved that the measures assigned to the objects provide a representation both for differences and for ratios. This procedure is illustrated in Fig. 3.13.

We are now ready to formulate and prove a representation theorem [20].

3.5.4 Representation for Intensive Structures

Let us start with a formal definition of an intensive structure.

Definition 3.19 (*[Finite] intensive structure*) Let A be a (not empty) set of objects carrying the property x. Let $(A, \succsim_d, \succsim_r)$ be such that $(A, \succsim_d)$ is a difference structure and let $S = \{s_0, s_1 \ldots s_n\}$ be a series of standards for it. Let $S_0 = \{a \in A | a \sim s_0\}$ and $B = A - S_0$. Let the following axioms hold true:

19.1 $\succsim_r$ is a weak order on $B \times B$;
19.2 for each $a, b, c \in B$, $\Delta_{ab} \succsim_d \Delta_{cb} \Leftrightarrow a/b \succsim_r c/b$;

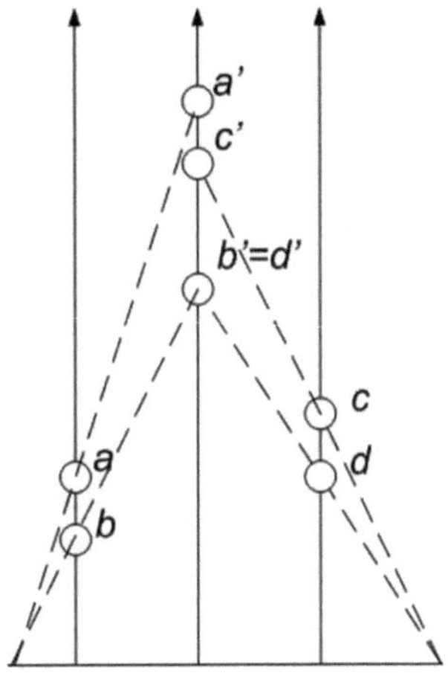

Fig. 3.13 How to obtain a representation

19.3 for $a, b, a', c' \in B$ and $i, j, k, l, k \cdot i, k \cdot j \leq n$, if $a \sim s_i$, $b \sim s_j$, $a' \sim s_{k \cdot i}$ and $b' \sim s_{k \cdot j}$, then $a/b \sim a'/b'$.

Then, we say that $(A, \succcurlyeq_d, \succcurlyeq_r)$ is an (finite empirical) intensive structure.

We thus obtain the following representation.

Theorem 3.20 (Representation for intensive structures) *Let A be a finite (not empty) set of objects carrying the property x, $(A, \succcurlyeq_d, \succcurlyeq_r)$ a (finite) intensive structure and $S = \{s_0, s_1 \ldots s_n\}$ be a series of standards for $(A, \succcurlyeq_d)$. Then, for each $a, b, c, d \in B$, with $a \sim s_i$, $b \sim s_j$, $c \sim s_k$, $d \sim s_l$, if $i, j, k, l, j \cdot l, i \cdot l, k \cdot j \leq n$, there exists a function $m: B \rightarrow \mathbb{R}$ such that*

$$\Delta_{ab} \succcurlyeq_d \Delta_{cd} \Leftrightarrow m(a) - m(b) \geq m(c) - m(d),$$

$$a/b \succcurlyeq_r c/d \Leftrightarrow \frac{m(a)}{m(b)} \geq \frac{m(c)}{m(d)}.$$

Proof Let $a \sim s_i$, $b \sim s_j$, $c \sim s_k$, $d \sim s_l$.

Let $a' \sim s_{l \cdot i}$ and $b' \sim s_{l \cdot j}$: then, for Axiom 19.3, $a/b \sim a'/b'$.

Let then $c' \sim s_{j \cdot k}$ and $d' \sim s_{j \cdot l}$: thus, similarly, $c/d \sim c'/d'$.

Then, for transitivity, $a/b \succcurlyeq c/d \Leftrightarrow a'/b' \succcurlyeq c'/d'$.

But, since $b' \sim d'$, for Axiom 19.2, $a'/b' \succcurlyeq_r c'/d' \Leftrightarrow \Delta_{a'b'} \succcurlyeq_d \Delta_{c'b'}$.

Considering the representation theorem for differences, $\Delta_{a'b'} \succcurlyeq_d \Delta_{c'b'} \Leftrightarrow m(a') - m(b') \geq m(c') - m(d') \Leftrightarrow li - lj \geq jk - jl \Leftrightarrow li \geq jk \Leftrightarrow \frac{i}{j} \geq \frac{k}{l}$.

Thus, we finally obtain both $\Delta_{ab} \succcurlyeq_d \Delta_{cd} \Leftrightarrow m(a) - m(b) \geq m(c) - m(d)$ and $a/b \succcurlyeq_r c/d \Leftrightarrow \frac{m(a)}{m(b)} \geq \frac{m(c)}{m(d)}$, which completes the proof. □

Let us now consider uniqueness.

Theorem 3.21 (Uniqueness for intensive structures) *Let A be a finite (not empty) set of objects carrying the property x, $(A, \succsim_d, \succsim_r)$ a (finite) intensive structure and $m: B \to \mathbb{R}$ a measure function for it. Then, any other measure function m' is such that, for each $a \in B$,*

$$m'(a) = \alpha m(a),$$

with $\alpha > 0$.

Proof We have first to prove that if m' satisfies $m'(a) = \alpha m(a)$, with $\alpha > 0$, then it satisfies the representation. This can be verified by substitution.

Then, we have to prove that any m' that satisfies the representation is in the form $m'(a) = \alpha m(a)$. Indeed, if m' satisfies the representation for difference, it must be of the form $m'(a) = \alpha m(a) + \beta$. Furthermore, for $a, b, c, d \in B$, such that $\frac{m(a)}{m(b)} \geq \frac{m(c)}{m(d)}$, also $\frac{\alpha m(a)+\beta}{\alpha m(b)+\beta} \geq \frac{\alpha m(c)+\beta}{\alpha m(d)+\beta}$ must hold true, which implies $\beta = 0$, which concludes the proof. □

Let us conclude with a simple example. Let $A = \{a, b, c, o\}$, with $a \succ b \succ c \succ o$. Assume also that

$$ao \succ_d ac \sim_d bo \succ_d ab \sim_d bc \sim_d co;$$
$$ac \succ_r bc \succ_r ab.$$

Then, a proper measure function is, e.g.,

$$m = \{(o, 0), (c, 1), (b, 2), (a, 3), \}.$$

We obtain, e.g.,

$$m(a) - m(c) = 2 > m(b) - m(c) = 1;$$
$$\frac{m(a)}{m(c)} = 3 > \frac{m(b)}{m(c)} = 2.$$

Another proper measure function would be:

$$m' = \{(o, 0), (c, 2), (b, 4), (a, 6), \}.$$

Note that the zero is the same.

Fig. 3.14 Comparison and addition of segments

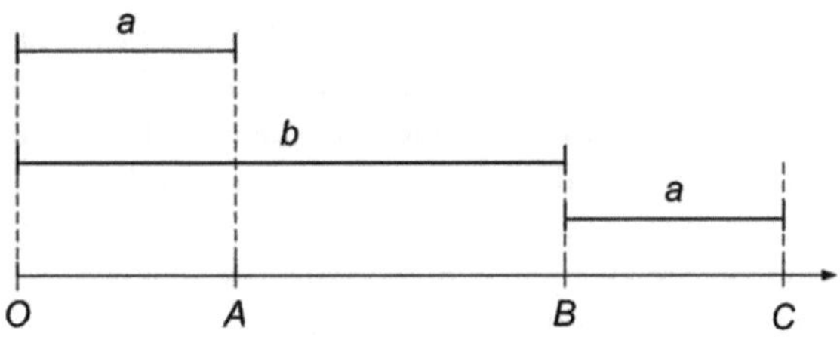

3.6 Ratio Scales for Extensive Structures

3.6.1 The Role of Additivity in Measurement

We are now at the core of measurement, as traditionally intended.

As we know, Helmholtz outlined that in many cases, a quantity can be regarded as the amount of something, resulting from the sum of a number of elementary parts, or units, of that something. So the empirical properties that allow measurement to take place are similar to those that hold amongst natural numbers and make counting possible and include order and addition.[23] Later on, Campbell highlighted the role of physical addition in allowing the construction of a reference scale. We investigate now the role of empirical addition in attaining a representation for extensive quantities. As noted in the previous section, they can be geometrically represented by segments: we will thus use this analogy for illustrating their properties. The key properties of segments, as well as of numbers, are order and addition, and we can visualise them as in Fig. 3.14.

We can compare two segments, a and b, by placing them on a semi-straight line, with one extremum coincident with the origin, O, of the line.[24] Then, we establish a correspondence between each segment and its second extremum on the line; in the figure, the segments a and b correspond to the points A and B respectively. In this example, $b \succ a$, since the point A is internal to the segment OB. This geometrical construction is similar to the real comparisons of blocks by an optical comparator.

Furthermore, we can add two segments by placing one adjacent to the other on a straight line and considering the segment delimited by the two non coincident extremes. For example, in the figure, we have $c = a \circ b$, where c is the segment OC.[25] Again this is similar to the real operation of piling two blocks.

We require that the empirical addition operation, is *commutative*, that is $a \circ b = b \circ a$, and *associative*, i.e., $a \circ (b \circ c) = (a \circ b) \circ c$, as happens with natural numbers. These properties are very important since they allow to introduce the *sum of a finite number of elements*. In fact, consider three elements, a, b, c, we can write $d = a \circ b \circ c$,

---

[23] Remember Sect. 1.2.

[24] Instead of actually placing them on the reference line, we have drawn them parallel to it, to make the figure more readable.

[25] Remember that the symbol "$\circ$" denotes addition of entities other than numbers.

where the second member can be interpreted as: "take two elements, sum them and then sum the result with the third element". This is possible since, thanks to the afore properties, any addition sequence will yield the same result. In this way, we can extend the addition operation, originally defined for two elements, to i elements, writing $b = a_1 \circ a_2 \circ \cdots \circ a_i$. In particular, we can consider the case where the elements are all equivalent to each other, that is $a_1 \sim a_2 \sim \cdots \sim a_i$. Denoting these elements, informally, with the same symbol, a, we can define the sum of i *perfect copies* of a as ia. A perfect copy a' of an element a is an element that can be safely substituted for a in any relation.[26]

Resembling the properties of natural numbers, we also require a property that links order and additivity: if we add an element to another, the resulting element is greater than the original. For example, in Fig. 3.14, c, which is the sum of a and b is greater than both of them. This also implies that all the elements are "positive", since summing any of them to another we will increase it, as happens with positive numbers. Thus, this property is called *positiveness* or *monotonicity*.[27]

Let us now investigate the role of additivity in attaining at the representation of extensive quantities. The basic idea is that any element a in A is equivalent to a sum of elementary elements, as in Helmholtz's original approach, and that the measure value can be assigned in accordance to this property. A measure assigned in this way will satisfy the representation theorem, that can be formulated in two, essentially equivalent, ways.

We can write either

$$m(a \circ b) = m(a) + m(b),$$

which means that the measure of the empirical sum of two objects equals the sum of the individual measures of the two objects, or

$$a \circ b \sim c \Leftrightarrow m(a) + m(b) = m(c),$$

which means that if an object is equivalent to the sum of two others, its measure will be equal to the sum of the measures of the other two. We will consider this latter formulation in the following.

A representation theorem was proved, for infinite sets, by Hölder in 1901 and constituted a fundamental result for the overall theory of measurement [3, 6]. In the case of infinite sets, a so-called Archimedean property is required, which can be formulated in this way: for every (whatever great) object a and for every (whatever small) object b, there exists a number i such that $ib \succcurlyeq a$. This is a very strong structural property which conveys, in a sense, the fundamental idea for a characteristic of being "quantitative", or extensive. It implies that there is no real gap between an

---

[26] The notion of "perfect copy" can be stated formally, but this results in a rather cumbersome mathematical framework [3]. We prefer to simply assume, when needed, that perfect copies, or replicas, of objects are available and we will often denote them with the same symbol of the original element.

[27] We have encountered a few monotonicity conditions so far. In fact, monotonicity concerns order conservation, and since order is a key property in measurement, monotonicity is also important.

object expressing a characteristic in a very high degree and another expressing it in a very low one, since you can always reach the former by summing a sufficient numbers of replicas of the latter. Note that this is not always true, e.g., in human sciences: the value of a human life, for example, in our vision, cannot be equalled by summing a number, whatever great, of valuable material objects.

Anyway, remaining within extensive systems, let us briefly review the basic idea under Hölder's Theorem. Consider an element a and an element u, conventionally chosen, to which we assign $m(u) = 1$. Then, for the Archimedean property, there will be an integer i such that $iu \preccurlyeq a \prec (i+1)u$. If the equivalence $iu \sim a$ holds true, we simply assign $m(a) = im(u) = i$. Otherwise, we can define a range of possible values for the measure of a, as $i \leq m(a) < i + 1$. Furthermore, we can refine this approximation, since either $2iu \preccurlyeq a \prec (2i+1)u$ or $(2i+1)u \preccurlyeq a \prec (2i+2)u$. Suppose, for example, that the former holds true. Then, we conclude that $i \leq m(a) < i + 1/2$, and so forth. Hölder basically proved that this procedure converges to a unique assignment of $m(a)$, thanks to the properties of the continuum.

So far for Hölder's Theorem. Yet, we want to consider *finite* sets, according to the general approach of this book. This makes some difference. First, we can no longer sum any pair of elements in A; since, if A is finite, it must have a maximum element, say z. We cannot sum z with another element because, for the monotonicity property, we would obtain a greater element, which is impossible, since we have assumed that z is the maximum. We can visualise this by thinking of summing rods in a room: only summing operations that give rise to rods not exceeding the maximum dimension of the room will be allowed. On the other hand, we will assume that all operations that satisfy this constraint are permitted.

The second important consequence of dealing with a finite set, is that the Archimedean property is no more necessary, although we have to assume a solvability condition instead. We require that if an element is greater than another, it is always possible to find elements that summed to the lesser one allow matching the greater.

If these properties hold true, the representation for finite sets can be achieved. The basic idea is that, given any two elements, a and b, it is always possible to find an element e and two integers, i and j,such that $a \sim ie$ and $b \sim je$. Consequently, for their sum $c = a \circ b$, we obtain $c \sim (i + j)e$, which is illustrated in Figs. 3.15 and 3.16.

In Fig. 3.14, $b \succ a$; then, for the solvability condition, there must be elements that summed to a match b. Let us sum a copy a' of a first. We see, in Fig. 3.15, that we still do not reach b. If we summed again another copy, a'', of a we would exceed b. So there must be another element, lower than a, that enables the matching. Call it e. Then, we can verify that, in this example, a is also matched by summing twice e, as shown in Fig. 3.16. So, in this example, $a \sim 2e$, $b \sim 5e$ and their sum, $c = a \circ b$, satisfies $c \sim 7e$.

Compare this with what happens with difference structures: there, a *concatenation of intervals* is possible, here a *concatenation of objects* is allowed. This latter is a stronger structural property, since it implies the former, whilst the converse is not true. In fact, in an extensive structure, it is possible to define difference as the inversion

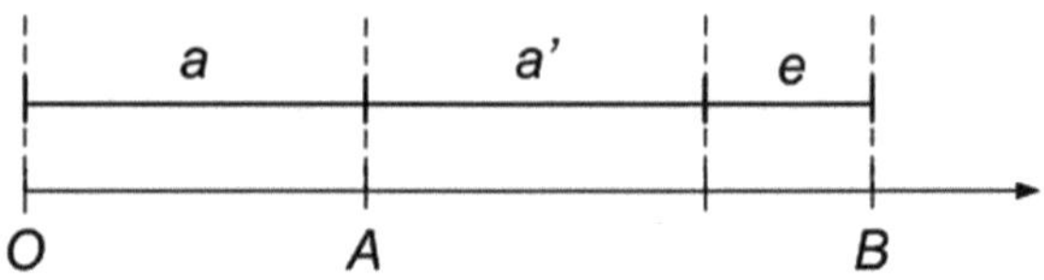

Fig. 3.15 Matching of element b, as in Fig. 3.14, which is greater than a, via the summing operation

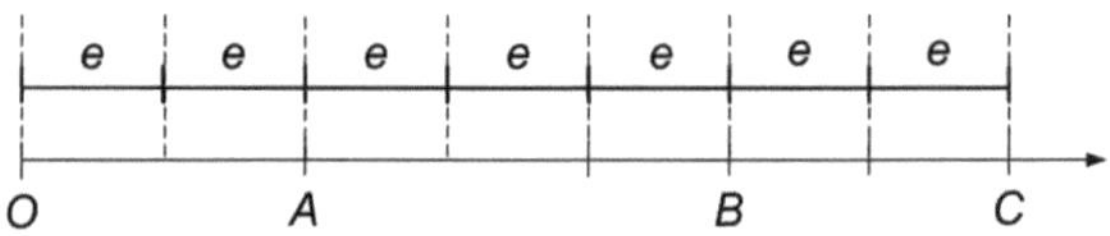

Fig. 3.16 Matching objects a and b by summing (replicas of) e repeatedly

of addition, that is

$$\Delta_{ab} = c \Leftrightarrow a = b \circ c. \tag{3.1}$$

Consequently, since in an extensive structure, the difference is an element of A, it is possible to concatenate differences, that is to concatenate intervals in respect of the difference between their extremes. On the other hand, in difference structure, there is no way of concatenating objects.

To sum up, for the representation to hold true in a finite extensive structure, we have to assume

- a weak order,
- an addition operation, commutative and associative, applicable to all the pairs that give rise to a result not greater than the maximum element of A,
- a monotonicity condition and
- a solvability condition.

3.6.2 Representation for Extensive Structures

For formally defining extensive structures, note that the addition operation, $c = a \circ b$, is a binary operation, since it applies to pairs, (a, b) of elements of A, and can also be formalised as a function from pairs of elements of A into elements of A as well. We can thus write $\circ\colon A \times A \to A$. If A is finite, we have seen that $\circ$ cannot be applied to all the pairs of elements of A. Its domain is instead a subset $B \subset A \times A$ and we will write $\circ\colon B \to A$. We can also associate to "$\circ$" the ternary relation $a \circ b \sim c$. With this premise, we are now ready for formally defining extensive structures.

Definition 3.22 (*Extensive structure*) Let A be a (not empty) set of objects carrying the property x. Then, $(A, \succcurlyeq, \circ)$ is an (finite empirical) extensive structure if $\succcurlyeq$ is a binary relation on A and $\circ$ is a function $\circ\colon B \to A$, where $B \subset A \times A$, that satisfy, for each $a, b, c, d \in A$, the following axioms:

22.1 $\succcurlyeq$ is a weak order;
22.2 if $(a, b) \in B$ and $(a \circ b, c) \in B$, then $(b, c) \in B$, $(a, b \circ c) \in B$ and $(a \circ b) \circ c \succcurlyeq a \circ (b \circ c)$;
22.3 if $(a, c) \in B$ and $a \succcurlyeq b$, then $(c, b) \in B$ and $a \circ c \succcurlyeq c \circ b$;
22.4 if $a \succ b$, then there exists $d \in A$ such that $(b, d) \in B$ and $a \succcurlyeq b \circ d$;
22.5 if $(a, b) \in B$, then $a \circ b \succ a$.

From the afore discussion, it is easy to recognise the role of each axiom. Axiom 22.1 concerns the order amongst the objects; Axioms 22.2 and 22.3 together define B and the associative and commutative property of addition; Axiom 22.4 is a solvability condition and Axiom 22.5 a monotonicity property.

Let us first establish a useful property of the sum ia .

Lemma 3.23 *Let $(A, \succcurlyeq, \circ)$ be an extensive structure and let ia be defined by induction as*

23.1 $1 \in I_a$ and $1a = a$;
23.2 if $(i-1) \in I_a$ and $((i-1)a, a) \in B$, then $i \in I_a$ and $ia = ((i-1)a) \circ a$;
23.3 if $(i-1) \in I_a$ and not $((i-1)a, a) \in B$, then for all $j \geq i$, j is not in I_a.

I_a is the set of consecutive positive integers for which ia is defined.
Then, if $i, j \in I_a$ and $(i + j) \in I_a$,

$$(ia) \circ (ja) = (i + j)a.$$

Proof We proceed by double induction.

Let us first assume $i = 1$ and let us prove that

$$a \circ (ja) = (j + 1)a.$$

We proceed by induction with respect to j. For $j = 1$, we obtain

$$a \circ (1a) = a \circ a = 2a,$$

according to the general definition of ja. Then, if the statement holds true for a generic value of j, that is if

$$a \circ (ja) = (j + 1)a,$$

then, it also holds true for $j + 1$, since

$$\begin{aligned} a \circ (j+1)a &= a \circ (a \circ ja) = a \circ (ja \circ a) \\ &= (a \circ ja) \circ a = (j+1)a \circ a = (j+2)a. \end{aligned}$$

Thus, the Lemma holds true for $i = 1$.

Let us now proceed by induction with respect to i.

We have already proved that things work for $i = 1$.

Let us now show that, if the statement holds true for a generic i, that is if

$$(ia) \circ (ja) = (i + j)a,$$

then, it also holds true for $j + 1$.

Indeed we obtain

$$\begin{aligned}(i + 1 + j)a &= (i + j + 1)a = (i + j)a \circ a = (ia) \circ (ja) \circ a \\ &= (ia) \circ a \circ (ja) = (i + 1)a \circ (ja),\end{aligned}$$

which completes the proof, by double induction, of the Lemma. □

And here the representation theorem comes.

Theorem 3.24 (Representation for extensive structures) *Let A be a finite (not empty) set of objects carrying the property x and let* $(A, \succsim, \circ)$ *be an extensive structure. Then there exists a function* $m: A \to \mathbb{R}$ *such that, for each* $a, b, c \in A$, $(a, b) \in B$,

$$a \sim b \circ c \Leftrightarrow m(a) = m(b) + m(c).$$

Proof Let us first consider the left-to-right implication, that is

$$a \sim b \circ c \Rightarrow m(a) = m(b) + m(c).$$

Since the structure includes an order amongst the objects, let $S = \{s_1, \ldots, s_n\}$ be a series of standards as per Definition 3.8. Then s_1 is a minimal element for A. Consider now any two successive elements $s_i, s_{i+1} \in S$. We want to prove that $s_{i+1} \sim s_i \circ s_1$.

Since $s_{i+1} \succ s_i$, for Axiom 22.4, there exists an element $e \in A$ such that $s_{i+1} \succsim s_i \circ e$. Now compare e with s_1. It cannot be $e \prec s_1$ since s_1 is minimal in A. But neither can we have $e \succ s_1$, since in that case, for Axiom 22.3, we would obtain $s_i \prec s_i \circ s_1 \prec s_i \circ e \precsim s_{i+1}$, which would imply the existence of an element, $s_i \circ s_1$, strictly in between s_i and s_{i+1}, which is impossible because of the way S has been constructed. Then we must have $e \sim s_1$, $s_{i+1} \sim s_i \circ e$, and, again for Axiom 22.3, $s_{i+1} \sim s_i \circ s_1$. So the series of standards is equally spaced, with "resolution" determined by the minimum element s_1. Then, for $i, j, i + j \leq n$, for Lemma 3.23, we obtain $is_1 \circ js_1 = (i + j)s_1$.

Let us now define, as usually, for $a \in A$, $s_i \in S$, the measure function as follows:

$$m(a) = m(s_i) = i \Leftrightarrow a \sim s_i.$$

Consider any $(b, c) \in B$. There will be $i, j < n$ such that $b \sim s_i$, $c \sim s_j$ and $i + j \leq n$. So $m(b) = i$ and $m(c) = j$. On the other hand, $b \circ c \sim is_1 \circ js_1 =$

$(i + j)s_1 \sim s_{i+j}$. Consequently, $m(b \circ c) = i + j = m(b) + m(c)$, which completes the proof of the left-to-right implication of the theorem.

For proving the converse, that is

$$m(a) = m(b) + m(c) \Rightarrow a \sim b \circ c, \tag{3.2}$$

consider again that for $(b, c) \in B$, there will be $i, j < n$ such that $b \sim s_i$, $c \sim s_j$, $i + j \leq n$ and $b \circ c \sim is_1 \circ js_1 = (i + j)s_1 \sim s_{i+j}$. So $m(b) = i$, $m(c) = j$, $m(b) + m(c) = i + j$. Then $m(a) = i + j$, which implies, $a \sim s_{i+j} \sim b \circ c$. This completes the proof of the theorem. □

Note that the measure function also provides a proper representation of order, since an extensive structure also includes an order structure.

We can also consider uniqueness conditions.

Theorem 3.25 (Uniqueness for extensive structures) *Let A be a finite (not empty) set of objects carrying the property x, $(A, \succsim, \circ)$ an extensive structure and $m: A \to \mathbb{R}$ a measure function for it. Then, any other measure function m' is such that, for each $a \in A$,*

$$m'(a) = \alpha m(a),$$

with $\alpha > 0$.

Proof We have to prove that

(a) if m is a valid measure function, them also $m' = \alpha m$, with $\alpha > 0$, is appropriate;
(b) if both m and m' are valid measure functions, there is $\alpha > 0$ such that $m' = \alpha m$.

Concerning statement (a), simply note that if $m(a) = m(b) + m(c)$, then also $\alpha m(a) = \alpha m(b) + \alpha m(c)$ holds true, which implies that $m'(a) = m'(b) + m'(c)$, as required.

Concerning statement (b), let $S = \{s_1, \ldots, s_n\}$ be a series of standards for the structure under consideration. We first prove that any valid measure function m must satisfy $m(s_1) > 0$. In fact, for each $a \in A$, such that $(a, s_1) \in B$, for Axiom 22.5, we obtain $a \circ s_1 \succ a$. Then, $m(a) + m(s_1) > m(a)$, and thus $m(s_1) > 0$.

Then, for each $a \in A$, there exists $s_i \in S$, such that $a \sim s_i$. Since S is equally spaced, $s_i \sim is_1$, which implies that also $a \sim is_1$. Thus, if both m and m' are valid measure functions, then both $m(a) = im(s_1)$ and $m'(a) = im'(s_1)$ hold true. Then, $m'(a) = \frac{m'(s_1)}{m(s_1)} m(a) = \alpha m(a)$, with $\alpha > 0$. □

Let us conclude by "constructing" a simple example. Let us start from the set $A' = \{a, b, c\}$, with $a \succ b \succ c$. Let us include an addition operation. Since c is the minimum element, $b \succ c$ must be reached by summing, at least, a copy of c to c itself. Let us call e such a copy and assume that this is exactly the case, that is $b \sim c \circ e$.

Then, in the simplest case, we can assume that also a can be reached by adding e to b, this time. That is $a \sim b \circ e$. If we want that a remain the maximum element, we can stop here. Then, we have

Table 3.5 The addition operation in the example

	b	c	e
b	/	a	a
c	a	b	b
e	a	b	b

Each element in the table is the result of the addition of the corresponding row and column elements

$$A = \{a, b, c, e\};$$
$$a \succ b \succ c \sim e;$$
$$B = \{(b, c), (c, b), (b, e), (e, b), (c, e), (e, c)\}.$$

Lastly, the addition operation is illustrated in Table 3.5.

3.7 Derived Scales

3.7.1 Derived Versus Fundamental Scales

Campbell, as we know, distinguished between fundamental and derived quantities. Fundamental, in his view, are those for which it is possible to construct a reference scale, thanks to the "physical addition" operation, and that can be measured by direct comparison with that scale. Derived instead are those that can be measured only indirectly, thanks to some physical law that relates them to other, independently measurable, quantities.

We can maintain, to some extent, this classification, provided that we generalise and update it on the basis of recent, widely accepted, results of the representational theory.

We thus call *fundamental* a scale that can be constructed on the basis of the *internal properties*, or *intra-relations*, of the quantity under consideration. Occasionally, we also call fundamental such a quantity and its measurement, when it is based on comparison with a reference scale.

On the other hand, we call *derived* a scale which is obtained on the basis of *relations linking the quantity under consideration with other quantities*, or, briefly, on the basis of *inter-relations*. Occasionally, we also call derived the quantity itself and its measurement obtained in this way.

For example, a scale for mass, constructed by generating the multiples and submultiples of an initial, arbitrary, unitary element, and by assigning their values in accordance with the additivity property, can be considered a fundamental scale, and fundamental can be called mass, as a quantity, and a measurement procedure for it, when it is based on a comparison with that scale.

Instead, a temperature scale based on a graduated mercury-in-glass thermometer can be considered derived from the length scale, through the property of thermal dilatation of fluids. In fact, this property is not "internal" to temperature (as would be the "warmer" relation), but rather relates two distinct quantities, temperature and length.

Another important example is loudness,[28] L, measured in *sone*, which, in the case of pure tones at, say 1 kHz, can be considered a perceptual characteristic derived from the corresponding physical quantity, *sound intensity*.

We will discuss these examples in the next sections, where we will provide a kernel axiomatic theory for derived quantities. We will consider only the case of one quantity derived from another; the case of one quantity derived from the combination of two or more others will be briefly addressed in Chap. 7, which is devoted to multidimensional measurement.

3.7.2 Representation for Derived Scales

3.7.2.1 Cross-Order Structures and Derived Ordinal Scales

Let us start, as we have done in the case of fundamental scales, from order relations. Consider two sets of objects, A and B, related to quantities y and x, respectively; we want to derive the measure of y from the measure of x. For example, y could be the temperature of an object, $a \in A$, and x the height of the mercury column in a mercury thermometer, $b \in B$. Note that there is a correspondence between a and b, since b is the height of the mercury column, *when the thermometer is in contact with object a*. We express this by a (cross-)equivalence relation, $a \sim b$. More precisely, we want firstly to make clear that that there is an order amongst temperatures, i.e., in A, then an order amongst the heights of mercury columns, that is in B, and lastly an equivalence between temperatures of objects and corresponding mercury-column heights, that is an equivalence *across* A and B. All this can be concisely expressed by the following notion, that we call "cross-order".[29]

Definition 3.26 (*Cross-order*) Let A and B be two sets. We call *cross-order* a weak order relation over $A \cup B$ and we denote it by the symbol $\succsim$.

Then, we introduce a cross-order structure on A and B. We further require that the set B is wide enough to express all the possible manifestations of the objects in A. That is, we require to have, e.g., enough mercury-column heights for representing all possible temperature states of our interest. Mathematically, this a kind of solvability condition.

---

[28] Additional details on loudness are provided in Chap. 8, Sect. 8.2.

[29] This notion of "cross-order" seems to be not standard in set theory, although it sounds quite "natural" from our standpoint. So, we provide a formal definition for it.

Definition 3.27 (*Cross-order structure*) Let A and B be two (not empty) sets of objects carrying, respectively, the properties y and x. We say that $(A, B, \succcurlyeq)$ is a *cross-order (empirical) structure*, if

27.1 $\succcurlyeq$ is a cross-order over $A \cup B$,
27.2 for each $a \in A$ there exists $b \in B$, such that $a \sim b$.

Suppose, for example, that $A = \{a_1, a_2\}$, $B = \{b_1, b_2, b_3\}$ and that $b_1 \succ b_2 \succ b_3$, $a_1 \sim b_1$ and $a_2 \sim b_3$ hold true. Then, by transitivity, we can conclude that $a_1 \succ a_2$, that is we can infer relations amongst elements of A *without comparing them directly*, but rather comparing them with elements of B and then *deriving* the searched relations from those holding amongst the corresponding elements of B.

All this can be formalised through the following representation theorem.

Theorem 3.28 (Representation for cross-order structures) *Let $(A, B, \succcurlyeq)$ be a cross-order structure for properties y and x. Then, for each $a_1, a_2 \in A$, there exist two elements $b_1, b_2 \in B$, a function $m_x \colon B \to \mathbb{R}$, a function $m_y \colon A \to \mathbb{R}$ and a monotone increasing function $g \colon \mathbb{R} \to \mathbb{R}$, such that for each $a_1, a_2 \in A$,*

$$a_1 \succcurlyeq a_2 \Leftrightarrow m_y(a_1) \geq m_y(a_2) \Leftrightarrow g(m_x(b_1)) \geq g(m_x(b_2)).$$

Proof Since $\succcurlyeq$ is a weak order for $A \cup B$, $(B, \succcurlyeq)$ is an order structure also. Then, there exists a measure function $m_x \colon B \to \mathbb{R}$ such that, for each $b_1, b_2 \in B$, $b_1 \succcurlyeq b_2 \Leftrightarrow m_x(b_1) \geq m_x(b_2)$. Then, let g be a monotone increasing function, and define m_y by $m_y(a) = g(m(b)) \Leftrightarrow a \sim b$ for each a. Note that, for Axiom 15.2, for each a, an element $b \in B$ equivalent to a always exists. Consider now any $a_1, a_2 \in A$: there will be $b_1, b_2 \in B$ such that $a_1 \sim b_1$ and $a_2 \sim b_2$. Then, $a_1 \succcurlyeq a_2 \Leftrightarrow b_1 \succcurlyeq b_2 \Leftrightarrow m_x(b_1) \geq m_x(b_2) \Leftrightarrow g(m_x(b_1)) \geq g(m_x(b_2)) \Leftrightarrow m_y(a_1) \geq m_y(a_2)$. □

For example, y could represent the temperature of objects that can be interfaced to a mercury thermometer and x the height of the mercury column of such a thermometer. $a \sim b$ means that object a produces height b when put in contact with the thermometer.

Coming back to the afore example, the theorem is satisfied by assuming $m_x(b_1) = 3$, $m_x(b_2) = 2$, $m_x(b_3) = 1$ and $g(x) = x$, which yields $m_y(a_1) = 3$ and $m_y(a_2) = 1$.

Suppose now that $S = \{s_i | i \in I\}$ is a series of standards for x and, consequently, $R_x = \{(s_i, m_x(s_i))\}$ is a reference scale for x. Then, $R_y = \{(s_i, g(m_x(s_i))\}$ can serve as a reference scale for y. That is to say, a series of heights of a mercury-in-glass thermometer, which in principle would constitute a height scale, can serve also as a temperature scale, thanks to their relation with temperature. This can perhaps give a concrete idea of what deriving a scale implies.

Lastly, note that A and B can denote both two distinct sets of objects, or the same set of objects but considered in respect of two properties, y and x, of each object. For example, we could consider the set of mercury columns in mercury thermometers: each of them is characterised both by its height, x, and by its temperature y.

3.7.2.2 Cross-Difference Structures and Derived Interval Scales

If cross-ordering provides a basic introduction to the empirical properties underlying derived scales, cross-differences allow an almost general treatment of this topic.

The idea is to have now also a cross-ordering amongst differences on the two sets, A and B, with properties closely resembling those of difference structures.

Let us first consider a weak order amongst intervals in A and in B. We will denote it by $\succsim_d$. Note that from this order amongst intervals it is possible to derive an order amongst objects, in much the same way as we have done with difference measurements (Sect. 3.4.3 and Lemma 3.13). Let us formulate this.

Definition 3.29 (*Cross-difference order*) Let A and B be two sets: we call cross-difference order a weak order relation over $A \times A \cup B \times B$ and we denote it by the symbol $\succsim_d$.

Lemma 3.30 *Let A and B be two sets and $\succsim_d$ a cross-difference order over them. Then, the relation $\succsim$, on $A \cup B$, defined, for each $a, b, c \in A \cup B$, by*

$$a \succsim b \Leftrightarrow \Delta_{ac} \succsim \Delta_{bc},$$

is a cross-order for A, B.

Proof As in Lemma 3.13 and according to Definition 3.26. □

We can now define a cross-difference structure.

Definition 3.31 (*Cross-difference structure*) Let A and B be two (not empty) sets of objects, carrying respectively properties y and x. Then, $(A, B, \succsim_d)$ is an (empirical) cross-difference structure if

31.1 $\succsim_d$ is a cross-difference order on $A \times A \cup B \times B$;
31.2 for each $a, b, c, d \in A \cup B$, if $\Delta_{ab} \succsim \Delta_{cd}$, then $\Delta_{dc} \succsim \Delta_{ba}$;
31.3 for each $a, b, c, a', b', c' \in A \cup B$, if $\Delta_{ab} \succsim \Delta_{a'b'}$ and $\Delta_{bc} \succsim \Delta_{b'c'}$, then $\Delta_{ac} \succsim \Delta_{a'c'}$;
31.4 for each $a, b, c, d', d'' \in B$, if $\Delta_{ab} \succsim \Delta_{cd} \succsim \Delta_{aa}$, then there exists $d', d'' \in A$ so that $\Delta_{ad'} \sim \Delta_{cd} \sim \Delta_{d''b}$ (solvability condition, for B only).
31.5 for each $a \in A$, there exist $b \in B$, such that $a \sim b$ (solvability condition, across A and B).

We are finally in a position as to formulate and prove the representation theorem.

Theorem 3.32 (Representation for cross-difference structures) *Let $(A, B, \succsim)$ be a cross-difference structure for properties y and x. Then, there exists a function $m_x\colon B \to \mathbb{R}$, a function $m_y\colon A \to \mathbb{R}$ and a linear positive function $g\colon \mathbb{R} \to \mathbb{R}$, that is $g(x) = \alpha x + \beta$, with $\alpha > 0$, such that for each $a, b, c, d \in A$, there exist $a', b', c', d' \in B$ such that*

$$\Delta_{ab} \succsim \Delta_{cd} \Leftrightarrow m_y(a) - m_y(b) \geq m_y(c) - m_y(d)$$
$$\Leftrightarrow g(m_x(a')) - g(m_x(b')) \geq g(m_x(c')) - g(m_x(d')).$$

Proof Note that $(B, \succsim_d)$ is a difference structure and then there is a function $m_x \colon B \rightarrow \mathbb{R}$ that satisfies a difference representation.

We now prove first that, for each $a, b, c, d \in A$, if $a', b', c', d' \in B$ are such that $a \sim a'$, $b \sim b'$, $c \sim c'$ and $d \sim d'$, then

$$\Delta_{ab} \succsim \Delta_{cd} \Leftrightarrow \Delta_{a'b'} \succsim \Delta_{c'd'}.$$

To see this, fix any element $o \in A \cup B$, then

$$\begin{aligned}
a \sim a' &\Rightarrow \Delta_{ao} \sim \Delta_{a'o} \Rightarrow \Delta_{oa} \sim \Delta_{oa'},\\
b \sim b' &\Rightarrow \Delta_{bo} \sim \Delta_{b'o} \Rightarrow \Delta_{ob} \sim \Delta_{ob'},\\
c \sim c' &\Rightarrow \Delta_{co} \sim \Delta_{c'o} \Rightarrow \Delta_{oc} \sim \Delta_{oc'},\\
d \sim d' &\Rightarrow \Delta_{do} \sim \Delta_{d'o} \Rightarrow \Delta_{od} \sim \Delta_{od'}.
\end{aligned}$$

But

$$\begin{aligned}
\Delta_{ab} &\sim \Delta_{ao} \circ \Delta_{ob},\\
\Delta_{a'b'} &\sim \Delta_{a'o} \circ \Delta_{ob'},\\
\Delta_{cd} &\sim \Delta_{co} \circ \Delta_{od},\\
\Delta_{c'd'} &\sim \Delta_{c'o} \circ \Delta_{od'}.
\end{aligned}$$

Then,

$$a \sim a', b \sim b', c \sim c', d \sim d' \Rightarrow \Delta_{ab} \sim \Delta_{a'b'}, \Delta_{cd} \sim \Delta_{c'd'},$$

and consequently,

$$\Delta_{ab} \succsim \Delta_{cd} \Leftrightarrow \Delta_{a'b'} \succsim \Delta_{c'd'}.$$

Lastly, since $(B, \succsim_d)$ is a difference structure,

$$\begin{aligned}
&\Delta_{a'b'} \succsim \Delta_{c'd'} \Leftrightarrow m_x(a') - m_x(b') \geq m_x(c') - m_x(d')\\
&\Leftrightarrow [\alpha m_x(a') + \beta] - [\alpha m_x(b') + \beta] \geq [\alpha m_x(c') + \beta] - [\alpha m_x(d') + \beta]\\
&\Leftrightarrow g(m_x(a')) - g(m_x(b')) \geq g(m_x(c')) - g(m_x(d'))\\
&\Leftrightarrow m_y(a) - m_y(b) \geq m_y(c) - m_y(d),
\end{aligned}$$

which completes the proof. □

Cross-difference structures are very important in science. A few examples can help to appreciate their wide scope.

In liquid-in-glass thermometry, a proper (first approximation) law is

$$h - h_0 = \alpha(T - T_0),$$

which can be put in the form

$$y = \alpha x + \beta,$$

with $y = h$, $x = T$, $\beta = h_0 - \alpha T_0$.

Fechner's law,

$$\psi = \alpha \ln \varphi + \beta,$$

can be immediately put in the afore form, by simply identifying $y = \psi$ and $x = \ln \varphi$.

Stevens's power law,

$$\psi = \eta \varphi^{\theta},$$

can also be put in the same form, by firstly taking the logarithm

$$\ln \psi = \theta \ln \varphi + \eta,$$

and then by identifying $y = \ln \psi$, $x = \ln \varphi$, $\alpha = \theta$ and $\beta = \eta$.

Additional useful results on derived measurement, as here considered, can be found in the literature on psychophysical laws [3, 21].

3.7.3 Systems of Quantities

In science and technology, quantities are not isolated but form a system, that includes fundamental and derived ones. To better understand the links amongst them consider this simple, didactic, example.

Suppose that we have firstly defined the measurement of mass through its properties of order and addition, through an equal-arm balance. Then, we choose an unitary element, u_m, and construct a reference scale. Suppose now that we have similarly defined a reference scale for length, based on plane-parallel blocks and a length comparator, and we have chosen a unitary element, u_l. We now consider an additional class of empirical facts, those related to elasticity of solid bodies: when we apply in certain conditions a mass to certain bodies, that henceforth we will call "springs", we observe a change in their length. Then, we have another way for defining the measurement of mass, based on its ability to produce length variations in springs (when properly coupled, details are inessential here). We could take the same unitary element u_m and measure the length variation, y_{u_m}, that it produces when applied to a selected spring, and then define the mass of an element a, that produces a length variation y_a, as

$$m(a) = \frac{y_a}{y_{u_m}} m(u_m).$$

Suppose now to observe that the two mass scales so far established coincide. Then, our knowledge has advanced since from now onwards mass can be regarded

- either as that property that has an empirical extensive structure that may be implemented by an equal-arm balance, or
- that property that causes, in a class of springs, proportional length variations.

In particular, we can decide, if convenient, to derive the mass unit (scale) from the length unit (scale), by choosing, as u_m, *that* mass that produces a length variation equivalent to u_l.

To sum up, if we have two characteristics that can be measured independently from each other, we can assume, on the basis of empirical facts, a law that links them. If this assumption is confirmed by experimentation, we can choose, if convenient, to derive the measurement of one of the two characteristics from the other.

Consider now another case. Suppose that we have already established the measurability of length and that we assume the law of elasticity holds true (on the basis of some qualitative observations). Then, we can define mass as that property that causes, in a class of springs, proportional length variations, and base its measurability on this definition. Perhaps we would be not totally satisfied with this solution (that anyway can be retained for some time, at least till a more satisfactory one is found) and we will probably look for some confirmation of the measurability of the second characteristic, mass in this example, which basically may be achieved in two ways, either

- by developing a direct measurement method, based on its "internal" properties (the equal-arm balance case), or
- by finding an additional law relating the mass with another measurable characteristic, say z, and checking whether we obtain consistent results.

What we have so far discussed can be generalised to the case of n quantities. This example also shows that it is not necessary to measure all the characteristics directly; note anyway that *at least one needs to be measured directly*.

The development of a system of quantities greatly helps the progress of science as well as the practical execution of measurement. This has been the case with the constitution and development of the International System of Metrology.

3.7.4 The International System of Metrology

The signing of the famous Metre Convention in Paris, in 1875, was just the culminating point of a process started by the end of the eighteenth century, at the time of the French revolution and in the mood of the Enlightenment thought ([22, 23]). Although the need of reference standards for measurement in trade, agriculture and construction, has been recognised by mankind from ancient times, it was only in that period that a rationalisation of the system of units was searched. In fact at that time, modern science had been firmly established and the need of accurate measurements for its development had been clearly recognised. Furthermore, the philosophers of the Enlightenment were in search of a rational foundation of knowledge, which has a natural counterpart, in science, in the search of universal reference standards, independent from place and time. In 1799, the decimal metric system was instituted and two platinum standards representing the metre and the kilogram were deposed in the Archives de la Republique in Paris.

By the end of the nineteenth century, the Metre Convention, signed by representatives of seventeen nations, created the Bureau International des Poids et Mesures (BIPM), the international reference body for metrology, and established a permanent organisational structure for coordinating metrology activities [24].[30] Such an organisation includes the International Committee for Weights and Measures (CIPM), a body supported by a number (currently ten) of technical committees for providing recommendations for the development of the different fields of metrology. The CIPM reports to the General Conference on Weights and Measures (CGPM), a biennial international meeting, the necessary decisions for the operation of the world metrology system are taken. A major output of such a coordination is the maintenance and development of the International System of Units (SI), that is a single, coherent system of measurements throughout the world, for physical—in a broad sense—sciences.

Thanks to such a system, a unique, stable, primary reference, recognised and accepted worldwide, is maintained for each quantity. Furthermore, quantities in the system are linked by a set of relations, the currently accepted physical laws, and consequently progress in one quantity influences other quantities as well. We can thus say that the measurability of each quantity is founded not only on its properties, but also on the overall-system's coherence, which is continually checked, both theoretically and experimentally.

The system evolves with time. It was firstly concerned with mechanical quantities (length, mass and time), and then it moved towards other fields in science. In 1946, a basic unit for electrical quantities, the ampere, was included, then the kelvin, for thermodynamic temperature and the candela, for luminous intensity, in 1954. In 1971, the mole was added, as the base unit for amount of substance, bringing the total number of base units to seven.

For our interdisciplinary perspective, the introduction of the candela is particularly relevant, since luminous intensity measures the human response to a physical stimulus and then extends the scope of the system from pure physical (or chemical) quantities to human-dependent properties. Such a system may perhaps undergo a noteworthy reorganisation in a few years, since there is a proposal to revise the definition of the base quantities through the natural constants that link them.

The cooperation in the system also includes the publication of documents addressing methodological issues. We have already mentioned the International vocabulary of terms in metrology (VIM)[31] and the Guide to the expression of uncertainty in

[30] There are now 53 members of the BIPM, including all the major industrialised countries.

[31] See footnote 1, in Chap. 1.

measurement (GUM).[32] The evolution of the VIM can be noted. From its first publication in 1984 [25], it underwent two substantial revisions, in 1993 [26] and 2007 [1]. These revisions were substantial and were necessary for two reasons. On the one hand, terminological issues are strictly related to the understanding of basic ideas in measurement, which evolves as long as measurement theory does. On the other hand, measurement science tends to become more and more interdisciplinary. In fact, the reason for the first and, especially, the second revision was to "take account of the needs of chemistry and related fields" and to "cover measurements in chemistry and laboratory medicine for the first time", since "it is taken for granted that there is no fundamental difference in the basic principles of measurement in physics, chemistry, laboratory medicine, biology or engineering". Instead the GUM, originally published in 1993 [27], only underwent minor modifications thenceforth. Yet, supplemental guides dealing with specific implementation or interpretation aspects were prepared and are still under development. We will discuss uncertainty evaluation and expression in some detail, in Chap. 9.

Recently, the cooperation amongst members of the BIPM has been reinforced and made more effective by the institution of a Mutual Recognition Agreement (CIPM MRA) [28], which specifies the organisational and technical requirements for the mutual recognition of measurements performed by National Metrological Institutes (NMIs). A major tool for such a recognition are *key comparisons*. In some, comparisons are performed directly against an international reference facility at the BIPM. In others, a stable travelling standard is circulated amongst several NMIs, which are asked to provide a measurement value for it, accompanied by an uncertainty statement. An international committee of NMI experts in the field evaluates the resulting data and provides practical information on the degree of comparability of the individual results. Similar exercises, called *inter-comparisons*,[33] are performed amongst laboratories at lower levels of the metrological structure, and they are very effective for guaranteeing the performance of the overall system of metrology. We will treat inter-comparisons in Chap. 10. Their use could perhaps be extended to measurement in behavioural sciences also, as we will mention in Chap. 8.

3.8 Summary

We have considered fundamental and derived measurement scales.

A scale for a quantity x is called fundamental if it is based on the internal properties of x (intra-relations).

A scale for a quantity y is called derived if it is obtained through the relations of y with another quantity x (inter-relations) (or with some other quantities).

We have studied three types of fundamental scales, ordinal, interval and ratio, referring to order, difference, intensive or extensive empirical structures. Since all

[32] The GUM was presented in Sect. 2.2.6.

[33] In fact key comparisons are just inter-comparisons performed at the highest level.

Table 3.6 Fundamental scales

Empirical structure	Representation	Scale type	Uniqueness
Order	$a \succcurlyeq b \Leftrightarrow m(a) \geq m(b)$	Ordinal	Monotone increasing
Difference	$\Delta_{ab} \succcurlyeq \Delta_{cd} \Leftrightarrow$ $m(a) - m(b) \geq$ $m(c) - m(d)$	Interval	$m'(a) = \alpha m(a) + \beta$
Intensive	$a/b \succcurlyeq c/d \Leftrightarrow$ $\frac{m(a)}{m(b)} \geq \frac{m(c)}{m(d)}$	Ratio	$m'(a) = \alpha m(a)$
Extensive	$a \sim b \circ c \Leftrightarrow$ $m(a) = m(b) + m(c)$	Ratio	$m'(a) = \alpha m(a)$

Table 3.7 Derived scales

Empirical structure	Representation	Derived scale	Functional relation
Cross-order	$a \succcurlyeq b \Leftrightarrow$ $g(m_x(a')) \geq g(m_x(b'))$	Ordinal	g is monotone increasing
Difference	$\Delta_{ab} \succcurlyeq \Delta_{cd} \Leftrightarrow$ $g(m_x(a')) - g(m_x(b')) \geq$ $g(m_x(c')) - g(m_x(d'))$	Interval	$g(x) = \alpha x + \beta$

these structures have an empirical weak order, we can always form a series of standards S, by selecting one element in each of the equivalence classes in A, with respect to the equivalence relation $\sim$:

$$S = \{s_i \in A | i \in I \text{ and, for } i, i+1 \in I, s_i \prec s_{i+1}\},$$

where I is a proper set of indices:

- $I = \{1, \ldots, n\}$, for order or extensive structures,
- $I = \{0, 1, \ldots, n\}$, for interval or intensive structures.

If we associate to each standard in the series its measure, we obtain a reference scale,

$$R = \{(s_i, m(s_i)), i \in I\}.$$

For each $a \in A$ and for $s \in S$, we can define a measure function m as *virtually* obtainable by direct comparison with the reference scale, as

$$m(a) = m(s) \Leftrightarrow a \sim s.$$

The measure function satisfies a representation theorem and a uniqueness conditions, which are summarised in Table 3.6. Note that each structure also satisfies the representation theorems of those that precede it in the table.

Concerning derived scales, considering two sets of objects, A and B, with associated characteristics y and x, their empirical cross structure, $\mathcal{C}$, now comes into play. If proper conditions are satisfied, the scale of y can be derived from the scale of x. The most important of such properties is cross-order, that is a weak order over $A \cup B$.

Representations for cross-order and cross-difference structures are summarised in Table 3.7, where $a, b, c, d \in A$ and $a', b', c', d' \in B$.

A system of quantities in general is made of both fundamental and derived quantities. It is required that at least one of them is fundamental.

References

1. ISO: ISO/IEC Guide 99:2007 International Vocabulary of Metrology—Basic and General Terms (VIM). ISO, Geneva (2007)
2. Finkelstein, L.: Theory and philosophy of measurement. In: Sydenham, P.H. (ed.) Handbook of Measurement Science, vol. 1, pp. 1–30. Wiley, Chichester (1982)
3. Roberts, F.S.: Measurement Theory, with Applications to Decision-making, Utility and the Social Sciences. Addison-Wesley, Reading, MA (1979). Digital Reprinting (2009) Cambridge University Press, Cambridge
4. Huler, S.: Defining the Wind: The Beaufort Scale. Crown, New York. ISBN 1-4000-4884-2 (2004)
5. Narens, L.: Abstract Measurement Theory. MIT Press, Cambridge (1985)
6. Krantz, D.H., Luce, R.D., Suppes, P., Tversky, A.: Foundations of Measurement, vol. 1. Academic Press, New York (1971)
7. Russell, B.: Introduction to the Mathematical Philosophy. George Allen and Unwin, London (1919)
8. Ferguson, A., Myers, C.S., Bartlett, R.J.: Quantitative estimates of sensory events. Final Report—British Association for the Advancement of Science, vol. 2, pp. 331–349 (1940)
9. Kant, I.: Critik der reinen Vernunft, Johann FriedrichHartknoch (Italian edition: Kant I (2004) Critica della ragion pura (trans: Esposito C).Bompiani, Milano). Riga (1781/1787)
10. Campbell, N.R.: Physics—The Elements. Reprinted as: foundations of science (1957). Dover, New York (1920)
11. Stevens, S.S.: The direct estimation of sensory magnitudes: loudness. Am. J. Psychol. **69**, 1–25 (1956)
12. Stevens, S.S.: On the theory of scales and measurement. Science **103**, 667–680 (1946)
13. Narens, L.: A theory of ratio magnitude estimation. J. Math. Psychol. **40**, 109–129 (1996)
14. Steingrimsson, R., Luce, R.D.: Evaluating a model of global psychophysical judgements (Part I and Part II). J. Math. Psychol. **50**, 290–319 (2005)
15. Berglund, B., Rossi, G.B., Townsend, J., Pendrill, L. (eds.): Measurement with Persons. Taylor and Francis, New York (2012)
16. Torgerson, W.S.: Distances and ratios in psychophysical scaling. Acta Psychol. **19**, 201–205 (1961)
17. Miyamoto, J.M.: An axiomatization of the ratio difference representation. J. Math. Psychol. **27**, 439–455 (1983)
18. Birnbaum, M.H.: Comparison of two theories of ratio and difference judgement. J. Exp. Psychol. **109**, 304–319 (1980)
19. Rule, S.J., Curtis, D.W.: Ordinal properties of subjective ratios and differences. J. Exp. Psychol. **109**, 296–300 (1980)
20. Rossi, G.B., Crenna, F.: On ratio scales. Measurement **46**, 29–36 (2013). doi:10.1016/j.measurement.2013.04.042

21. Luce, R.D.: On the possible psychophysical laws. Psychol. Rev. **66**, 81–95 (1959)
22. Balducci, E.: Storia del pensiero umano. Edizioni Cremonese, Citt di Castello (1987)
23. Reale, G., Antiseri, D.: Storia della filosofia. Bompiani, Milano (2008)
24. BIPM: The International System of Units, 8th edn. STEDI, Paris (2006)
25. BIPM, CEI, ISO, OIML: International Vocabulary of Basic and General Terms in Metrology. ISO, Genéve (1984)
26. BIPM, IEC, IFCC, ISO, IUPAC, IUPAP, OIML: International Vocabulary of Basic and General Terms in Metrology, 2nd edn (1993)
27. BIPM, IEC, IFCC, ISO, IUPAC, IUPAP, OIML: Guide to the Expression of Uncertainty in Measurement. ISO, Geneva. Corrected and reprinted (1995), ISBN 92-67-10188-9 (1993)
28. BIPM: Mutual Recognition. STEDI, Paris (2008)

Chapter 4
The Measurement Scale: Probabilistic Approach

4.1 Working with Probability

4.1.1 The Nature of Probability

The discussion on the nature of probability was part of nineteenth century epistemological debate. According to Hacking, probability may be understood either as a *relative frequency* or as a *degree of belief* [1]. This is also related to seeing the objects of probability as *events* or *statements*, and the two perspectives are often convertible: if I am interested in the probability that tomorrow it will rain, I may see "tomorrow it will rain" as an event, which may happen or not, or as a statement, which may be true or false.

Historically, both positions were pursued up to their extreme consequences [2]. Von Mises, for example, fully investigated the frequentistic approach which admitted the possibility of a probabilistic estimation only when it refers to a "collective" of realisations. On the other hand, De Finetti reached a subjectivistic vision. Pursued up to its extreme consequences, this approach leads to considering probabilistic statements as reflecting the vision of a single person, and, as such, having limited scientific value, since scientific knowledge is understood to be inter-subjective.

In my opinion, these two views are related to the general question in science of whether the properties that we attribute to objects have an *ontic* or an *epistemic* character. Ontic means that they are inherent in the object, epistemic means that they are strongly dependent on our cognitive categories. In the former perspective, we consider the *randomness* of some phenomena, that is the fact that they happen or evolve in a way that is not completely predictable. Randomness is a property of a system under observation. In the epistemic perspective, we rather consider *uncertainty*, which is a property of the observer. We may be uncertain in the description of a phenomenon because of its randomness, or due to our ignorance about it, or for a combination of both.

G. B. Rossi, *Measurement and Probability*, Springer Series in Measurement Science and Technology, DOI: 10.1007/978-94-017-8825-0_4,

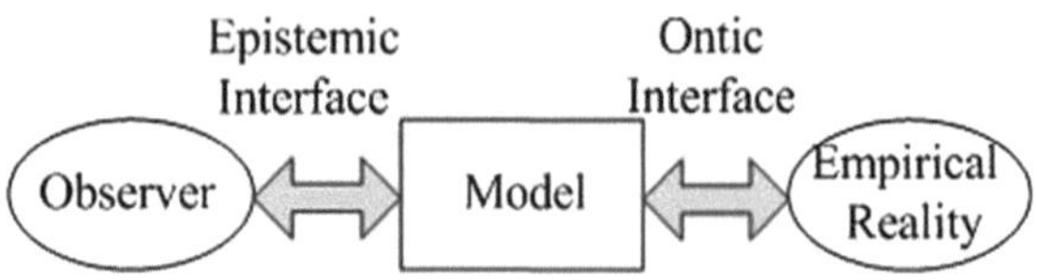

Fig. 4.1 The role of the model in the scientific method of observing the world

In my opinion, the notion of *model* can be of great help for settling this question. In quite general terms, a (scientific) model can be understood as an abstract system intended to represent, to some extent and from a certain standpoint, a real system. Modern science basically stands on models; they are a lens through which we look at reality, as shown in Fig. 4.1.

The model necessarily interfaces with us on one side and with the empirical reality on the other. As long as it is a model of ours, it must comply with our cognitive categories and hence its epistemic character. Yet, as it "bites into reality", as Barone used to say [3], it must take on some ontic aspects.

So these two perspectives are not, in my opinion, irreconcilable; rather, they are two faces of the same coin, in general not separable. Coming back to probability, as this is used to develop probabilistic models, it belongs to this pattern. Probabilistic models are those that are expressed in terms of probabilistic relations and/or variables, and as models, they usually give a description of things which is both epistemic and to some extent, ontic. Thus, concerning the "nature" of probability, I basically regard it as a primitive notion—everyone intuitively understands the term "probability" as they do terms such as "line" or "plane"—mathematically characterised by a set of axioms. In science, probability is used as a kind of "logic" that allows models to be developed and inferences made, whose validity is subjected to evaluation by the scientific community, as is the case for any scientific construct. The adherence of the scientist who formulated the model to some credo, Bayesian, frequentistic or any other, does not, in my opinion, add to or subtract from the validity of the model.

4.1.2 The Rules of Probability

Probability concerns either events or statements. The corresponding mathematical structure is based either on set theory or on mathematical logics. The two approaches are formally equivalent as it is, in many cases their interpretation, as aforementioned. We will use mainly a set theoretic approach, since we believe it may sound more familiar to people involved in measurement, yet, we will also briefly mention the other.

Consider a finite set, $\Omega = \{\omega_1, \ldots, \omega_n\}$, of elementary events or atoms. For example, $\Omega = \{f_1, f_2, f_3, f_4, f_5, f_6, \}$ in die rolling or $\Omega = \{h, t\}$ (h for heads and t for tails) in coin tossing. In general, an event, A, is a subset of Ω. For example, the event "odd" in die rolling corresponds to the set $\{f_1, f_3, f_5\}$. In general, suppose we wish to model an *experiment*. A *trial* is a single performance of it. At each trial, we observe a single outcome, $\omega \in \Omega$. Then, we say that an event A occurs during a trial if the outcome of the trial is included in it, i.e. if $\omega \in A$. In this perspective, Ω is the *certain event* and the empty set $\varnothing$ is the *impossible event*.

Let us denote by $\mathcal{F}$ the class of all the events. Since they are sets, for working with them, we need to be able to perform the basic operations of union and intersection. Note the meaning of such operations: if $C = A \cup B$, C is satisfied whenever either A or B happens; $D = A \cap B$ instead means that both A and B must occur in order that D is verified. We thus require that if A and B belong to $\mathcal{F}$, then also their union and intersection belong to it. Furthermore, let $\bar{A}$ be the complement (the complementary set) of A, that is, $\bar{A} = \Omega - A$. Then, we also require that $\bar{A} \in \Omega$. All this is expressed by saying that $\mathcal{F}$ is *a boolean algebra of sets*. For a finite Ω, $\mathcal{F}$ often is taken as the set of all the subsets of Ω, usually denoted by 2^{Ω}, since it has 2^n elements. In the case of coin tossing, $\Omega = \{h, t\}$ and $\mathcal{F} = 2^{\Omega} = \{\varnothing, \{h\}, \{t\}, \Omega\}$. Note that $\varnothing$ and Ω must always be included in $\mathcal{F}$.

Lastly, we can introduce probability, P, as a function of the events with values in the interval $[0, 1]$, $P : \mathcal{F} \rightarrow [0, 1]$, that satisfies the following properties [4]:

1. for each $A \in \mathcal{F}$, $P(A) \geq 0$;
2. $P(\Omega) = 1$; and
3. for each $A, B \in \mathcal{F}$, if $A \cap B = \varnothing$, then $P(A \cup B) = P(A) + P(B)$.

If, say, $A = \{\omega_1, \omega_2, \omega_3\}$, its probability can be denoted either by $P(A)$, by $P(\{\omega_1, \omega_2, \omega_3\})$ or, as a shorthand notation, by $P\{\omega_1, \omega_2, \omega_3\}$.

To sum up, the basic structure for working with probability is called a *probability space*, $\mathcal{S} = (\Omega, \mathcal{F}, P)$, where

- Ω is a (finite) set;
- $\mathcal{F}$ is a boolean algebra of sets on Ω; and
- P is a probability function, satisfying the above properties.

Probability so far introduced is called *absolute* or *categorical*; another key notion is now needed, that of *relative* or *conditional* probability. Let A and B be two events, with $P(B) > 0$. Then, the probability of A given (conditioned on) B is

$$P(A|B) = \frac{P(A \cap B)}{P(B)}.$$

Its meaning is the following: $P(A|B)$ is the probability of A calculated not in the whole set Ω, but in the set of the outcomes that satisfy B (Fig. 4.2). In fact, suppose that Ω has n elements, that they are all equally probable, each with probability $1/n$, that A has n_A elements, $B\, n_B$, $(A \cap B)\, n_{AB}$. Then, $P(A) = n_A/n$, $P(B) = n_B/n$, $P(A \cap B) = n_{AB}/n$ and

$$P(A|B) = \frac{n_{AB/n}}{n_B/n} = \frac{n_{AB}}{n_B}.$$

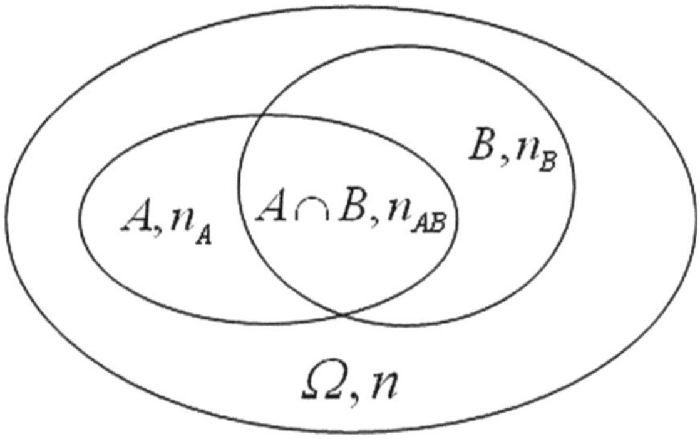

Fig. 4.2 The notion of conditional probability

Conditional probability expresses the way an event influences another. If $P(A|B) \neq P(A)$, B influences A, since its occurence changes the probability of A; in particular, if $P(A|B) > P(A)$, the occurence of B facilitates that of A; if $P(A|B) < P(A)$, the occurrence of B hinders that of A. On the other hand, if $P(A|B) = P(A)$, the two events are independent.

With these premises, the basic rules for probability calculus can be derived [5]. They concern the probability of the union and of the intersection of events:

- $P(A \cup B) = P(A) + P(B) - P(A \cap B)$,
- $P(A \cap B) = P(A|B)P(B) = P(B|A)P(A)$.

Note that A and B are independent, then

- $P(A \cap B) = P(A)P(B)$.

Note that if Ω is finite, if we assign a probability value to each of its elements,

- $P(\omega_i) = p_i$, for $i = 1, 2, \ldots, n$,

since each event, A, is a subset of Ω, it is possible to calculate its probability as

- $P(A) = \sum_{\omega_i \in A} p_i$.

Two additional rules play a central role in probability. Let $\{A_1, A_2, \ldots, A_n\}$ be a *Partition* of the sample space Ω and B an event. The *principle of total probability* states that

- $P(B) = \sum_{i=1}^{n} P(B|A_i)P(A_i)$.

From this formula, it is also possible to obtain $P(A_j|B)$, $1 \leq j \leq n$, which yields the famous *Bayes–Laplace rule*:

- $P(A_j|B) = \frac{P(B|A_j)P(A_j)}{\sum_{i=1}^{n} P(B|A_i)P(A_i)}$.

These two rules are amongst the most basic and well-established principles in the theory of probability. The former is possibly the oldest rule, as it can be traced back to Bernoulli himself; the latter is one century younger as it was developed, almost simultaneously and independently, by Bayes and Laplace [6]. From an epistemological standpoint, let us interpret each A_i as *a possible cause* of B, *the effect*. Then, the principle of total probability allows us *to predict the effect from the causes*, which, in physics and, generally, in science, is called a *direct problem*, whilst the Bayes–Laplace rule allows us *to identify the cause of a given observed effect*, which is called an *inverse problem*. In medicine, for instance, a diagnosis is an inverse problem whilst a prognosis is a direct one. A noteworthy variation in the Bayes–Laplace rule occurs when all the causes are equally probable. This yields:

- $P(A_j|B) = \frac{P(B|A_j)}{\sum_{i=1}^{n} P(B|A_i)}$,

Which is indeed the original form of the rule.

4.1.3 An Illustrative Example

Consider a system made by a box with two white balls, w_1 and w_2, and two red ones, r_1 and r_2.

Consider firstly a simple experiment, E_0, where a ball is extracted from the box. Let the balls be undistinguishable at touch and let the box be shaken before the extraction. Then, we can assume that all the possible outcomes, forming the set $\Omega = \{w_1, w_2, r_1, r_2\}$, have the same probability, thus equal to $1/4$. If we define the following events,

A = "white ball",
$\bar{A}$ = "red ball",

we simply obtain

$$P(A) = P\{w_1, w_2\} = 1/2,$$
$$P(\bar{A}) = P\{r_1, r_2\} = 1/2.$$

Consider now another experiment, E_1, where one ball is extracted from the box, then returned; the box is shaken; and a ball is extracted again. Here, the result is a sequence of two extracted balls, and there are 16 such possible sequences, as shown in Fig. 4.3. Define, similarly to above, the following events:

A = "white ball at the first extraction",
$\bar{A}$ = "red ball at the first extraction",
B = "white ball at the second extraction" and
$\bar{B}$ = "red ball at the second extraction".

Fig. 4.3 Experiment E_1

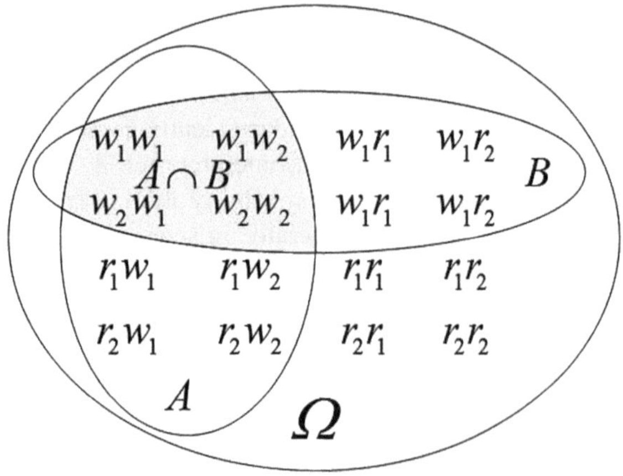

Note that events A and B are here independent, since the result of the first extraction does not influence the second. So, we obtain

$$P(A) = P(B) = 1/2,$$
$$P(B|A) = P(B) = 1/2 \text{ and}$$
$$P(A \cap B) = P(A)P(B) = 1/4.$$

These results can be checked by Venn diagrams, as in Fig. 4.3.

Consider now another, more interesting, experiment, E_2. Here, we extract one ball, without looking at it, we do not return it to the box, and we make a second extraction. Let the events A, $\bar{A}$, B, and $\bar{B}$ be defined as above. What changes is now that A and B are no longer independent, since what happens in the first extraction now does influence the second. So the possible results now include only those sequences where the outcomes of each extraction are different, as shown in Fig. 4.4. There are 12 such possible results.

Consider two problems.

(a) We have not yet made the experiment, and we want to predict what will happen. In particularly, we are interested in the probability of B, $P(B)$, based on what we know about the system and the way the experiment is performed. This is a typical *direct* problem.
(b) We have just performed the experiment; we do not know the colour of the first extracted ball but we do know that the second was white; that is, B occurred. Here, we want to make a statement on the unknown result of the first extraction, based on what we know—i.e. that B occurred—that is, we look for $P(A|B)$. This is a typical *inverse* problem.

As we know, problem (a) can be solved by the principle of total probability. We obtain

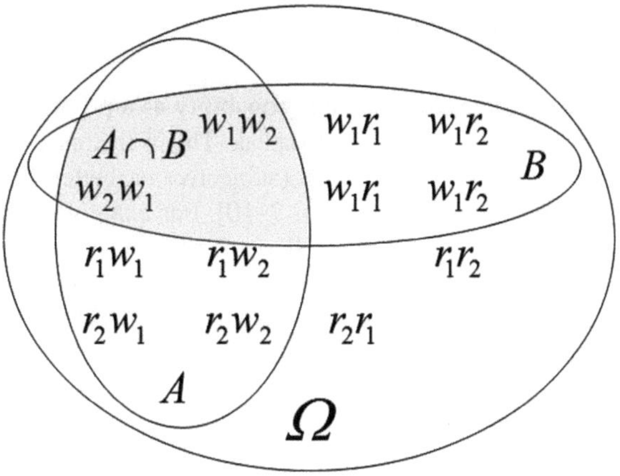

Fig. 4.4 Experiment E_2

$$P(B) = P(B|A)P(A) + P(B|\bar{A})P(\bar{A}),$$
$$P(A) = P(\bar{A}) = 1/2,$$
$$P(B|A) = 1/3,\ P(B|\bar{A}) = 2/3,$$
$$P(B) = \frac{1}{3} \cdot \frac{1}{2} + \frac{2}{3} \cdot \frac{1}{2} = \frac{1}{2}.$$

Problem (b) can be solved instead by means of the Bayes–Laplace rule, obtaining

$$P(A|B) = \frac{P(B|A)P(A)}{P(B|A)P(A) + P(B|\bar{A})P(\bar{A})} = \frac{\frac{1}{3} \cdot \frac{1}{2}}{\frac{1}{3} \cdot \frac{1}{2} + \frac{2}{3} \cdot \frac{1}{2}} = \frac{1}{3}.$$

Again these results can be checked by Venn diagrams, as shown in Fig. 4.4.

As a final comment, suppose that we are actually performing experiment E_2. In case (a), we are uncertain about the occurrence of B due to the intrinsic randomness of the experiment. We can think that the probability we assign to B is of an ontic character.

On the other hand, in case (b), the result of first extraction is perfectly defined: we could simply look at the first extracted ball and we will exactly know whether A occurred or not. Our uncertainty now is due to from a lack of information and is thus rather of an epistemic nature.

Yet note that in both cases, we have obtained the correct result, by simply applying the rules of probability, without being specially concerned about the nature of probabilities we were working with. This seems to confirm our previous statement that a scientifically good result can be obtained provided that good probabilistic models are developed, fairly independently from the nature of the involved probabilities.

4.1.4 Probability as a Logic

So far we have considered probability as a property of events, accounting for their "propensity to occur", so to speak. This is the ontic view. Alternatively, it is possible to interpret probability as a (subjective or, better, inter-subjective) degree of belief in some statement [1, 2, 4, 7–10]. For example, consider again the sample space $\Omega = \{\omega_1, \ldots, \omega_n\}$ as the collection of all the possible outcomes of some experiment, E. It is easy to establish a one-to-one correspondence between each ω_i and the statement $\omega_i' =$ "ω_i occurs (in experiment E)". Furthermore, the event $\{\omega_i\} \cup \{\omega_j\}$ will correspond to the statement $\omega_i' \vee \omega_j'$, where $\vee$ is the logical "or" operator. Similarly, the event $\{\omega_i\} \cap \{\omega_j\}$ will correspond to the statement $\omega_i' \wedge \omega_j'$, where $\wedge$ is the logical "and" operator. More generally, if events A and B correspond to statements A' and B', then

- $A \cup B$ corresponds to $A' \vee B'$, and
- $A \cap B$ corresponds to $A' \wedge B'$.

Note also that the certain event Ω corresponds to the identically true statement "one of the ω_i occurs", and the impossible event to the false statement "none of the ω_i occurs".

If we now assign the same probability to the events and to the statements, we can see the two different, although here equivalent, interpretations of probability.

In general, it is possible to define the probability space with P being a function of a boolean algebra of statements rather then a boolean algebra of sets, and these two approaches are, at least formally, mutually convertible. In fact, there is a famous theorem, due to Stone [8], that ensures that this correspondence can be always established.

From an epistemological standpoint, this means, in simple words, that it is possible both to start from observations of events and then make statements (probabilistic models) about them, or, conversely, to start from a model and then to check it by observations.

For the purpose of this book, it is good to keep an open mind towards these two approaches.

4.1.5 Probabilistic Variables

Scientific and technical models usually describe a piece of reality by a set of variables and of relations amongst them. A (numerical) *variable*, say x, is a mathematical entity that can assume values in a set of numbers, X.

In a deterministic approach, when the model is applied to a specific *situation*, the variable assumes a specific defined value, say $x = \xi, \xi \in X$. In fact, from a scientific standpoint, a situation (context, state of things,...) can be defined by the set of values that the related variables assume in it.

Thus, a *deterministic variable* is one that assumes a specific value in any specific context.

In contrast, in a probabilistic approach, generally speaking, a variable in a specific context can only be characterised by a plurality of values with an associated probability distribution.

Recalling discussion in Sect. 4.1.1, this can be due to two main reasons:

- either the variable describes a *population* of individuals, rather than a single individual; for example, the (modulus of the) velocity of the molecules in a gas, in a given thermodynamic condition, varies from one molecule to another and can only be described by a probability distribution;
- or the variable describes a single individual, but its value is uncertain, as it happens in measurement;
- or by a combination of them.

Thus, the notion of *probabilistic* (*or random*) *variable* comes into play. For formalising it, we need a way for assigning a probability to the event/statement $x = \xi, \xi \in X$, that says that the variable assumes a specific value.

To do that, we introduce a probability space $\mathcal{S} = (\Omega, \mathcal{F}, P)$ and we establish a correspondence (a function) between the points of the sample space and the possible values of the variable. In fact, a probabilistic variable can be defined as one such function, $x : \Omega \rightarrow X$. For example in the case of coin tossing, we establish such a function by simply printing the number from 1 to 6 on the faces of the die. In this way, the probability that, say, $x = 3$, equals the probability of the event $\{f_3\}$, that is, for a fair die, $P(x = 3) = P\{f_3\} = 1/6$. Note anyway that another variable, y, could be defined by printing the number 1 on the even faces and the number 0 on the odd. In this case, we obtain, e.g., $P(y = 1) = P(\{f_2, f_4, f_6\}) = 1/2$. In general, *the probability that a variable takes one specific value equals the probability of the set of points of the sample space that yield that value.*

A (discrete and finite) probabilistic variable, is fully characterised by its *probability distribution*, $P_x : X \rightarrow [0, 1]$, defined, for each $\xi \in X$, by

$$P_x(\xi) = P(x = \xi) = P\left(\{\omega \in \Omega | x(\omega) = \xi\}\right).$$

Note that occasionally we will denote the probability distribution of x by $P(x)$ for short, when this causes no ambiguity. Variables so far considered are individual or scalar. A vector variable, instead, is an ordered collection of scalar variables, $\mathbf{x} = (x_1, x_2, \ldots, x_n)$, and is fully characterised by the joint probability distribution

$$P_{\mathbf{x}}(\xi_1, \xi_2, \ldots, \xi_n) = P\left((x_1 = \xi_1) \cap (x_2 = \xi_2) \cap \cdots \cap (x_n = \xi_n)\right).$$

Concerning notation, the second member of this expression will also be written as $P(x_1 = \xi_1, x_2 = \xi_2, \ldots, x_n = \xi_n)$ and the probability distribution will also be denoted by $P(\mathbf{x})$, as a shorthand notation.

4.1.6 Probabilistic Functions

Consider now a set of objects $A = \{a_1, a_2, \ldots, a_m\}$ having a property described by a variable x with values in a set $X = \{\xi_1, \xi_2, \ldots, \xi_n\}$. In a deterministic context, a situation is defined by assigning a specific value to the property as expressed by each object: $x_a = \xi = f(a)$, for each a, with $\xi \in X$. Here, x_a is a (deterministic) variable, associated to element a, that acts as a parameter, and f a deterministic function $f : A \to X$. These two perspectives, the parametrical variable and the function, although formally equivalent, are somewhat complementary as it will appear in the following. Again, we look for a way for giving a precise meaning to the expression $P(\xi = f(a))$ that intuitively accounts for uncertainty. Similarly to what we have done above, we consider a collection of functions from A to X, a probability space $\mathcal{S} = (\Omega, \mathcal{F}, P)$ and we establish a correspondence between such functions and the points of the sample space, Ω. This time, it is general enough to assume a one-to-one correspondence, so that the functions can be directly labelled by the point ω to which they correspond, thus forming the set $\{f_\omega | \omega \in \Omega\}$. Then, the statement $\xi = f(a)$ is verified for all those functions in the set that associate ξ to a. Consequently, we obtain:

$$P(\xi = f(a)) = P(\{\omega | f_\omega(a) = \xi\}) = \sum_{\omega | f_\omega(a) = \xi} P(\omega).$$

Note that for each fixed a, this yields the definition of the probabilistic variable x_a. So *a probabilistic function is equivalent to a collection of probabilistic variables*, $\{x_a | x_a : \omega \to X, a \in A\}$. The complete characterisation of such a collection of probabilistic variables is provided by the joint probability distribution

$$P_{x_{a_1}, x_{a_2}, \ldots, x_{a_n}}(\xi_1, \xi_2, \ldots, \xi_n) = P((x_{a_1} = \xi_1) \cap (x_{a_2} = \xi_2) \cdots \cap (x_{a_n} = \xi_n)),$$

where each ξ_i spans X. Each point, $\boldsymbol{\xi} = (\xi_1, \xi_2, \ldots, \xi_n)$, of this distribution corresponds to the statement/event $(x_{a_1} = \xi_1) \cap (x_{a_2} = \xi_2) \cdots \cap (x_{a_n} = \xi_n)$, which in turn corresponds to the statement/event $(f(a_1) = \xi_1 \cap f(a_2) = \xi_2 \cap \cdots \cap f(a_n) = \xi_n)$, which defines the function f. Therefore, each point of the domain of the joint probability distribution of the collection of probabilistic variables

$$\{x_a | x_a : \omega \to X, a \in A\}$$

corresponds to a function of the collection of functions

$$\{f_\omega | f_\omega : A \to X, \omega \in \Omega\}$$

and it has the same probability. Thus, we obtain:

Table 4.1 A simple example of a probabilistic function

	a	b	c	$P(f_i)$
f_1	0	0	0	0.01
f_2	0	0	1	0.05
f_3	0	1	0	0.05
f_4	0	1	1	0.10
f_5	1	0	0	0.05
f_6	1	0	1	0.10
f_7	1	1	0	0.10
f_8	1	1	1	0.54
	x_a	x_b	x_c	

$$\begin{aligned} P_{x_{a_1},x_{a_2},\ldots,x_{a_n}}(\xi_1, \xi_2, \ldots, \xi_n) &= P((x_{a_1} = \xi_1) \cap (x_{a_2} = \xi_2) \cdots \cap (x_{a_n} = \xi_n)) \\ &= P(f_\omega | (f_\omega(a_1) = \xi_1 \cap f_\omega(a_2) \\ &= \xi_2 \cap \cdots \cap f_\omega(a_n) = \xi_n)). \end{aligned}$$

This concept is very important and not so easy to understand. A simple numerical example may help. Let $A = \{a, b, c\}$ and $X = \{0, 1\}$. Then, there are eight functions from A to X, which are listed in Table 4.1. Each row of numbers represents both a point of the joint probability distribution and a function, with its associated probability. For example,

$$P_{x_a x_b x_c}(0, 1, 1) = P((x_a = 0) \cap (x_b = 1) \cap (x_c = 1)) = P(f_4) = 0.1.$$

The same structure may serve also as a model for another class of situations, where there is just one object, with a property, x, that can take different values at different times. This can be modelled by interpreting A as a set of time instants, $A = T = \{t_1, t_2, \ldots, t_n\}$. The resulting structure is often called a *stochastic process* and comes into play in dynamic measurement, to be treated in Chap. 12.

4.1.7 Probabilistic Relations

Consider now the case of a set of objects A, where a weak order relation is defined, as we have amply discussed in Chap. 3. In a deterministic context, for each pair of objects, say a, b, we can definitely assess one, and only one, of the three possibilities: $a \succ b$, $a \sim b$ or $a \prec b$. Yet, as we have discussed in Chap. 2, in general there are good reasons for looking for a probabilistic formulation. This requires us to give a precise meaning to, say, the expression $P(a \succ b) = \xi$, with $\xi \in [0, 1]$.

To achieve that we should firstly note that an order relation may be understood in two ways:

Table 4.2 An illustrative example of a probabilistic order structure on $A = \{a, b, c\}$

$\succsim_\omega$	Ordering	$m(a)$	$m(b)$	$m(c)$	P	$\mathcal{A}_\omega$
$\succsim_1$	$a \succ b \succ c$	3	2	1	0.2	$\mathcal{A}_1 = (A, \omega_1)$
$\succsim_2$	$a \succ c \succ b$	3	1	2	0.2	$\mathcal{A}_2 = (A, \omega_2)$
$\succsim_3$	$b \succ a \succ c$	2	3	1	0.0	$\mathcal{A}_3 = (A, \omega_3)$
$\succsim_4$	$b \succ c \succ a$	1	3	2	0.0	$\mathcal{A}_4 = (A, \omega_4)$
$\succsim_5$	$c \succ a \succ b$	2	1	3	0.0	$\mathcal{A}_5 = (A, \omega_5)$
$\succsim_6$	$c \succ b \succ a$	1	2	3	0.0	$\mathcal{A}_6 = (A, \omega_6)$
$\succsim_7$	$a \sim b \succ c$	2	2	1	0.1	$\mathcal{A}_7 = (A, \omega_7)$
$\succsim_8$	$a \sim c \succ b$	2	1	2	0.1	$\mathcal{A}_8 = (A, \omega_8)$
$\succsim_9$	$b \sim c \succ a$	1	2	2	0.0	$\mathcal{A}_9 = (A, \omega_9)$
$\succsim_{10}$	$a \succ b \sim c$	2	1	1	0.3	$\mathcal{A}_{10} = (A, \omega_{10})$
$\succsim_{11}$	$b \succ a \sim c$	1	2	1	0.0	$\mathcal{A}_{11} = (A, \omega_{11})$
$\succsim_{12}$	$c \succ a \sim b$	1	1	2	0.0	$\mathcal{A}_{12} = (A, \omega_{12})$
$\succsim_{13}$	$a \sim b \sim c$	1	1	1	0.1	$\mathcal{A}_{13} = (A, \omega_{13})$

For the sake of simplicity, we have taken $\Omega = \{1, 2, \ldots, 13\}$

- in a specific meaning, when we write $a \succsim b$, we intend that the relation $\succsim$ *holds for the pair* $(a, b) \in A$,
- in a general meaning, when we consider the relation $\succsim$ on A, *we refer to the set of all the pairs of elements of A for which it holds.*

Consider, for example, the set $A = \{a, b, c\}$. A weak order on A, in the general meaning, is defined once that have listed all the pairs of elements that satisfy it. For example $\{(a, b), (b, c), (a, c)\}$ is one such an order, in which a is greater than b and b is greater than c. The same ordering can be displayed, more explicitly, as $a \succ b \succ c$. In Table 4.2, all the possible orders/orderings for A are listed and (conventionally) numbered.

Similarly to what we have done, in the previous section, with probabilistic functions, we introduce a probability space, $\mathcal{S} = (\Omega, \mathcal{F}, P)$, and we establish a one-to-one correspondence between each order and each point, ω, of the sample space. So each order will obtain the same probability of the associated sample point:

$$P(\succsim_\omega) = P\{\omega\}, \quad \text{for each } \omega \in \Omega.$$

A very simple example of one such probability assignment is provided in Table 4.2. This probability assignment is applied to relations intended in the "general sense" defined above. What about relations in the specific meaning? Very simple: consider for example $a \succ b$: what is its probability? If we look at the table, considering only orderings with a non-null probability, we see that it holds true in orderings 1, 2, 8 and 10, whilst it does not in 7 and 13. So its probability is:

$$P(a \succ b) = P(\succsim_1) + P(\succsim_2) + P(\succsim_8) + P(\succsim_{10}) = 0.8.$$

As a general rule, we can simply state that, for each $a, b \in A$,

$$P(a \succcurlyeq b) = P\{\omega | a \succcurlyeq_\omega b\}.$$

So far for probabilistic relations. As a further straightforward generalisation, we can consider *probabilistic relational structures*. Look again at Table 4.2, it is apparent that there is a correspondence between each order, $\succcurlyeq_\omega$ and the order structure $\mathcal{A}_\omega = (A, \succcurlyeq_\omega)$, that appears in the last column. So the above rule for calculating the probability of a generic relation can be restated as

$$P(a \succcurlyeq b) = P\{\mathcal{A}_\omega | a \succcurlyeq_\omega b\}.$$

This is immaterial if only one relation is involved, but it allows applying this procedure also to structures that have more than one associated relation.

Lastly, before concluding, let us briefly consider some possible interpretations of such probability assignments. A pretty straightforward one arises in the case of jury testing. Suppose that the elements of A are sounds and that the jury of 100 people has been asked to rate them according to their pleasantness. Suppose that 30 of them have provided the ordering $\succcurlyeq_{10}$, 20 the ordering $\succcurlyeq_1$, 20 the ordering $\succcurlyeq_2$, 10 the ordering $\succcurlyeq_7$, 10 the ordering $\succcurlyeq_8$ and 10 the ordering $\succcurlyeq_{13}$. Then, the probability assignment may be interpreted as reflecting the preferences of the judges. In the case of physical measurement, we may think to have 100 equally reliable comparators and that again they have rated the objects as above.

4.1.8 Continuity

The notion of continuum was one of the great achievements of modern science. Such thinkers as Newton and Leibiniz competed for the priority in developing the mathematics of infinitesimal calculus [11]. Once scientists gained that idea and mastered its mathematics, they became used to adopting continuity as a standard assumption for representing reality. In spite of the experimental evidence of just noticeable thresholds in perception, psychologists did not hesitate to speak of perceptual continua for representing a person's inner world.

On the other hand, in science, a parsimony principle is generally accepted: models should be as simple as possible and unnecessary assumptions should be avoided [12]. As we have discussed in Chap. 3, in measurement, there seems to be no real need for a continuous representation and we develop the theory in this book for discrete and finite structures. Yet, since using continuous variables is common practice in measurement also, we will re-formulate some of the results of the next chapter in terms of continuous probabilistic (or random) variables. We do not examine in detail such a notion, which is otherwise amply treated in all good textbooks in probability and mathematical statistics [5, 13]. We simply mention that dealing with continua requires us to assume countably infinite additivity, amongst the axioms of probability, that the domain of probability must be a sigma-algebra (which again means that an infinite union of events must be well defined) and a continuous probabilistic variable is now a mapping $x : \Omega \rightarrow \mathbb{R}$. The probability distribution now becomes a *probability density function*, $p_x(\xi)$, that allows calculating the probability that the variable takes value in each interval, finite or infinite, or reals numbers. That is

$$P(x_1 \leq x < x_2) = \int_{x_1}^{x_2} p_x(\xi)\mathrm{d}\xi.$$

Furthermore, the important principles of total probability and of the Bayes–Laplace rule can now be restated as follows:

$$p_y(\xi) = \int_X p_{y|x}(\xi, \zeta)p_x(\zeta)\mathrm{d}\zeta,$$

and

$$p_{x|y}(\xi, y_0) = \frac{p_{y|x}(y_0, \xi)p_x(\xi)}{\int_X p_{y|x}(y_0, \zeta)p_x(\zeta)\mathrm{d}\zeta}.$$

Note that the following shorthand notation can be used instead:

$$p(y) = \int_X p(y|x)p(x)\mathrm{d}x,$$

and

$$p(x|y) = \frac{p(y|x)p(x)}{\int_X p(y|x)p(x)\mathrm{d}x},$$

where arguments and variables are denoted by the same symbols.

Other useful results on continuous probabilistic variables will be recalled in the next chapter, when needed.

4.1.9 Non-probabilistic Approaches to Measurement Uncertainty

If we look back to the definition of probability, we note that additivity is its main feature. Note that both the principle of total probability and the Bayes–Laplace rule rely on additivity. Is it really needed? If we remember our discussion on the aspects of probability, ontic and epistemic, we recognise that this assumption is mainly required under an ontic perspective. If we favour the epistemic side and we essentially look for a logic to express uncertainty, we may be tempted to drop this assumption. This is what happens, for instance, in the remarkable *evidence theory*. This theory includes probability as a special case but, in its most general formulation, does not require additivity. As an example, if we look back at the Bayes–Laplace rule, we may interpret B as a symptom, such as a headache, and the A_j as the possible causes, such as high blood pressure, visual stress, etc; in this perspective, the rule provides an excellent procedure for a rational diagnosis! But see the problem: the additivity property requires that *all possible causes* be considered, which is perhaps not always the case in practical application! So evidence theory may perhaps provide an interesting alternative approach to a general theory of measurement. Interesting results have been obtained, for example, by Ferrero and Salicone [14], as regards the expression of measurement uncertainty, or by Benoit [15], mainly with reference to the scale aspects. Studies in this direction are still in their infancy however and, in my opinion, should be encouraged.

4.2 Probabilistic Representations

We are now ready to provide a probabilistic counterpart of the representation theorems presented in Chap. 3, as anticipated in Chap. 2.

Consider again the deterministic representation for weak orders,

$$a \succsim b \Leftrightarrow m(a) \geq m(b).$$

As we have just seen, it is possible to regard $a \succsim b$ as an event/statement, to which a probability can be assigned, $P(a \succsim b)$. On the other hand, in a probabilistic environment, the assignment of numbers to objects will no longer be unique. The function m will now be probabilistic rather than deterministic. Now, what can we require is that the probability of observing a relation between two objects is the same as the probability of obtaining the same kind of relation between the associated probabilistic functions. So, in the case of order, we now write

$$P(a \succsim b) = P(m(a) \geq m(b)).$$

Let us now show, in an informal way, how can this work. Looking at Table 4.2, we observe that, for each of the possible orders, it is possible to find a number assignment to the objects, $m(a)$, $m(b)$ and $m(c)$, that correctly represents such ordering. Actually, there are other possible assignments, but suppose that we adopt some conventional rules in order to make this assignment always unique. Such assignments are shown in columns 3–5 of the table: note that each number assignment has now the same probability of the associated order. So it is also easy to check that the representation above holds: whenever $a \succsim b$ is satisfied, a numerical assignment satisfying $m(a) \geq m(b)$ is adopted and consequently, the two probabilities are the same.

Furthermore, as we have discussed in Sect. 1.6, it is also possible to assign probabilistic variables to the objects, x_a, x_b and x_c, by simply setting, for each structure, $x_a = m(a), x_b = m(b)$ and $x_c = m(c)$. The probability distributions for those probabilistic variables can be calculated and they appear as in Fig. 4.5 [16].

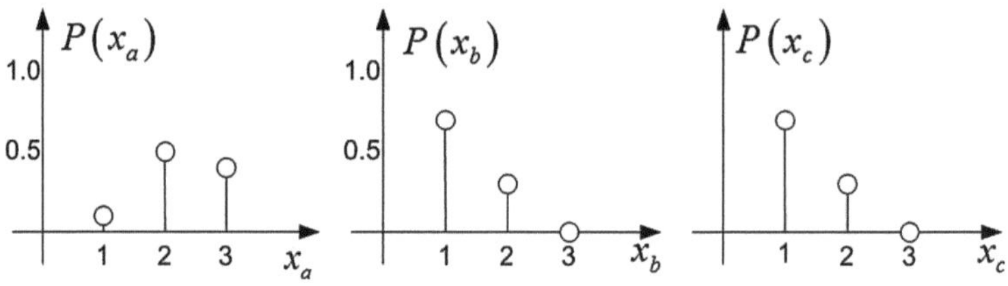

Fig. 4.5 Probability distributions for the probabilistic variables x_a, x_b and x_c

In this example, qualitatively, object a is on average greater than the others, and b is "equivalent" to c, in that it has the same probability distribution. It is important to note that the measure value, which in the deterministic case was uniquely identifiable as a function of the object, is now a probabilistic variable associated to it. This is a big change in perspective. In fact, in the deterministic case, each object manifests the characteristic under investigation in one (and only one) way. We can call *state* (of the object) the way in which an object manifests a characteristic. In the deterministic case, there is a one-to-one correspondence between objects and states and between states and (measure) values. In the probabilistic representation, due to the fuzziness of empirical relations, we can either say that there is a one-to-many correspondence between objects and states, whilst there still is a one-to-one correspondence between states and values (or that there still is a one-to-one correspondence between objects and states, but that each state is describable by a probabilistic distribution).[1]

We are now ready to consider probabilistic representations in a formal way, for order, interval, intensive and extensive structures, respectively.

4.3 Probabilistic Fundamental Scales

4.3.1 Order Structures

Consider a formal definition of a probabilistic order structure.

Definition 4.1 Let A be a finite (not empty) set of objects manifesting the property x. Let $\mathcal{A} = (A, \succsim)$ denote a generic (empirical) order structure on A, and let $\mathcal{E}$ be a finite collection of distinct order structures on A. Then, a(n) (empirical) probabilistic order structure is a probability space $\mathcal{S}_{\mathcal{E}} = (\Omega, \mathcal{F}, P)$, where the elements of Ω are in a one-to-one correspondence with the elements of $\mathcal{E}$, $\mathcal{F}$ is an algebra on Ω, and $P : \mathcal{F} \rightarrow [0, 1]$ is a probability function.

Note that this definition resembles exactly what we have done in the illustrative example. Then, the representation theorem may be stated and proved as follows.

[1] The former seems to be the perspective of statistical mechanics [6], the latter that of quantum mechanics [17].

Theorem 4.2 *Let A be a finite (not empty) set of objects manifesting the property x, and let* $\mathcal{S}_{\mathcal{E}} = (\Omega, \mathcal{F}, P)$ *be a probabilistic order structure on A. Then, there is a probabilistic function* $m = \{m_\omega : A \to \mathbb{N}, \omega \in \Omega\}$*, and a vector probabilistic variable* $\mathbf{x} = (x_a | x_a : \Omega \to \mathbb{N}, a \in A)$*, such that, for each* $a, b \in A$*,*

$$P(a \succsim b) = P(m(a) \geq m(b)) = P(x_a \geq x_b).$$

Proof For each $\omega \in \Omega$, there is one and only one structure, $\mathcal{A}_\omega = (A, \succsim_\omega) \in \mathcal{E}$ that corresponds to it. Let N be the number of elements in A, $n_\omega \leq N$ the number of equivalence classes in $\mathcal{A}_\omega$, $n = max\{n_\omega | \omega \in \Omega\}$, $X = \{1, 2, \ldots, n\}$.

Let $m_\omega : A \to X \subset \mathbb{N}$ be a measure function, constructed as in the proof of Theorem 3.8, that satisfies the representation of the weak order $\succsim_\omega$, associated to $\mathcal{A}_\omega$. We now define the probabilistic function m as the set of all such functions,

$$m = \{\{m_\omega | m_\omega : A \to X, \omega \in \Omega\}\}$$

with their associated probabilities:

$$P(m_\omega) = P(\omega).$$

Similarly, we introduce the vector probabilistic variable

$$\mathbf{x} = (x_a | x_a \colon \Omega \to \mathbb{N}, a \in A),$$

where each component is defined by

$$x_a(\omega) = m_\omega(a),$$

with

$$P(\mathbf{x}_\omega) = P(\omega).$$

(In order for $\mathbf{x}$ to be well defined on X^N, we assign null probability to the points of X^N not included in the previous assignment.)

We thus obtain

$$P(a \succsim b) = P\{\omega | a \succsim_\omega b\} = P\{\omega | m_\omega(a) \geq m_\omega(b)\} = P(m(a) \geq m(b)).$$

On the other hand, we also obtain

$$P(a \succsim b) = P\{\omega | a \succsim_\omega b\} = P\{\omega | x_a(\omega) \geq x_b(\omega)\} = P(x_a \geq x_b),$$

which completes the proof. □

4.3.2 Difference Structures

The treatment of difference structures and related interval scales essentially follows the same pattern. We firstly define a probabilistic difference structure.

Definition 4.3 Let A be a finite (not empty) set of objects manifesting the property x. Let $\mathcal{A} = (A, \succsim_d)$ denote a generic (empirical) difference structure on A, and let $\mathcal{E}$ be a finite collection of distinct difference structures on A. Then, a(n) (empirical) probabilistic difference structure is a probability space $\mathcal{S}_\mathcal{E} = (\Omega, \mathcal{F}, P)$, where the elements of Ω are in a one-to-one correspondence with the elements of $\mathcal{E}$, $\mathcal{F}$ is an algebra on Ω, and $P : \mathcal{F} \to [0, 1]$ is a probability function.

Then, the representation theorem immediately follows.

Theorem 4.4 *Let A be a finite (not empty) set of objects manifesting the property x, and let $\mathcal{S}_\mathcal{E} = (\Omega, \mathcal{F}, P)$ be a probabilistic difference structure on A. Then, there is a probabilistic function $m = \{m_\omega : A \to \mathbb{N}, \omega \in \Omega\}$, and vector probabilistic variable $\mathbf{x} = (x_a | x_a : \Omega \to \mathbb{N}, a \in A)$, such that, for each $a, b, c, d \in A$,*

$$P(\Delta_{ab} \succsim_d \Delta_{cd}) = P(m(a) - m(b) \geq m(c) - m(d)) = P(x_a - x_b \geq x_c - x_d).$$

Proof follows closely the same pattern of Theorem 4.2, and thus, it is omitted.

4.3.3 Intensive Structures

The definition of a probabilistic intensive structure follows the usual pattern.

Definition 4.5 Let A be a finite (not empty) set of objects manifesting the property x. Let $\mathcal{A} = (A, \succsim_d, \succsim_r)$ denote a generic (empirical) intensive structure on A, and let $\mathcal{E}$ be a finite collection of distinct intensive structures on A. Then, a(n) (empirical) probabilistic intensive structure is a probability space $\mathcal{S}_\mathcal{E} = (\Omega, \mathcal{F}, P)$, where the elements of Ω are in a one-to-one correspondence with the elements of $\mathcal{E}$, $\mathcal{F}$ is an algebra on Ω, and $P : \mathcal{F} \to [0, 1]$ is a probability function.

The formulation of the representation theorem is not so immediate, since the representation does not hold for all the quadruples, $a, b, c, d \in B$, but only for those that satisfy a special condition that, in practice, ensures that their ratios are comparable on a finite scale. This condition has little practical importance, since we can often assume, in a practical application, that the reference scale is wide enough. Yet, from a theoretical perspective, this aspect must be managed. We will simply reserve the representation to those quadruples of elements that satisfy the requirement in all the structures under consideration. The reader is invited to not overestimate the importance of this assumption, that is, instead expected to have little impact in practical application.

Theorem 4.6 *Let A be a finite (not empty) set of objects manifesting the property x, and let* $\mathcal{S}_{\mathcal{E}} = (\Omega, \mathcal{F}, P)$ *be a probabilistic intensive structure on A. For each* $\omega \in \Omega$, *let* $\mathcal{A}_\omega$ *be the structure associated to it,* S_ω *the corresponding series of standards,* $S_{0\omega}$ *and* B_ω *defined as in Definition 3.19,* $B = \bigcap_{\omega \in \Omega} B_\omega$. *Then, there is a probabilistic function* $m = \{m_\omega : B \to \mathbb{N}, \omega \in \Omega\}$ *and a vector probabilistic variable* $\mathbf{x} = (x_a | x_a : \Omega \to \mathbb{N}, a \in B)$, *such that, for* $a, b, c, d \in B_\omega$, *for each* $\omega \in \Omega$, *with* $a \sim s_i, b \sim s_j, c \sim s_k, d \sim s_l$, *for* $s_i, s_j, s_k, s_l \in S_\omega, i, j, k, l, j \cdot l, i \cdot l, k \cdot j \leq n_\omega$,

$$P(\Delta_{ab} \succeq_d \Delta_{cd}) = P(m(a) - m(b) \geq m(c) - m(d)) = P(x_a - x_b \geq x_c - x_d),$$

$$P(a/b \succeq_r c/d) = P\left(\frac{m(a)}{m(b)} \geq \frac{m(c)}{m(d)}\right) = P\left(\frac{x_a}{x_b} \geq \frac{x_c}{x_d}\right).$$

With the above assumptions, proof runs as in Theorem 4.2.

4.3.4 Extensive Structures

Extensive structures run smoothly.

Definition 4.7 Let A be a finite (not empty) set of objects manifesting the property x. Let $\mathcal{A} = (A, \succeq, \circ)$ denote a generic (empirical) extensive structure on A, and let $\mathcal{E}$ be a finite collection of distinct extensive structures on A. Then, a(n) (empirical) probabilistic extensive structure is a probability space $\mathcal{S}_{\mathcal{E}} = (\Omega, \mathcal{F}, P)$, where the elements of Ω are in a one-to-one correspondence with the elements of $\mathcal{E}$, $\mathcal{F}$ is an algebra on Ω, and $P : \mathcal{F} \to [0, 1]$ is a probability function.

Theorem 4.8 *Let A be a finite (not empty) set of objects manifesting the property x, and let* $\mathcal{S}_{\mathcal{E}} = (\Omega, \mathcal{F}, P)$ *be a probabilistic extensive structure on A. Then, there is a probabilistic function* $m = \{m_\omega : A \to \mathbb{N}, \omega \in \Omega\}$ *and a vector probabilistic variable* $\mathbf{x} = (x_a | x_a : \Omega \to \mathbb{N}, a \in A)$, *such that, for each* $a, b, c \in A$, $(a, b) \in B$,

$$P(a \sim b \circ c) = P(m(a) = m(b) + m(c)) = P(x_a = x_b + x_c).$$

Again proof follows closely that of Theorem 4.2.

4.4 Probabilistic Derived Scales

4.4.1 An Introductory Example

The probabilistic formulation for derived scales is not so straightforward as for fundamental ones. Let us see what happens by a simple but interesting numerical example.

Table 4.3 An illustrative example of a probabilistic cross-order structure on $A = \{a_1, a_2\}$, $B = \{b_1, b_2\}$, where g has been taken as the identity function, e.g. $g(1) = 1, g(2) = 2$

C_ω	Ordering	$m_y(a_1)$	$m_y(a_2)$	$m_x(b_1)$	$m_x(b_2)$	P	$\varphi_\omega = \varphi^{(i)}$
C_1	$a_1 \sim b_1 \succ b_2 \sim a_2$	2	1	2	1	0.6	$\varphi_1 = \varphi^{(1)}$
C_2	$a_1 \sim b_2 \succ b_1 \sim a_2$	2	1	1	2	0.1	$\varphi_2 = \varphi^{(4)}$
C_3	$a_1 \sim b_1 \sim a_2 \succ b_2$	2	2	2	1	0.025	$\varphi_3 = \varphi^{(2)}$
C_4	$a_1 \sim b_2 \sim a_2 \succ b_1$	2	2	1	2	0.025	$\varphi_4 = \varphi^{(3)}$
C_5	$b_1 \succ b_2 \sim a_1 \sim a_2$	1	1	2	1	0.025	$\varphi_5 = \varphi^{(3)}$
C_6	$b_2 \succ b_1 \sim a_1 \sim a_2$	1	1	1	2	0.025	$\varphi_6 = \varphi^{(2)}$
C_7	$a_1 \sim b_1 \sim b_2 \sim a_2$	1	1	1	1	0.1	$\varphi_7 = \varphi^{(1)}$
C_8	$a_2 \sim b_1 \succ b_2 \sim a_1$	1	2	2	1	0.05	$\varphi_8 = \varphi^{(4)}$
C_9	$a_2 \sim b_2 \succ b_1 \sim a_1$	1	2	1	2	0.05	$\varphi_9 = \varphi^{(1)}$

Let $A = \{a_1, a_2\}$ and $B = \{b_1, b_2\}$. The list of possible cross-orders is presented in Table 4.3, where each line corresponds to one structure, C_ω. There are nine such structures; for notation, simplicity we have taken $\omega_i = i$, for $i = 1, \ldots, 9$.

One major problem is that whilst in the deterministic case for each pair of elements $a_1, a_2 \in A$, we can find two elements $b_1, b_2 \in B$, where b_1 corresponds to a_1 and b_2 to a_2, for which the representation holds true, here, for each $a_1, a_2 \in A$ the elements of B that correspond to them in general change from one structure to the other, as can be checked in the table. For example, in C_1, b_1 corresponds to a_1 and b_2 to a_2, whilst in C_2, b_1 corresponds to a_2 and b_2 to a_1.

The way out of this is to introduce a set of functions, $\varphi^{(i)} : A \to B$, that define all possible such correspondences. In our example, there are four such functions:

$$\begin{aligned}
\varphi^{(1)} &= \{(a_1, b_1), (a_2, b_2)\}, \\
\varphi^{(2)} &= \{(a_1, b_1), (a_2, b_1)\}, \\
\varphi^{(3)} &= \{(a_1, b_2), (a_2, b_2)\}, \\
\varphi^{(4)} &= \{(a_1, b_2), (a_2, b_1)\}.
\end{aligned}$$

If we now denote by φ_ω the function appropriate to each structure, C_ω, we have the situation that appears in the last column of the table. Note that the same function, $\varphi^{(i)}$, can appear in more than one structures; so, the functions $\varphi^{(i)}$ *are not* in a one-to-one correspondence with the structures C_ω; the correspondence is rather as shown in the table.

Yet the set of functions $\Phi = \{\varphi^{(i)} | i \in \Omega' = \{1, 2, 3, 4\}\}$, with their associated probabilities, $P(\varphi^{(i)})$, define a probabilistic function, although the reference sample space is not $\Omega = \{1, 2, 3, 4, 5, 6, 7, 8, 9\}$, but $\Omega' = \{1, 2, 3, 4\}$. Let us practice calculating their probabilities: the probability of each function equals, as usually, the sum of the probabilities of the structures where it takes place. We obtain:

$$P(\varphi^{(1)}) = 0.6 + 0.1 + 0.05 = 0.75,$$
$$P(\varphi^{(2)}) = 0.025 + 0.025 = 0.05,$$
$$P(\varphi^{(3)}) = 0.025 + 0.025 = 0.05,$$
$$P(\varphi^{(4)}) = 0.1 + 0.05 = 0.15.$$

We can thus see that, for each $a_1, a_2 \in A$, in each structure, $\mathcal{C}_\omega$, there is a function, φ_ω, that maps them in a pair $b_1, b_2 \in B$, whose measure values in that structure are $m_{x,\omega}(b_1)$ and $m_{x,\omega}(b_2)$. Such values are mapped, by the function g into the corresponding measure values for a_1 and a_2, that is, $m_{y,\omega}(a_1)$ and $m_{y,\omega}(a_2)$, that satisfy, in that structure, the representation. For example, in structure $\mathcal{C}_1$ (first line after the heading in the table), $a_1 \succ a_2$, a_1, a_2 correspond to b_1, b_2, respectively, $m_x(b_1) = 2$, $m_x(b_2) = 1$, $m_y(a_1) = g(m_x(b_1) = 2$, $m_y(a_2) = g(m_x(b_2) = 1$, and $m_y(a_1) > m_y(a_2)$.

Thus, it is possible to calculate the probabilities of the empirical relations and of the numerical representations, as the sum of the probabilities of the structures in which they are verified, and to check that they match.

For example,

$$P(a_1 \succ a_2) = 0.6 + 0.1 = 0.7,$$
$$P(m_y(a_1) = g(m_x(b_1) > m_y(a_2) = g(m_x(b_2)) = 0.6 + 0.1 = 0.7.$$

We are now ready to formalise these ideas.

4.4.2 Probabilistic Cross-Order Structures

The probabilistic framework developed so far can be extended, at least formally, to derived scales as well. Consider a formal definition of a probabilistic cross-order structure.

Definition 4.9 (*Probabilistic cross-order structure*) Let A and B be two finite (not empty) sets of objects manifesting, respectively, the properties y and x. Let $\mathcal{C} = (A, B, \succeq)$ denote a generic (empirical) cross-order structure on $C = A \cup B$, and let $\mathcal{E}$ be a finite collection of distinct such cross-orders. Then, a(n) (empirical) probabilistic cross-order structure is a probability space $\mathcal{S}_\mathcal{E} = (\Omega, \mathcal{F}, P)$, where the elements of Ω are in a one-to-one correspondence with the elements of $\mathcal{E}$, $\mathcal{F}$ is an algebra on Ω, and $P : \mathcal{F} \to [0, 1]$ is a probability function.

A representation theorem follows.[2]

Theorem 4.10 (Representation for probabilistic cross-order structures) *Let $\mathcal{S}_\mathcal{E} = (\Omega, \mathcal{F}, P)$ be a probabilistic cross-order structure for properties y and x. Then, for*

[2] Proof is quite technical and can be omitted, at a first reading, without loss of continuity.

each $a_1, a_2 \in A$, there exist a probabilistic function φ, from A to B, a probabilistic function m_x, from B to $\mathbb{N}$, and a probabilistic function m_y, from A to $\mathbb{N}$, such that, for any monotone increasing (deterministic) function $g : \mathbb{N} \to \mathbb{N}$,

$$P(a_1 \succsim a_2) = P(m_y(a_1) \geq m_y(a_2)) = P(g(m_x(\varphi(a_1))) \geq g(m_x(\varphi(a_2)))).$$

Proof For each $\omega \in \Omega$, there is one and only one structure, $\mathcal{C}_\omega$ that corresponds to it, and two order substructures associated to it, namely $\mathcal{A}_\omega = (A, \succsim_\omega)$ and $\mathcal{B}_\omega = (B, \succsim_\omega)$.

Let N be the number of elements in A, L the number of elements in B, $n_\omega \leq N$ the number of equivalence classes in $\mathcal{A}_\omega$, $l_\omega \leq M$ the number of equivalence classes in $\mathcal{B}_\omega$, $n = \max\{n_\omega | \omega \in \Omega\}$, $l = \max\{n_\omega | \omega \in \Omega\}$, $X = \{1, 2, \ldots, l\}$, and $Y = \{1, 2, \ldots, q\}$, with $q \geq n$.

Consider the set $\Phi = \{\varphi | \varphi : A \to B\}$ of all the functions from A to B. For any structure $\mathcal{C}_\omega$, for each $a \in A$, there exists $b \in B$, such that $a \sim b$. Then, for any a just *fix* one such b; then, there is a function $\varphi \in \Phi$ that maps each a into the fixed corresponding equivalent b: call it φ_ω.

In this way, it is possible to define a probabilistic function φ by establishing a one-to-one correspondence between the elements of Φ and a sample space Ω'. Call, as we have done in the informal discussion above, $\varphi^{(\omega')}$ a generic element (function) in Φ, which corresponds to $\omega' \in \Omega'$. Then, its probability is

$$P(\varphi^{(\omega')}) = P\{\omega | \varphi_\omega = \varphi^{(\omega')}, \omega \in \Omega\}.$$

Consider now any monotonic increasing function $g : X \to Y$ and any $a_1, a_2 \in A$. For each structure $\mathcal{C}_\omega$, there is $b_{1,\omega} = \varphi_\omega(a_1)$ and $b_{2,\omega} = \varphi_\omega(a_2)$, $b_{1,\omega}, b_{2,\omega} \in B$, a function $m_{x,\omega} : B \to X$, a function $m_{y,\omega} : A \to Y$, defined, for each $a \in A$, by

$$m_{y,\omega}(a) = g(m_x(\varphi(a))),$$

such that, if $a_1 \succsim a_2$, then

$$\begin{aligned} &b_{1,\omega} = \varphi_\omega(a_1) \succsim b_{2,\omega} = \varphi_\omega(a_2), \\ &m_{y,\omega}(b_{1,\omega}) \geq m_{y,\omega}(b_{2,\omega}), \text{ and} \\ &m_{y,\omega}(a1) \geq m_{y,\omega}(a1). \end{aligned}$$

Then,

$$\begin{aligned} P(a_1 \succsim a_2) &= P\{\omega | a_1 \succsim_\omega a_2\} \\ &= P\{\omega(g(m_{x,\omega}(\varphi(a_1))) \geq g(m_{x,\omega}(\varphi(a_2))))\} \\ &= P(g(m_x(\varphi(a_1))) \geq g(m_x(\varphi(a_2)))). \end{aligned}$$

□

Table 4.4 Probabilistic representation for fundamental scales

Empirical structure	Probabilistic representation
Order	$P(a \succcurlyeq b) = P(m(a) \geq m(b)) = P(x_a \geq x_b)$
Difference	$P(\Delta_{ab} \succcurlyeq_d \Delta_{cd})$
	$= P(m(a) - m(b) \geq m(c) - m(d))$
	$= P(x_a - x_b \geq x_c - x_d)$
Intensive	$P(\Delta_{ab} \succcurlyeq_d \Delta_{cd}) = P(m(a) - m(b) \geq m(c) - m(d))$
	$= P(x_a - x_b \geq x_c - x_d)$,
	$P(a/b \succcurlyeq_r c/d) = P(\frac{m(a)}{m(b)} \geq \frac{m(c)}{m(d)}) = P(\frac{x_a}{x_b} \geq \frac{x_c}{x_d})$
Extensive	$P(a \sim b \circ c) = P(m(a) = m(b) + m(c)) = P(x_a = x_a + x_a)$

4.4.3 Probabilistic Cross-Difference Structures

Let us formally define cross-difference structures.

Definition 4.11 (*Probabilistic cross-difference structure*) Let A and B be two finite (not empty) sets of objects manifesting, respectively, the properties y and x. Let $\mathcal{C} = (A, B, \succcurlyeq_d)$ denote a generic (empirical) cross-difference structure on $C = A \times A \cup B \times B$, and let $\mathcal{E}$ be a finite collection of distinct such structures. Then, a(n) (empirical) probabilistic cross-difference structure is a probability space $\mathcal{S}_{\mathcal{E}} = (\Omega, \mathcal{F}, P)$, where the elements of Ω are in a one-to-one correspondence with the elements of $\mathcal{E}$, $\mathcal{F}$ is an algebra on Ω, and $P : \mathcal{F} \rightarrow [0, 1]$ is a probability function.

Here is the corresponding representation theorem.

Theorem 4.12 (Representation for probabilistic cross-order structures) *Let $\mathcal{S}_{\mathcal{E}} = (\Omega, \mathcal{F}, P)$ be a probabilistic cross-difference structure for properties y and x. Then, for each $a, b, c, d \in A$, there exist a probabilistic function φ, from A to B, a probabilistic function m_x, from B to $\mathbb{N}$, and a probabilistic function m_y, from A to $\mathbb{N}$, such that, for any linear positive function $g : \mathbb{N} \rightarrow \mathbb{N}$,*

$$\begin{aligned} P(\Delta_{ab} \succcurlyeq \Delta_{cd}) &= P(m_y(a) - m_y(b) \geq m_y(c) - m_y(d)) \\ &= P(g(m_x(\varphi(a))) - g(m_x(\varphi(b))) \geq g(m_x(\varphi(c))) \\ &\quad - g(m_x(\varphi(d)))). \end{aligned}$$

Proof develops along the same line as in the previous theorem and is thus omitted.

4.5 Summary

We have presented a probabilistic representation for both fundamental and derived measurement scales. The corresponding main results are presented in Tables 4.4 and 4.5, respectively.

Table 4.5 Probabilistic representation for derived scales

Empirical structure	Probabilistic representation
Cross-order	$P(a_1 \succsim a_2) = P(m_y(a_1) \geq m_y(a_2))$ $= P(g(m_x(\varphi(a_1))) \geq g(m_x(\varphi(a_2))))$
Cross-difference	$P(\Delta_{ab} \succsim \Delta_{cd}) = P(m_y(a) - m_y(b) \geq m_y(c) - m_y(d))$ $= P(g(m_x(\varphi(a))) - g(m_x(\varphi(b))) \geq g(m_x(\varphi(c))) - g(m_x(\varphi(d))))$

So far for the measurement scale, which ensures that measurement can be made. We have now to consider how to make measurement, that is, we have to consider the measurement process.

References

1. Hacking, I: An introduction to probability and inductive logic. Cambridge Press, Cambridge (2001) (Italian edition: Il Saggiatore, Milano, 2005)
2. Costantini, D.: I fondamenti storico-filosofici delle discipline statistico probabilistiche. Bollati Boringhieri, Torino (2004)
3. Barone, F.: I problemi epistemologici della misurazione. In: Cunietti, M., Mari, L. (eds.) Atti della X Giornata della Misurazione. CLUP, Milano (1992)
4. Narens, L: Theories of Probability: An Examination of Logical and Qualitative Foundations. World Scientific (2007)
5. Papoulis, A.: Probability, Random Variables and Stochastic Processes, 2nd edn. McGraw-Hill, Singapore (1984)
6. Costantini, D.: Verso una visione probabilistica del mondo. GEM, Padova (2011)
7. Kemeny, J.G. In: Schilpp (ed.) The Philosophy of Rudolf Carnap, p. 711. Cambridge University Press, London (1963)
8. Rigamonti, G.: Corso di logica. Bollati Boringhieri, Torino (2005)
9. Garibaldi, U., Scala, E.: Finitary Probabilistic Methods in Econophysics. Cambridge University Press, Cambridge (2010)
10. Haenni, R., Romeijn, J.W., Wheeler, G., Williamson, J.: Probabilistic Logics and Probabilistic Networks. Springer, Dordrecht (2011)
11. Balducci E (1987) Storia del pensiero umano. Edizioni Cremonese, Cittá di Castello
12. Reale, G., Antiseri, D.: Storia della filosofia. Bompiani, Milano (2008)
13. Monti, M., Pierobon, G.: Teoria della probabilità. Zanichelli, Bologna (2000)
14. Ferrero, A., Salicone, S.: Uncertainty: only one mathematical approach to its evaluation and expression? IEEE Trans. Instrumentation and Measurement **61**, 2167–2178 (2012)
15. Benoit, E.: Uncertainty in fuzzy scales based measurements. Paper presented at the 14th Joint Int. IMEKO TC1+TC7+TC13 Symposium, Jena, 31 Aug–2 Sept 2011
16. Rossi, G.B.: A probabilistic theory of measurement. Measurement **39**, 34–50 (2006)
17. Ghirardi, G.C.: Un'occhiata alle carte di Dio. Il Saggiatore, Milano (2003)

Chapter 5
The Measurement Process

5.1 How Can We Measure?

In order to measure something, we have first to construct a reference scale for the quantity of interest. This may be done, at least in principle, by selecting a subset of objects which is representative of all the possible states of the quantity and by assigning them a measure, in order to constitute a reference scale. Suppose, for example, that we have a finite set A of N objects and that there are $n < N$ distinct states. Then, we may select a *series of standards*,

$$S = \{s_i | i = 1, \ldots, n\}, \tag{5.1}$$

and assign a measure to them to form a reference scale,

$$R = \{(s_i, m(s_i)) | i = 1, \ldots, n\}. \tag{5.2}$$

This is just a reference pattern that may correspond to different practical implementations. In psychophysics, the implementation may follow closely the reference pattern: this is called "scaling" in that area. In physics, if the scale is additive, one may take advantage of the summing properties for an efficient implementation. In mass measurement, it is possible to realise a large number of elements of the scale by means of a limited number of real objects. In length measurement, a sophisticated optical standard may be used, consisting in a light beam produced by a high-precision laser. Here, the elements of the scale are the propagation planes of the electromagnetic field associated with the laser beam.

In any case, independently of its actual realisation, we may think, at this stage, to have a reference scale, as defined by formula 1, at our disposal. Then, the point is how to measure objects not included in the scale: this is the aim of the *measurement process*.

Basically, we may do that in two ways, *direct* and *indirect*, as we have already mentioned in Chap. 1: see Fig. 1.5 as a reminder. A direct measurement procedure

G. B. Rossi, *Measurement and Probability*, Springer Series in Measurement Science and Technology, DOI: 10.1007/978-94-017-8825-0_5,

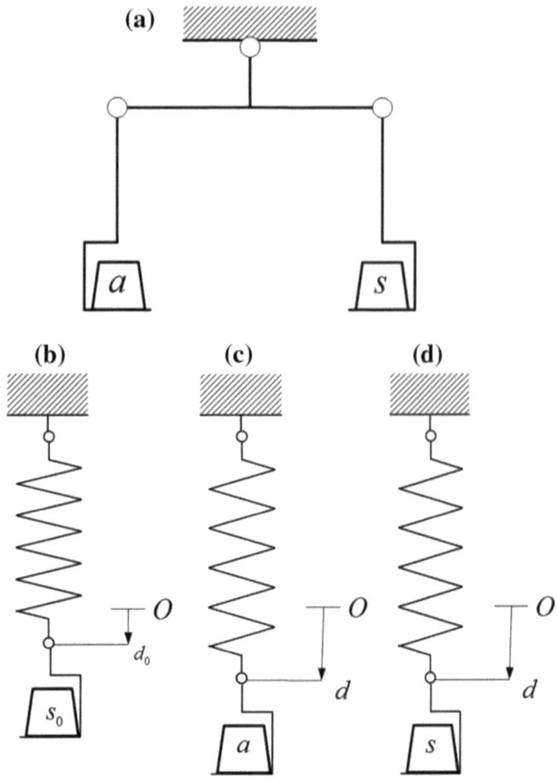

Fig. 5.1 Mass measurement: **a** direct versus **b–d** indirect

consists in comparing an unknown object, a, with the reference scale, in order to find a standard, s, which is equivalent to it. If the scale properly represents all the possible states of the quantity, we may assume that we will always find one such a standard. This is illustrated, for the case of mass, in Fig. 5.1a, where an equal-arm balance is used as the comparator.

Globally, the measurement process may be seen as a mapping from objects into measurement values. If we denote the measurement value by $\hat{x}$, we may describe the measurement process by the function γ,

$$\hat{x} = \gamma(a). \tag{5.3}$$

Let us briefly see how this function is defined in the case of direct measurement. The empirical evidence provided by the mass comparator is the equivalence between a and a standard, s, of the reference scale. In symbols: $a \sim s$. Let $x = m(a)$ be the *unknown* measure value[1] of a, $\hat{x} = m(s)$ the *known* value of the standard s. Then,

[1] Note that the function m does not represent an empirical operation, as γ instead does, but rather the (mathematical) existence of a correspondence between objects and numbers, as ensured by the

the function γ associates with the object a the value of the standard that has been selected by the comparator as equivalent to a. In formal terms,

$$\hat{x} = \gamma(a) \Leftrightarrow (a \sim s \text{ and } m(s) = \hat{x}). \tag{5.4}$$

In the indirect approach, instead of comparing directly the object with the reference scale, we use a calibrated measuring system (or instrument). *Calibration* is a fundamental operation in measurement, in which the behaviour of a measuring device is assessed by inputting it with the standard objects of a reference scale, whose values are known, and by recording the corresponding outputs of the device, which we will call (instrument) *indications*. In this way, the behaviour of the instrument can be described by a *calibration function* (or *calibration curve*).[2] We denote such a function by f, and we write $y = f(x)$, where x is the generic value of a standard object and y is the corresponding instrument indication. It is quite obvious to assume that the instrument will behave in the same way during the measurement process, and thus, after observing the indication y, it is possible to make an inference on the value of the measurand, since f is known, thanks to the preliminary calibration operation. Since the calibration operation involves a comparison of the instrument with the scale, using the instrument is equivalent to comparing the object with the scale *indirectly*, that is, through the intermediation of the instrument. Let us illustrate this procedure in the case of mass measurement, as shown in Fig. 5.1b, c. Here, the measuring system includes a spring, oriented according to the gravity field. Let us first discuss its calibration. We assume that the spring behaves like a linear elastic mechanical device and is thus governed by the equation

$$F = mg = k_s d, \tag{5.5}$$

where F is the force due to the weight of the loaded mass, m the loaded mass, k_s the stiffness of the spring and d the displacement of the free extreme of the spring when the mass is loaded. We may rewrite this equation as

$$d = \frac{g}{k_s} m = km, \tag{5.6}$$

where k is now the *sensitivity* of the measuring device. The goal of calibration is to determine experimentally the sensitivity k, since once k is known, the behaviour of the instrument is completely defined. This may be done by applying a standard s_0, whose known value is x_0 , and by recording the corresponding displacement of the free end of the spring, d_0 (Fig. 5.1b). Then, we may estimate k by

---

representation theorems that we have discussed in Chap. 3. In performing measurement, therefore, x has to be regarded as an unknown value, whilst $\hat{x}$ is the value that we *actually obtain* as the result of the measurement process.

[2] Calibration will be treated in greater detail in Chap. 11.

$$k = \frac{d_0}{x_0}. \tag{5.7}$$

After performing calibration, and having thus obtained a proper value for k, we are ready for making measurements. When measuring an object a, if we obtain the indication d, as shown in Fig. 5.1c, we can assign to a the measured mass value by solving Eq. (5.6) in respect of m, that is, trivially,

$$\hat{m} = d/k. \tag{5.8}$$

Compare now Fig. 5.1c, d. If s is the standard equivalent to a, s produces the same displacement d as the element a, and so through the indirect procedure, we assign to a the same value of the standard which is equivalent to it, as in direct measurement case, as expressed by formula (5.4). The difference is in the way we perform the comparison. In the direct procedure, both the object and the standard are inputted to the comparator at the same time: we call this a synchronous comparison or a comparison by *opposition*. In the indirect case, instead, they are inputted in different moments: this can be called an asynchronous comparison or by *substitution*. This latter procedure is often more practical and thus more often applied. Yet they are conceptually equivalent, at least at the present level of abstraction, and they can thus be treated in a unified way.

Let us now generalise what we have seen in this example. Let a be a generic element, $x = m(a)$ its (unknown) measure value (in respect of some quantity of our interest)[3] and y the instrument indication. Then, the behaviour of the instrument may be characterised by the calibration function

$$y = f(x). \tag{5.9}$$

Once that calibration has been performed and f is thus known, the measurement of an unknown object, a, can be performed by inputting it to the measuring system. If y is the indication obtained by the instrument, the measurement value

$$\hat{x} = f^{-1}(y) \tag{5.10}$$

can be assigned.

It is now time to formalise a little bit the above considerations in a general model of the measurement process, deterministic first and then probabilistic.

[3] Note that here m denotes the "measure" function, amply discussed in Chap. 3, and not "mass", as in the previous example.

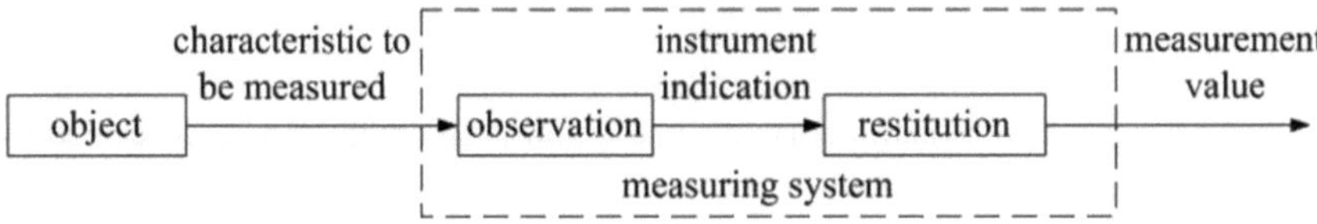

Fig. 5.2 Scheme of the general measurement process

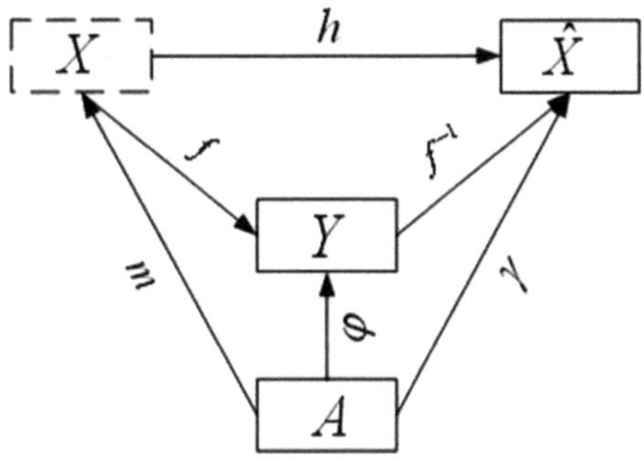

Fig. 5.3 Scheme of the deterministic model

5.2 Deterministic Model of the Measurement Process

Looking for a general model, unique for direct and indirect measurements, we propose to parse the measurement process into two subprocesses that we call *observation* and *restitution*, respectively: see Fig. 5.2.

In the observation phase, the measurand is inputted to the measuring system that produces an indication. In the restitution phase, the indication of the measuring system is interpreted on the basis of the calibration function and the measurement value is obtained. Measurement is the concatenation of observation and restitution. The two phases are conceptually distinct, since observation is where information is produced, thanks to the physical (or psychological) interaction between the object and the measuring system, whilst restitution is an information processing phase.[4]

Let us now present and discuss the general measurement model, illustrated in its various aspects, in Fig. 5.3.

Consider then a generic object a in regard of the property of our interest. In the observation phase, we obtain an indication for it: this can be described by a function, φ:

[4] Restitution may be sometimes very simple, as in the case of direct measurement, where it just consists in assigning to the measurand the same value of the standard that has been recognised as equivalent to it. In other cases, instead, it may be very complicated and challenging, as it happens in image-based measurement, where it involves sophisticated image processing procedures. Anyway, it is conceptually always present, since the instrument indication is, in general, a sign that needs to be interpreted for obtaining the measurement value, and restitution constitutes such an interpretation.

$$y = \varphi(a). \tag{5.11}$$

From the representational theory, we know that each object can be described by a value that, in the deterministic model, describes its relations with all the other elements that carry the same property:

$$x = m(a). \tag{5.12}$$

Then, observation can also be described, from a different standpoint, by the function,

$$y = f(x), \tag{5.13}$$

as we have seen in the previous section, where the function f may be called *calibration function*, inasmuch it is obtained by calibration, or *observation function*, since it describes observation. The link between φ and f is

$$\varphi(a) = f[m(a)]. \tag{5.14}$$

During measurement, x is unknown, whilst the function f is known, since we assume that we have previously calibrated the instrument. For assigning a proper value to the object, we can thus invert f. Restitution thus consists in the inversion of the observation function,

$$\hat{x} = f^{-1}(y). \tag{5.15}$$

The overall measurement process can thus be described by

$$\hat{x} = \gamma(a) \Leftrightarrow (y = \varphi(a) \text{ and } \hat{x} = f^{-1}(y)), \tag{5.16}$$

as shown in Fig. 5.3. As it appears from the figure, it is possible to decompose the overall measurement transformation, $\hat{x} = \gamma(a)$, in a few ways. Particularly significant is the path $a \mapsto y \mapsto \hat{x}$, which describes the way in which measurement is actually performed. It also corresponds to the model presented by Mari and colleagues in a few papers [1, 2] and also mentioned in Ref. [3]. Unfortunately, such a mapping, being from objects to numbers, does not allow an overall analytical representation. An alternative (sub) path can thus be considered, $x \mapsto y \mapsto \hat{x}$, that is, instead from numbers to numbers, and thus allows analytical representations [4–7].

In this way, we describe the measurement process by the function h, as

$$\hat{x} = h(x) = f^{-1}(f(x)) = x, \tag{5.17}$$

where the measurement function, denoted by h, in this ideal case reduces to an identity. This is what we have done, in the example in the previous section, and it is what we will do in the introductory part of the next section. The comparison between representations from objects to numbers, $\gamma : A \to \hat{X}$, and from numbers

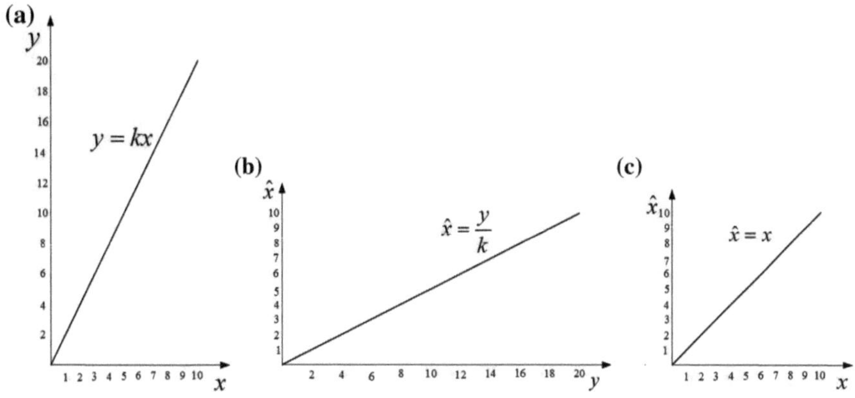

Fig. 5.4 Simple numerical example of a measurement process. **a** observation; **b** restitution; **c** measurement

to numbers, $h : X \rightarrow \hat{X}$, will be discussed in detail in Sect. 5.4 to follow, where the overall theoretical framework will be discussed in detail in its final, probabilistic, version.

This framework, although introduced by the indirect measurement example, may also be applied to the direct measurement case. Here, the observation phase consists in the comparison of the object with the scale and the "indication" is now the value of the selected standard, that is,

$$y = m(s) \Leftrightarrow a \sim s. \tag{5.18}$$

Moreover, having assumed, as usual, $m(a) = x$, since

$$a \sim s \Leftrightarrow m(s) = m(a). \tag{5.19}$$

We trivially obtain

$$y = f(x) = x, \tag{5.20}$$

and, consequently, again formula 5.17 holds true.

To conclude, let us consider a very simple illustration of the functions f, f^{-1} and h, provided by Fig. 5.4. Here, we have assumed $y = kx = 2x$.

Considering observation (a), if, for example, the value of the measurand is $x = 6$, the output, that is, the instrument indication, will be $y = 12$. Conversely, in restitution (b), if we obtain $y = 12$, we may infer that the value that has caused such an indication is $\hat{x} = 6$. The overall measurement transformation (c) reduces to an identity: if the value of the measurand is $x = 6$, the resulting measurement value will be $\hat{x} = 6$ as well.

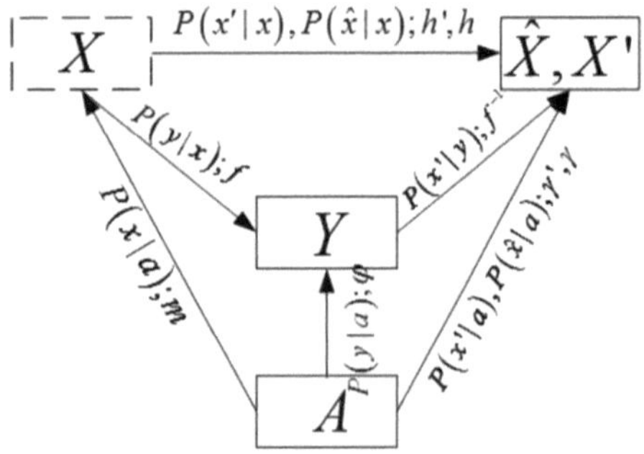

Fig. 5.5 Scheme of the probabilistic model

5.3 Probabilistic Model of the Measurement Process

As we have discussed in Chap. 2, a deterministic model is unable to account for uncertainty which, instead, is an inherent feature of real measurement. So in this and in the next sections, we will developed a probabilistic counterpart of the model presented in Fig. 5.3, summarised in Fig. 5.5.

In particular, in this section, we will focus on the path $x \mapsto y \mapsto \hat{x}$ that is on measurement considered as a mapping from numbers to numbers.

Concerning observation, what we expect now is that when we input a measurand whose value is x, and we repeat the experiment more times, we do not always obtain the same indication. Rather, the result varies at random, usually with some indications that are more frequent than others. We may describe this behaviour by assigning a probability distribution to the indications, conditioned on the value of the measurand, that is,

$$P(y|x), \tag{5.21}$$

which replaces formula (5.13) in describing observation. This is illustrated in Fig. 5.6a, to be compared with Fig. 5.4a.

In this example, if, say, $x = 6$, the indication may assume three values, $y = 10$, $y = 12$ or $y = 14$, with the following probabilities:

$$P(y = 10|x = 6) = 0.2,$$
$$P(y = 12|x = 6) = 0.6,$$
$$P(y = 14|x = 6) = 0.2.$$

This conditional distribution may still be obtained by calibration, by properly accounting for uncertainty.[5]

Restitution, in the deterministic model, constituted the inversion of observation. Here, it may be understood as the *probabilistic inversion* of observation. See Fig. 5.6b, and compare it with Fig. 5.4b. Now if we observe, e.g. $y = 12$, we may understand

[5] Calibration will be treated in some detail in Chap. 11.

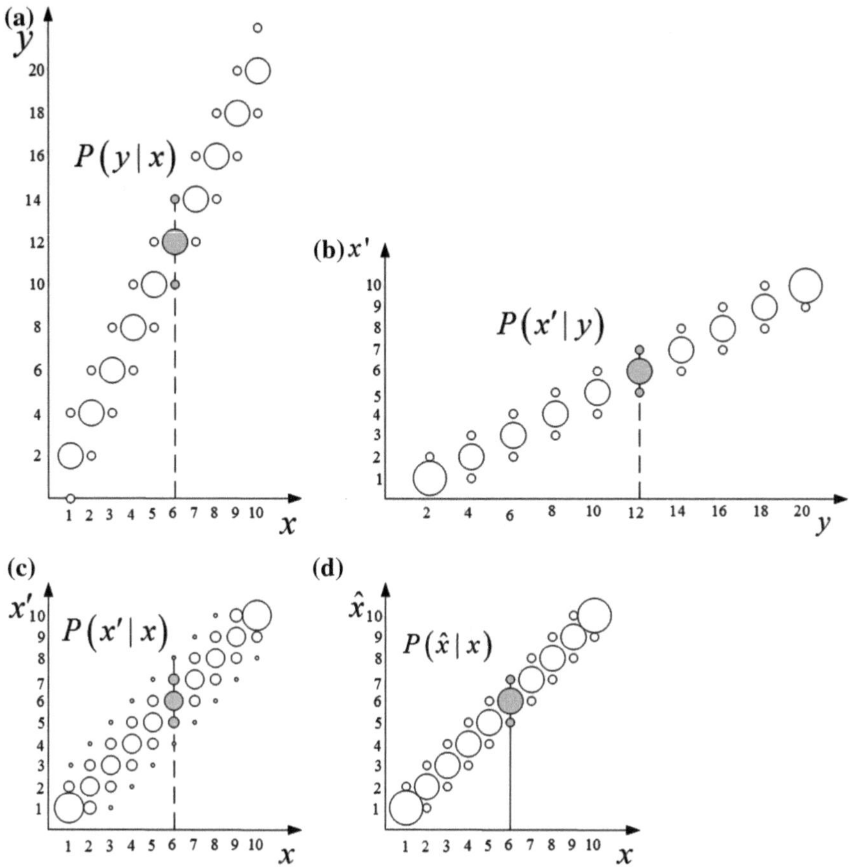

Fig. 5.6 Illustration of a probabilistic model of a measurement process: **a** observation; **b** restitution; **c** measurement, yielding the probabilistic measurement value; and **d** measurement, yielding the expected measurement value. The circles represent the probabilities under consideration, which are proportional to the diameters of the associated circles

(see again Fig. 5.6a) that such an indication may have been "caused" either by $x' = 5$ or by $x' = 6$ or by $x' = 7$. These three possibilities—"possible causes" in Laplace language—have the following probabilities:

$$P(x' = 5|y = 12) = 0.2,$$
$$P(x' = 6|y = 12) = 0.6,$$
$$P(x' = 7|y = 12) = 0.2.$$

In general, it is possible to perform this probabilistic inversion by the Bayes–Laplace ruleBayes-Laplace rule that, in this case, reads

$$P(x'|y) = \left[\frac{P(y|x)}{\sum_x P(y|x)} \right]_{x=x'}, \tag{5.22}$$

and by substituting x' for x.[6]

Measurement may be still viewed as the combination of observation and restitution. Yet we have two possibilities, depending on the way we define the measurement value.

In the deterministic framework, for each object, a, having a measure value, $x = m(a)$, we obtain a (unique) measurement value, $\hat{x}$, which is a single value and is equal to the measure value.

In the probabilistic case, instead, for each possible value of the measurand, x, we obtain a probabilistic variable, x', which we will call *(probabilistic) measurement value*.

Yet we can still recover a single value, as it is customary in practice, by taking the expected value of x':

$$\hat{x} = E(x'|y), \tag{5.23}$$

where E is the expectation operator. We will call $\hat{x}$ *(expected) measurement value.*

When we simply say "measurement value", we will often refer to $\hat{x}$, for conformity with the usual practice. Anyway, when it will be clear from the context, we will also call x' measurement value, for short.

Then, the deterministic mapping, $x \mapsto \hat{x}$, is now replaced either by the probabilistic mapping, $x \mapsto x'$, from the measure value to the probabilistic measurement value, or by the mapping, $x \mapsto \hat{x}$, from the measure value to the expected measurement value.

In the former case, we obtain

$$P(x'|x) = \sum_y P(x'|y)P(y|x). \tag{5.24}$$

This overall transformation is illustrated in Fig. 5.6c. For example, for $x = 6$, we can obtain five values, with the following probabilities:

$$\begin{aligned} P(x' = 4|x = 6) &= 0.04, \\ P(x' = 5|x = 6) &= 0.24, \\ P(x' = 6|x = 6) &= 0.44, \\ P(x' = 7|x = 6) &= 0.24, \\ P(x' = 8|x = 6) &= 0.04. \end{aligned}$$

In the latter case, we obtain instead:

[6] This substitution may sound odd to some readers familiar with Bayesian statistics. The reason for this substitution is that x and x' describe what happens in two distinct stages in the measurement process. Additional reasons for this distinction will appear in the next sections and in Chap. 6.

$$P(\hat{x}|x) = \sum_y \delta[\hat{x} - E(x'|y)]P(y|x), \tag{5.25}$$

presented in Fig. 5.6d.

The observation process (a) is the same, and restitution (b) is also the same, but here, once we obtain the distribution $P(x'|y)$, we take the expected value, $\hat{x} = E(x'|y)$ as the final result. So, for example, if we observe $y = 12$, we will provide $\hat{x} = 6$ as the final result. The resulting description of the overall measurement process is then provided in Fig. 5.6d and is different from that of Fig. 5.6c. Here, if the value of the measurand is $x = 6$, we obtain

$$P(\hat{x} = 5|x = 6) = 0.2,$$
$$P(\hat{x} = 6|x = 6) = 0.6,$$
$$P(\hat{x} = 7|x = 6) = 0.2.$$

In both cases, we obtain an identity transformation *in the average*: then, for $x = 6$, the expected value of x' is still 6, which means that the overall process is unbiased. On the other hand, some dispersion appears, because of uncertainty, as expected.

5.4 Probability Space of the Measurement Process

So far, we have presented the main idea under the probabilistic model of the measurement process, in an intuitive, informal way. What we have presented is enough for addressing most applications, yet here we want to probe further some theoretical aspects. Anyway, the reader more interested in application aspects may skip this section at a first reading, whilst he/she should continue reading the next one, where we provide additional indications that are very important for application.

Consider then again the graph in Fig. 5.5 and the two main paths through it. As already noted, one connects measure values, in X, with measurement values, in X' or $\hat{X}$, and thus links different *numerical representations*. Let us denote by h' or h the corresponding mappings. Here, we do not consider the specific measured object, but just its value, and we study how it is transformed into the final measurement value. This is the perspective usually followed in most applications.

Yet we can consider objects also, starting from box A, for reaching again X' or $\hat{X}$. Let us denote by γ' or γ the corresponding mappings. We already noted that this constitutes a more complete description, since measurement is to be intended as mapping from objects to numbers and not just from numbers to numbers.

This holds true from a theoretical standpoint. But what is the difference between starting from the measure value (alternative 1) or from the object (alternative 2), in practical application?

Consider an example. Suppose that we want to measure the temperature of the room where the reader is reading this book. Suppose that there is a thermometer somewhere on a wall of the room. So a reasonable way of acting is to obtain the room temperature and its uncertainty by the thermometer. Yet this is, strictly speaking, correct, only if the room temperature can be properly represented by a single constant temperature value, independently from where, in the room, it is actually detected. If this is the case, alternative 1 correctly describes this situation. The reader may also note that this is what we usually do.

Yet it may also be the case that there are some small variations in room temperature, for example, it may be higher close to the heater and lower close to the window. In this case, we should explicitly consider the "object" to be measured, i.e. "the room", and the possible values that can be associated with it. Remember that, in a probabilistic approach, as we have amply discussed in Chap. 4, each object is no longer completely described by a single measure value, but rather by a probability distribution on the set of the possible values. Then, how to account for this additional uncertainty, which can not be revealed by the single thermometer on the wall? One possibility could be to put a few thermometers in the room and then from their indications, perhaps with the aid of some thermal simulation programme, obtain a thermal map of the room and, consequently, the probability distribution of "room temperature". This may be too costly and not necessary for an ordinary lecture room, but can be required for some sophisticated laboratory room, where some critical experiment is carried out. Our theory, for being complete, must allow treating both cases.

5.4.1 From Numbers to Numbers

Let us consider now the first alternative, that is, the path from X to X' or to $\hat{X}$, with the corresponding mappings, h' or h. Let us first discuss, in greater depth, its rationale. We have seen that given a set of objects A and a quantity x associated with them, a measuring system is an empirical system capable of interacting with every element $a \in A$ and to produce an indication y, on the basis of which it is possible to assign a value x' to a.

But what is the essential feature of the measuring system? It is that its output depends on the state of the object to be measured but not on its individuality: in other words, two objects that manifest the same state should be *exchangeable* for the measuring system. Now, remembering that a measuring system is characterised by the conditional probability distribution $P(y|x)$, this property may be expressed by requiring that, for each $a, b \in A$,

$$P(y|x_a = x) = P(y|x_b = x). \tag{5.26}$$

In reality, in the previous section, we have already implicitly assumed this property when we have characterised the measuring system by the distribution $P(y|x)$, without mentioning any specific object. With this premise, we may now specify the *probability space* that underlies the transformation that links the value of the measurand, x, to the measurement value, x', as a *vector probabilistic variable*,

$$\mu = (x, y, x'), \tag{5.27}$$

characterised by the joint probability distribution[7] (in shorthand notation)

$$P(x, y, x') = P(x)P(y|x)P(x'|x, y). \tag{5.28}$$

So it is essential to discuss the meaning and the characteristics of these three factors in our model.

$P(x)$ is the "prior" distribution of the measurand, and two main cases merit mentioning.

In a generic measurement process, we may assume a "vague" distribution, that is to say, for example, a uniform distribution over the set of all possible values of the measurand, the measuring range. In fact, when we use a measuring device, we implicitly assume that the value of the measurand will fall into the measuring range and the probabilistic expression of this knowledge is just a uniform distribution over that range.

Another important case, which occurs mainly in conformity assessment, is when x represents a parameter that characterises some production process, such as the length of a workpiece of the amount of some pollutant. In these latter cases, the distribution $P(x)$ may be known on the basis of the historical records of the process under consideration.

The second distribution appearing in formula (5.28), $P(y|x)$, is the distribution that characterises the measurement process. It must satisfy the property expressed by formula (5.26), and it must express the link holding between x and y. So we should require[8] further that, for each $x \in X$,

$$P(y|x) \neq P(y). \tag{5.29}$$

Lastly, for what we have discussed in the previous section about restitution, we have

$$P(x'|x, y) = P(x'|y) = \left[\frac{P(y|x)}{\sum_x P(y|x)}\right]_{x=x'}. \tag{5.30}$$

This completes the searched probabilistic framework.

Let us now illustrate it by a very simple numerical example, reported in Table 5.1 and in Figs. 5.7 and 5.8.

Suppose that $X = \{1, 2\}$, that the distribution $P(x)$ is

---

[7] This factorisation in three distributions simply results from the application of a rule of probability calculus; see e.g. [8].

[8] If instead, for some x, $P(y|x) = P(y)$, the indication would be, in those cases, independent from x, and measurement would be thus impossible, since we would not obtain any information from the instrument.

Table 5.1 An example of a probabilistic mapping from measure values to measurement values

ω	x	y	x'	$\hat{x}$	P
1	1	0	1	1	0.12
2	1	0	2	1	0.00
3	1	1	1	1	0.27
4	1	1	2	1	0.09
5	1	2	1	2	0.03
6	1	2	2	2	0.09
7	1	3	1	2	0.00
8	1	3	2	2	0.00
9	2	0	1	1	0.00
10	2	0	2	1	0.00
11	2	1	1	1	0.06
12	2	1	2	1	0.02
13	2	2	1	2	0.06
14	2	2	2	2	0.18
15	2	3	1	2	0.00
16	2	3	2	2	0.08

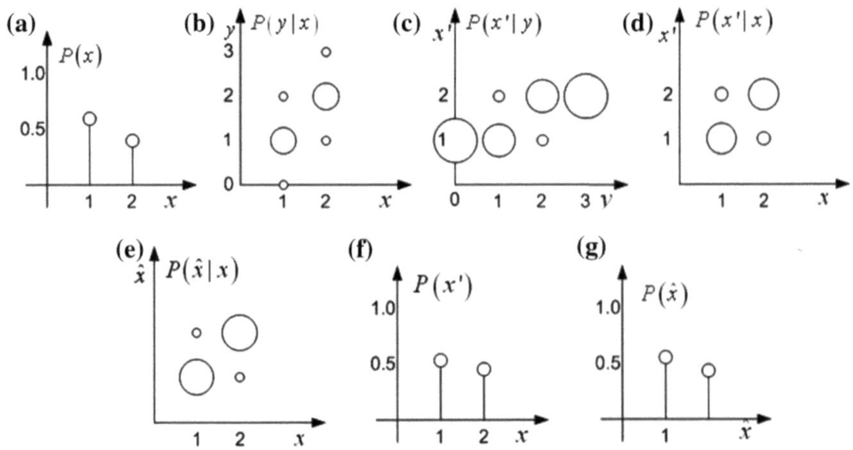

Fig. 5.7 Scheme of the probabilistic model: from numbers to numbers, showing all involved probability distributions: **a** $P(x)$; **b** $P(y|x)$; **c** $P(x'|y)$; **d** $P(x'|x)$; **e** $P(\hat{x}|y)$; **f** $P(x')$; **g** $P(\hat{x})$

$$P(x = 1) = 0.6,$$
$$P(x = 2) = 0.4,$$

as shown in Fig. 5.7a and that $P(y|x)$ is

$$P(y = 0|x = 1) = 0.2,$$
$$P(y = 1|x = 1) = 0.6,$$
$$P(y = 2|x = 1) = 0.2,$$
$$P(y = 1|x = 2) = 0.2,$$
$$P(y = 2|x = 2) = 0.6,$$
$$P(y = 3|x = 2) = 0.2.$$

as shown in Fig. 5.7b. Then, we obtain, for $P(x'|y)$,

$$P(x' = 1|y = 0) = 1.0,$$
$$P(x' = 2|y = 0) = 0.0,$$
$$P(x' = 1|y = 1) = 0.75,$$
$$P(x' = 2|y = 1) = 0.25,$$
$$P(x' = 1|y = 2) = 0.25,$$
$$P(x' = 2|y = 2) = 0.75,$$
$$P(x' = 1|y = 3) = 0.0,$$
$$P(x' = 2|y = 3) = 1.0,$$

as shown in Fig. 5.7c. Let us consider now both the probabilistic and the expected measurement value.

For the former, x', we obtain (Fig. 5.7d),

$$P(x' = 1|x = 1) = 0.7,$$
$$P(x' = 2|x = 1) = 0.3,$$
$$P(x' = 1|x = 2) = 0.3,$$
$$P(x' = 2|x = 2) = 0.7.$$

All the possible combinations of the involved variables are listed in the table; they are 16 in total. For each of them, corresponding to a point ω of the sample space Ω, the associated probability is calculated. For example, for $\omega = 1,\ x = 1,\ y = 0$ and $x' = 1$. The probability of this combination of values is, according to formulae 5.28 and 5.30,

$$P(x = 1, y = 0, x' = 1) = P(x = 1)P(y = 0|x = 1)P(x' = 1|y = 0) =$$
$$0.6 \times 0.2 \times 1.0 = 0.12,$$

and so forth. The distribution of the probabilistic measurement value can be easily obtained from the table and is

$$P(x' = 1) = 0.54,$$
$$P(x' = 2) = 0.46,$$

as shown in Fig. 5.7f.

Consider now the expected measurement value. Note that the expected value would be in general in between 1 and 2. Here, we round it to the closer integer value in order to have $\hat{X}$ still coincident with X. Note a big difference between x' and $\hat{x}$: for each value of y, we obtain an entire distribution for x', here consisting of two possible values, whilst we obtain just one value for $\hat{x}$:

$$\text{if } y = 0, \text{ then } \hat{x} = 1,$$
$$\text{if } y = 1, \text{ then } \hat{x} = 1,$$
$$\text{if } y = 2, \text{ then } \hat{x} = 2,$$
$$\text{if } y = 3, \text{ then } \hat{x} = 2.$$

Again these results appear in the table. Concerning the distribution $P(\hat{x}|x)$, we obtain (Fig. 5.7e)

$$P(\hat{x} = 1|x = 1) = 0.8,$$
$$P(\hat{x} = 2|x = 1) = 0.2,$$
$$P(\hat{x} = 1|x = 2) = 0.2,$$
$$P(\hat{x} = 2|x = 2) = 0.8.$$

The final distribution for $\hat{x}$ is

$$P(\hat{x} = 1) = 0.56,$$
$$P(\hat{x} = 2) = 0.44,$$

presented in Fig. 5.7g.

Both final distributions can be compared with the distribution of the value of the measurand, $P(x)$, Fig. 5.7a. In both cases, we note a greater "dispersion" due to the measurement process: in fact, both of them are closer to the maximum uncertainty distribution that would consist in assigning a probability of 0.5 to both values. The distribution of $\hat{x}$ has a smaller dispersion than that of x', due to the operation of expectation, which constitutes a kind of averaging.

5.4.2 *From Things to Numbers*

Consider now the second alternative, the path from A to X' or to $\hat{X}$, in Fig. 5.5, corresponding to the function γ' and γ, respectively. For the sake of simplicity, we will discuss mainly the first case, from A to X', since the second is essentially

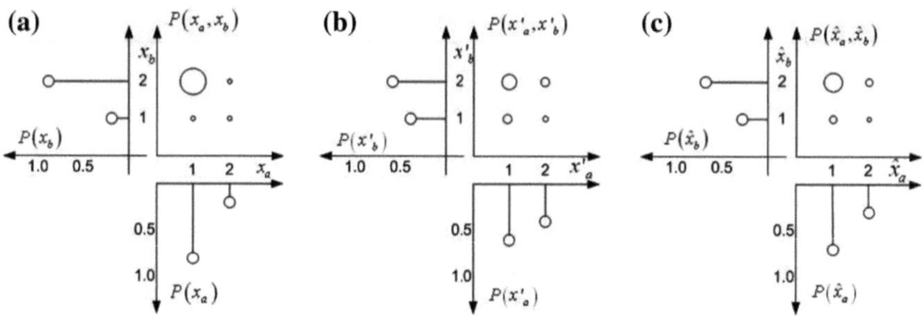

Fig. 5.8 Scheme of the probabilistic model: from things to numbers

similar, the main difference being that it involves, as we already know, an additional expectation operation.

The involved mathematical structure is thus the collection of probabilistic variables

$$(x'_a)_{a \in A}, \tag{5.31}$$

where the parameter a denotes the object under consideration.

Let us briefly explicate the related probabilistic structure.

Let $X' = X = \{1, 2, \ldots, n\}$ and $A = \{a_1, a_2, \ldots, a_N\}$, $n \leq N$. A complete probabilistic description involves the individual probability distributions of the variables, $P(x'_a)$, $P(x'_b)$, ..., and their joint distributions of any order up to N. Second-order distributions are, for example, $P(x'_a, x'_b)$, $P(x'_c, x'_d)$, $P(x'_a, x'_c)$, and the overall probability distribution is $P(x'_{a_1}, x'_{a_2}, \ldots x'_{a_N})$.

Individual (first-order) distributions, for each $a \in A$, $j \in X'$, $i \in X$, are given by

$$P(x'_a = j) = \sum_{i \in X} P(x' = j | x = i) P(x_a = i). \tag{5.32}$$

Second-order distributions, for each $a, b \in A$, $i, j \in X'$, $h, k \in X$, can be obtained by

$$P(x'_a = i, x'_b = j) = \sum_{h,k \in X} P(x' = i | x = h) P(x' = j | x = k) P(x_a = h, x_b = k). \tag{5.33}$$

Lastly, the overall joint distribution, for $j_1, \ldots, j_N \in X'$, $i_1, \ldots, i_N \in X$, is defined by

$$P(x'_{a_1} = j_1, \ldots, x'_{a_N} = j_N) = \sum_{i_1, \ldots, i_N \in X} P(x' = j_1 | x = i_1), \ldots, P(x' = j_N | x = i_N)$$
$$P(x_{a_1} = i_1, \ldots, x_{a_N} = i_N). \tag{5.34}$$

Table 5.2 An illustrative example with $A = \{a, b\}$

Order	x_a	x_b	P
$a \prec b$	1	2	0.7
$a \sim b$	1	1	0.1
$b \prec a$	2	1	0.1
$b \sim a$	2	2	0.1

Let us demonstrate the application of these formulae in a very simple example. Suppose we have just two objects, a and b, whose relations are as in Table 5.2.

Suppose that we measure them with the same measuring system considered in the previous section. So let the distribution $P(y|x)$, $P(x'|y)$ and $P(x'|x)$ be as above. The goal of this example is to understand well the links amongst empirical relations, numerical representations and numerical relations.

Let us consider firstly, the probabilistic variables x_a and x_b, that provide a numerical representation of the empirical relations. From the table, it is immediate to note that their joint probability distribution is

$$P(x_a = 1, x_b = 1) = 0.1,$$
$$P(x_a = 1, x_b = 2) = 0.7,$$
$$P(x_a = 2, x_b = 1) = 0.1,$$
$$P(x_a = 2, x_b = 2) = 0.1,$$

and, consequently, the marginal distributions are

$$P(x_a = 1) = 0.8,$$
$$P(x_a = 2) = 0.2,$$
$$P(x_b = 1) = 0.2,$$
$$P(x_b = 2) = 0.8.$$

Such distributions are illustrated in Fig. 5.8a.

Consider now the (probabilistic) measurement values x'_a and x'_b. In this example, the overall distribution coincides with the second-order one, since we have just two objects. Before calculating them, note that—in another perspective—we may consider that a collection of probabilistic variables is equivalent to a probabilistic function, as discussed in Chap. 4, Sect. 4.1.6. In our case, we can consider the set of functions, $\gamma'_i : A \to X'$, that, in our example, are

$$\gamma'_1 = \{(a, 1), (b, 1)\},$$
$$\gamma'_2 = \{(a, 1), (b, 2)\},$$
$$\gamma'_3 = \{(a, 2), (b, 1)\},$$
$$\gamma'_4 = \{(a, 2), (b, 2)\}.$$

Table 5.3 Complete development of the illustrative example, in terms of probabilistic measurement value

ω	Ordering	x_a	x_b	x'_a	x'_b	γ'_i	P
1	$a \prec b$	1	2	1	1	γ'_1	0.147
2	$a \prec b$	1	2	1	2	γ'_2	0.343
3	$a \prec b$	1	2	2	1	γ'_3	0.063
4	$a \prec b$	1	2	2	2	γ'_4	0.147
5	$a \sim b$	1	1	1	1	γ'_1	0.049
6	$a \sim b$	1	1	1	2	γ'_2	0.021
7	$a \sim b$	1	1	2	1	γ'_3	0.021
8	$a \sim b$	1	1	2	2	γ'_4	0.009
9	$a \succ b$	2	1	1	1	γ'_1	0.021
10	$a \succ b$	2	1	1	2	γ'_2	0.009
11	$a \succ b$	2	1	2	1	γ'_3	0.049
12	$a \succ b$	2	2	2	2	γ'_4	0.021
13	$b \sim a$	2	2	1	1	γ'_1	0.009
14	$b \sim a$	2	2	1	2	γ'_2	0.021
15	$b \sim a$	2	2	2	1	γ'_3	0.021
16	$b \sim a$	2	2	2	2	γ'_4	0.049

With this in mind, we obtain

$$\begin{aligned}
P(x'_a = 1, x'_b = 1) &= P(\gamma'_1) = 0.226,\\
P(x'_a = 1, x'_b = 2) &= P(\gamma'_2) = 0.394,\\
P(x'_a = 2, x'_b = 1) &= P(\gamma'_3) = 0.154,\\
P(x'_a = 2, x'_b = 2) &= P(\gamma'_4) = 0.226,
\end{aligned}$$

from which it is possible to calculate the marginal distributions

$$\begin{aligned}
P(x'_a = 1) &= 0.62,\\
P(x'_a = 2) &= 0.38,\\
P(x'_b = 1) &= 0.38,\\
P(x'_b = 2) &= 0.62.
\end{aligned}$$

The results so far obtained are illustrated in Fig. 5.8b and in Table 5.3.

A similar treatment can be developed in terms of expected measurement value and is presented in Table 5.4.

Note that the functions $\gamma_i : A \to \hat{X}$ can be defined in the same way as the γ'_i:

$$\begin{aligned}
\gamma_1 &= \{(a, 1), (b, 1)\},\\
\gamma_2 &= \{(a, 1), (b, 2)\},\\
\gamma_3 &= \{(a, 2), (b, 1)\},\\
\gamma_4 &= \{(a, 2), (b, 2)\}.
\end{aligned}$$

Table 5.4 Complete development of the illustrative example, in terms of expected measurement value

ω	Ordering	x_a	x_b	$\hat{x}_a$	$\hat{x}_b$	γ_i	P
1	$a \prec b$	1	2	1	1	γ_1	0.112
2	$a \prec b$	1	2	1	2	γ_2	0.448
3	$a \prec b$	1	2	2	1	γ_3	0.028
4	$a \prec b$	1	2	2	2	γ_4	0.112
5	$a \sim b$	1	1	1	1	γ_1	0.064
6	$a \sim b$	1	1	1	2	γ_2	0.016
7	$a \sim b$	1	1	2	1	γ_3	0.016
8	$a \sim b$	1	1	2	2	γ_4	0.004
9	$a \succ b$	2	1	1	1	γ_1	0.016
10	$a \succ b$	2	1	1	2	γ_2	0.004
11	$a \succ b$	2	1	2	1	γ_3	0.064
12	$a \succ b$	2	2	2	2	γ_4	0.016
13	$b \sim a$	2	2	1	1	γ_1	0.004
14	$b \sim a$	2	2	1	2	γ_2	0.016
15	$b \sim a$	2	2	2	1	γ_3	0.016
16	$b \sim a$	2	2	2	2	γ_4	0.064

The only difference is that now, the image is the set X instead of the set X', but the two sets have the same elements, even if their interpretation is different.

We can thus calculate, similarly to above, the following probabilities:

$$\begin{aligned}
P(\hat{x}_a = 1, \hat{x}_b = 1) &= P(\gamma_1) = 0.196,\\
P(\hat{x}_a = 1, \hat{x}_b = 2) &= P(\gamma_2) = 0.484,\\
P(\hat{x}_a = 2, \hat{x}_b = 1) &= P(\gamma_3) = 0.124,\\
P(\hat{x}_a = 2, \hat{x}_b = 2) &= P(\gamma_4) = 0.196,
\end{aligned}$$

and also the marginal distribution:

$$\begin{aligned}
P(\hat{x}_a = 1) &= 0.68,\\
P(\hat{x}_a = 2) &= 0.32,\\
P(\hat{x}_b = 1) &= 0.32,\\
P(\hat{x}_b = 2) &= 0.68,
\end{aligned}$$

as shown in Fig. 5.8c. This last representation allows us to give a precise meaning to formula (2.34), which we presented in Chap. 2 by intuitive arguments only. Let us reproduce it here:

$$P(\hat{x} = \gamma(a)). \tag{5.35}$$

In our example, it corresponds to the following statements:

$$P(1 = \gamma(a)) = P(\gamma_1) + P(\gamma_2) = 0.68,$$
$$P(2 = \gamma(a)) = P(\gamma_3) + P(\gamma_4) = 0.32,$$
$$P(1 = \gamma(b)) = P(\gamma_1) + P(\gamma_3) = 0.32,$$
$$P(2 = \gamma(b)) = P(\gamma_2) + P(\gamma_4) = 0.68.$$

Finally, we may also calculate the relational probabilities associated with the measurement values, $\hat{x}$, by their joint probability distribution. We obtain

$$P(\hat{x}_a < \hat{x}_b) = P(\hat{x}_a = 1, \hat{x}_b = 2) = 0.484,$$
$$P(\hat{x}_a > \hat{x}_b) = P(\hat{x}_a = 2, \hat{x}_b = 1) = 0.124,$$
$$P(\hat{x}_a = \hat{x}_b) = P(\hat{x}_a = 1, \hat{x}_b = 1) + P(\hat{x}_a = 2, \hat{x}_b = 2) = 0.392,$$

that can be compared with the relational probabilities for the measure value, x,

$$P(x_a < x_b) = P(x_a = 1, x_b = 2) = 0.7,$$
$$P(x_a > x_b) = P(x_a = 2, x_b = 1) = 0.1,$$
$$P(x_a = x_b) = P(x_a = 1, x_b = 1) + P(x_a = 2, x_b = 2) = 0.2,$$

and with the relational probabilities amongst the objects (Table 5.2):

$$P(a \prec b) = 0.7,$$
$$P(a \succ b) = 0.1,$$
$$P(a \sim b) = 0.2.$$

Now, we know from the representation, and we may check in this example that, e.g.,

$$P(b \succcurlyeq a) = 0.9 = P(x_b \geqslant x_a).$$

But what happens with the measurement values? Here, we find that $P(\hat{x}_b \geq \hat{x}_a) = 0.876$, which is no longer equal to $P(b \succcurlyeq a)$.

What happens now is that

$$P(b \succcurlyeq a) \geq P(b \prec a) \Leftrightarrow P(\hat{x}_b \geq \hat{x}_a) \geq P(\hat{x}_b < \hat{x}_a).$$

Although we do not have, at present, a formal proof, it may be that this is what happens with measurement values in general.

5.5 Systematic Effects

In Chap. 2, we mentioned two main classes of uncertainty sources: those giving rise to random variations and those causing systematic effects. The probabilistic framework presented in the previous section was introduced in reference mainly to random variations. Yet we will show now that it is capable of accounting for systematic effects also, provided that the distribution $P(y|x)$ is understood in a more sophisticated way. Remember the way we introduced systematic effects: if we measure the same object with two equally reliable measuring systems, R and S, it may happen that we *repeatedly* obtain two different values, one for each system,

$$\hat{x}_R = \gamma_R(a), \hat{x}_S = \gamma_S(a). \tag{5.36}$$

We can still express this situation in probabilistic terms, especially if we regard probability as a logic, as mentioned in Sect. 4.1.4. If the two instruments are *equally reliable*, we can assign the same probability to them:

$$P(R) = 0.5,$$
$$P(S) = 0.5.$$

Then, we assign, in a natural way,

$$P(\gamma(a) = \hat{x}_R|R) = 1.0,$$
$$P(\gamma(a) = \hat{x}_S|R) = 0.0,$$
$$P(\gamma(a) = \hat{x}_R|S) = 0.0,$$
$$P(\gamma(a) = \hat{x}_S|S) = 1.0,$$

and we obtain

$$P(\gamma(a) = \hat{x}_R) = P(\gamma(a) = x'_R|R)P(R) + P(\gamma(a) = \hat{x}_R|S)P(S) = 1.0 \times 0.5 + 0.0 \times 0.5 = 0.5,$$
$$P(\gamma(a) = \hat{x}_S) = P(\gamma(a) = \hat{x}_S|R)P(R) + P(\gamma(a) = \hat{x}_S|S)P(S) = 0.0 \times 0.5 + 1.0 \times 0.5 = 0.5.$$

This is the rationale for treating systematic deviations in probabilistic terms as well.

Along this line of thought, let us discuss the typical case of having a single instrument, whose output depends systematically on some influence quantity, such as temperature for example. See Fig. 5.9.

The idea is somewhat similar to the previous example. Although here we have just one instrument, what happens can be modelled as a blind random sampling from a set of three instruments, each having a different value of θ. So we may assign a probability of 1/3 to each of these possibilities. Now, for any fixed value of θ, one of the curves holds true, let call it f_θ. Then, after introducing the discrete Dirac impulse operator δ, defined by

$$\delta(i) = \begin{cases} 1, \text{ if } i = 0, \\ 0, \text{ if } i \neq 0, \end{cases} \tag{5.37}$$

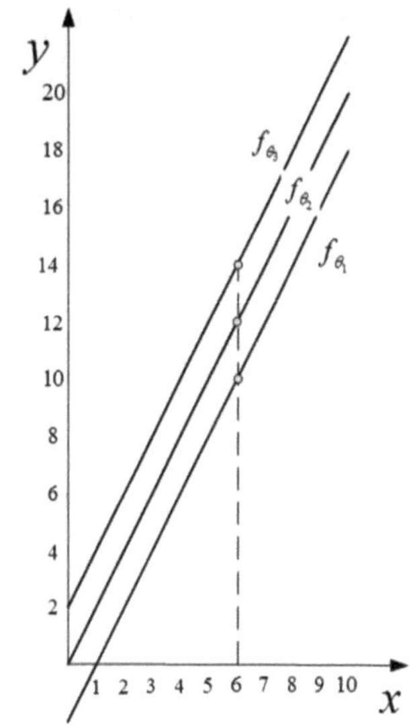

Fig. 5.9 Systematic effect on a measuring instrument

we can define the related conditional probability by

$$P(y|x,\theta) = \delta(y - f_\theta(x)). \tag{5.38}$$

The overall conditional probability is

$$\begin{aligned} P(y|x) &= \sum_\theta P(y|x,\theta)P(\theta) \\ &= 1/3\delta(y - f_{\theta_1}(x)) + 1/3\delta(y - f_{\theta_2}(x)) + 1/3\delta(y - f_{\theta_3}(x)). \end{aligned} \tag{5.39}$$

For example, in the case of Fig. 5.9, we obtain, for $x = 6$,

$$\begin{aligned} P(y = 10|x = 6) &= 1/3\delta(y - f_{\theta_1}(6)) = 1/3, \\ P(y = 11|x = 6) &= 1/3\delta(y - f_{\theta_2}(6)) = 1/3, \\ P(y = 12|x = 6) &= 1/3\delta(y - f_{\theta_3}(6)) = 1/3. \end{aligned}$$

Concerning restitution, we obtain, for each θ,

$$P(x'|y,\theta) = \delta(x' - f_\theta^{-1}(y)), \tag{5.40}$$

and, globally,

$$P(x'|y) = \sum_\theta P(x'|y,\theta)P(\theta). \tag{5.41}$$

This is illustrated in Fig. 5.10. For example, for $y = 12$, we obtain

Fig. 5.10 Restitution when a systematic effect is present

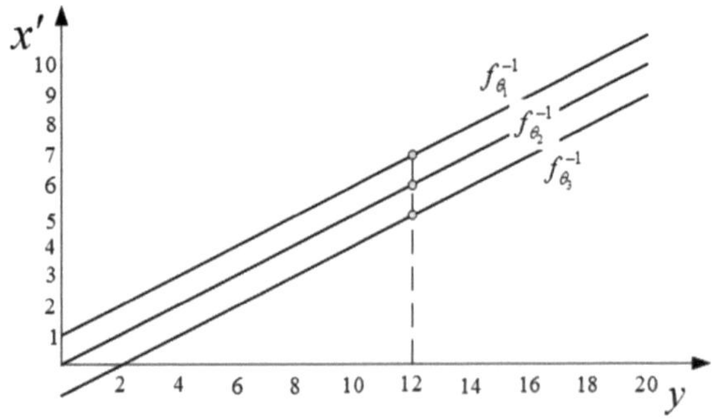

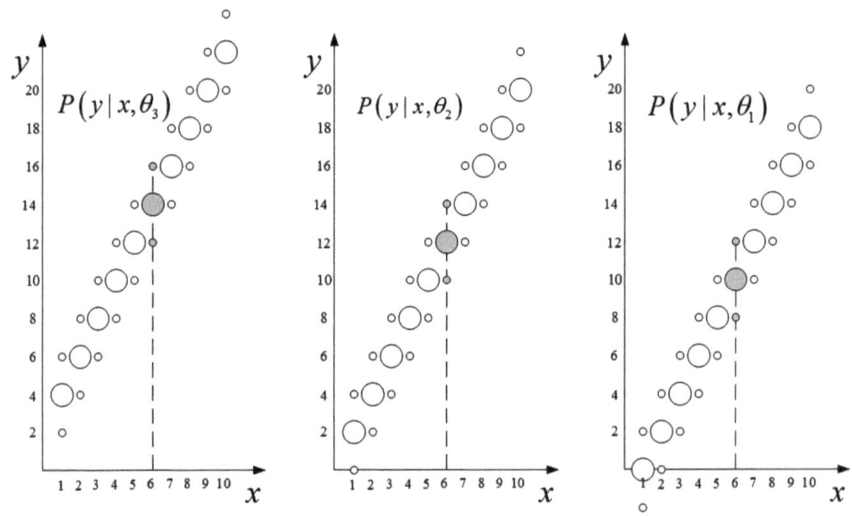

Fig. 5.11 Observation affected by both random variations and systematic effects

$$
\begin{aligned}
P(x' = 5|y = 12) &= 1/3\delta(x' - f_{\theta_3}^{-1}(12)) = 1/3, \\
P(x' = 6|y = 12) &= 1/3\delta(x' - f_{\theta_2}^{-1}(12)) = 1/3, \\
P(x' = 7|y = 12) &= 1/3\delta(x' - f_{\theta_1}^{-1}(12)) = 1/3.
\end{aligned}
$$

It is interesting to see now how random variations and systematic effects combine together.

If we consider our previous example, and suppose that there is also a systematic effect, we will have now a different probability distribution, $P(y|x)$, for each value of θ, as illustrated in Fig. 5.11.

For each fixed value of θ, we may perform restitution as before, according to formula (5.22), obtaining a result that is still conditioned on θ:

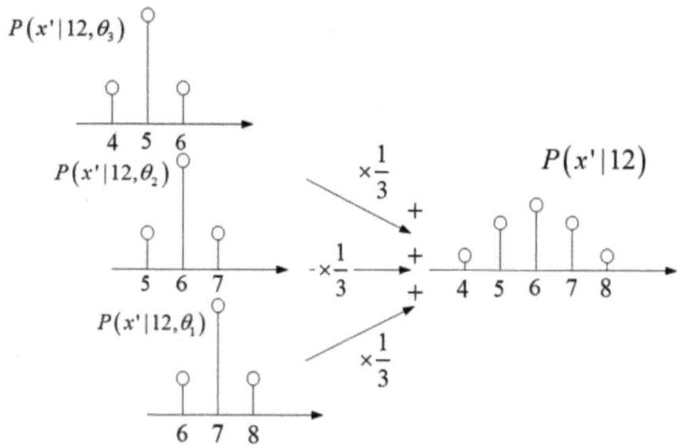

Fig. 5.12 Graphical illustration of restitution, for $y = 12$

$$P(x'|y, \theta) = \left[\frac{P(y|x, \theta)}{\sum_x P(y|x, \theta)}\right]_{x=x'}. \tag{5.42}$$

We have then to *de-condition* this result in respect of θ:

$$P(x'|y) = \sum_\theta P(x'|y, \theta)P(\theta). \tag{5.43}$$

This procedure is graphically illustrated in Fig. 5.12, for the case of $y = 12$.

To sum up, the proper formulae, in the presence of some influential quantity, are, for observation

$$P(y|x) = \sum_\theta P(y|x, \theta)P(\theta), \tag{5.44}$$

for restitution

$$P(x'|y) = \sum_\theta \left[\frac{P(y|x, \theta)}{\sum_x P(y|x, \theta)}\right]_{x=x'} P(\theta), \tag{5.45}$$

and for measurement, in terms of probabilistic measurement value,

$$P(x'|x) = \sum_y P(x'|y)P(y|x), \tag{5.46}$$

and in terms of expected measurement value:

$$P(\hat{x}|x) = \sum_y P(\hat{x}|y)P(y|x). \tag{5.47}$$

Lastly, note that

$$P(\hat{x}|y) = \delta(\hat{x} - E(x'|y)), \tag{5.48}$$

and thus, by substituting, we finally obtain

$$P(\hat{x}|x) = \sum_{y} \delta(\hat{x} - E(x'|y))P(y|x). \tag{5.49}$$

It is very interesting to consider what happens when we perform a measurement based on repeated observations of the same measurand in the same conditions. It is common practice, in such a case, to average the result, and it is commonly accepted that such a result is better than that obtained by a single observation. How can we treat this case in the present probabilistic framework, and what will we obtain? The reader is invited to try and figure out this case: we will treat it in Chap. 6.

5.6 Continuous Versus Discrete Representations

So far, we have represented the measure and measurement values as *discrete* probabilistic variables. As we have mentioned in Sects. 3.1 and 4.1.8, it is common in the professional practice, in education and in standardisation, to use *continuous* probabilistic variables instead. So what relation is there between these two alternatives? In particular, is a discrete representation a real limitation?

A discrete and finite representation basically implies to assume a finite number, n, of equivalence classes, since it is immaterial how many objects are there in any class. This in turn is equivalent to having a finite resolution, Δx_r, and an absolute maximum value, x_{max}.

But these two assumptions are indeed very reasonable. In fact, as we have discussed at the beginning of Chap. 3, we cannot attain, experimentally, an infinite resolution nor an infinite measurement range.

Furthermore, at present, we have developed a complete probabilistic theory for discrete representations, whilst we do not have any for continuous ones. So we will adopt, in this book, the, rather unusual but epistemologically justified, perspective that measurement values are *inherently discrete*, although they may be regarded as continuous, provided that some suitable interpolation is applied. Let us then define such a continuous representation. A generic discrete value, x_i, can be expressed as

$$x_i = i \Delta x_r, \tag{5.50}$$

with i being an integer number, and Δx_r the measurement resolution, referred to the proper measurement unit.

Then, we may introduce a continuous measure value, x, related to the discrete one in this way: the probability density function, $p(x)$ of x, is related to the probability distribution $P(x_i)$ of the discrete variable x_i by

$$p(x) = \begin{cases} P(x_i)/\Delta x_r, & \text{if there exists } i \text{ such that } x_i - \Delta x_r \le x < x_i + \Delta x_r; \\ 0, & \text{elsewhere.} \end{cases} \tag{5.51}$$

Note that this implies, for each i,

$$P(x_i) = \int_{x_i - \Delta x_r}^{x_i + \Delta x_r} p(x)\mathrm{d}x. \tag{5.52}$$

This latter is a system of integral equations that describes $p(x)$, albeit not uniquely. Formula (5.51) provides a simple solution. The probability density function, $p(x)$, obtained in this way is non-negative and has unitary integral over its domain. If we impose the additional constraint of continuity, a challenging curve-fitting problem arises, which anyway has been amply discussed in the scientific literature [9, 10].

We do not pursue this further here, since what is important here is to note that both a discrete representation and a continuous representation make sense and that if a discrete one is more directly attained through the theory, a correspondent continuous one may always be obtained, at least through formula (5.51).

For a continuous representation, formulae similar to (5.44–5.49) may be easily obtained, basically by substituting integrals to sums. We obtain, for observation

$$p(y|x) = \int_{\Theta} p(y|x,\theta)p(\theta)\mathrm{d}\theta, \tag{5.53}$$

for restitution

$$p(x'|y) = \int_{\Theta} \left[\frac{p(y|x,\theta)}{\int_X p(y|x,\theta)} \right]_{x=x'} p(\theta)\mathrm{d}\theta, \tag{5.54}$$

and for measurement, in terms of probabilistic measurement value, x',

$$p(x'|x) = \int_{Y} p(x'|y)p(y|x)\mathrm{d}y, \tag{5.55}$$

or, in terms of expected measurement value, $\hat{x}$,

$$p(\hat{x}|x) = \int_{Y} \delta(\hat{x} - E(x'|y))p(y|x)\mathrm{d}y. \tag{5.56}$$

where δ now denotes the usual (continuous) Dirac delta operator: see Ref. [5] for a thorough discussion.

Table 5.5 Synopsis of the probabilistic framework

The measurement scale		
Select type	Representation	
Ordinal	$P(a \succcurlyeq b) = P(x_a \geq x_b)$	
Interval	$P(\Delta_{ab} \succcurlyeq \Delta_{cd}) = P(x_a - x_b \geq x_c - x_d)$	
Ratio (intensive)	$P(a/b \succcurlyeq c/d) = P(\frac{x_a}{x_b} \geq \frac{x_c}{x_d})$	
Ratio (extensive)	$P(a \sim b \circ c) = P(x_a = x_b + x_c)$	
The measurement process		
Process	Discrete representation	Continuous representation
Observation	$P(y\|x) = \sum_\theta P(y\|x, \theta)P(\theta)$	$p(y\|x) = \int_\Theta p(y\|x, \theta)p(\theta)\mathrm{d}\theta$
Restitution	$P(x'\|y) = \sum_\theta \left[\frac{P(y\|x,\theta)}{\sum_x P(y\|x,\theta)}\right]_{x=x'} P(\theta)$	$p(x'\|y) = \int_\Theta \left[\frac{p(y\|x,\theta)}{\int_X p(y\|x,\theta)}\right]_{x=x'} p(\theta)\mathrm{d}\theta$
Measurement	$P(x'\|x) = \sum_y P(x'\|y)P(y\|x)$	$p(x'\|x) = \int_Y p(x'\|y)p(y\|x)\mathrm{d}y$
	$P(\hat{x}\|x) = \sum_y \delta\left[\hat{x} - E(x'\|y)\right] \times \left[\sum_\theta P(y\|x, \theta)P(\theta)\right]$	$p(\hat{x}\|x) = \int_Y \delta\left[\hat{x} - E(x'\|y)\right] \times \left[\int_\theta p(y\|x, \theta)p(\theta)\mathrm{d}\theta\right] \mathrm{d}y$

5.7 Overall Probabilistic Framework and Generalisations

It is now time to sum up the probabilistic framework so far developed. It consists of two parts: the measurement scale and the measurement process. For the measurement process, both a discrete representation and a continuous representation are possible. Instead, this distinction does not apply to the scale issue. A summary of the main results is presented in Table 5.5.

In respect of this, some further generalisations are possible that basically consist in moving from scalar to vectorial representations.

Firstly, it is possible that influence quantities are more than one: this may be easily handled by simply replacing the scalar θ with the vector $\boldsymbol{\theta}$. Yet there is another, more substantial point. Influence quantities, in the wide sense, may be both physical and environmental parameters, such as temperature, humidity, vibration or rather parameters of the model, such has, typically, the standard deviation of a probability distribution. Well, in both cases, the same model may be applied, yet an important difference outcomes between quantities about which it is possible to learn from the measurement process and others for which it is impossible. We will discuss this important distinction in Chap. 6 and, in greater depth, in Chap. 9.

Another generalisation concerns instrument indications: again, they may form a vector, $\mathbf{y}$, in place of y, and again, this may correspond to two cases:

- we may repeat several time the observation of the same measurand or
- we may have an indirect measurement.

In the first case, vector $\mathbf{y}$ is simply a collection of observations of the same kind, $\mathbf{y} = (y_1, y_2, \ldots, y_m)$. In the second, it may include observations from different instruments. For example, in the measurement of density, $\rho = m/V$, we may have an indication related to mass and one to volume: $\mathbf{y} = (y_m, y_V)$. Again, these topics will be addressed in Chaps. 6 and 9.

Table 5.6 Continuous vectorial representation of the measurement process

The measurement process	
Process	Representation
Observation	$p(\mathbf{y}\|\mathbf{x}) = \int_{\Theta} p(\mathbf{y}\|\mathbf{x}, \boldsymbol{\theta})p(\boldsymbol{\theta})\mathrm{d}\boldsymbol{\theta}$
Restitution	$p(\mathbf{x}'\|\mathbf{y}) = \int_{\Theta}[\frac{p(\mathbf{y}\|\mathbf{x},\boldsymbol{\theta})}{\int_{\mathbf{X}} p(\mathbf{y}\|\mathbf{x},\boldsymbol{\theta})}]_{\mathbf{x}=\mathbf{x}'} p(\boldsymbol{\theta})\mathrm{d}\boldsymbol{\theta}$
Measurement	$p(\mathbf{x}'\|\mathbf{x}) = \int_{\mathbf{Y}} p(\mathbf{x}'\|\mathbf{y})p(\mathbf{y}\|\mathbf{x})\mathrm{d}\mathbf{y}$
	$p(\hat{\mathbf{x}}\|\mathbf{x}) = \int_{\mathbf{Y}} \delta[\hat{\mathbf{x}} - E(\mathbf{x}'\|\mathbf{y})][\int_{\Theta} p(\mathbf{y}\|\mathbf{x}, \boldsymbol{\theta})p(\boldsymbol{\theta})\mathrm{d}\boldsymbol{\theta}]\mathrm{d}\mathbf{y}$

Lastly, it may be that the measurand is itself a vector, as when we measure the position of a point in the space, $\mathbf{y} = (y_x, y_y, y_z)$, or in the case of dynamic measurement, where the measurand may be a time-sampled signal. We will treat the first case in Chap. 7, devoted to multidimensional measurement, and the second in Chap. 12, concerning dynamic measurement.

By now let us simply show the generalised formulae in Table 5.6, considering continuous representations only, for the sake of brevity.

References

1. Mari, L.: Measurement in economics. In: Boumans, M. (ed.) Measurability, pp. 41–77. Elsevier, Amsterdam (2007)
2. Frigerio, A., Giordani, A., Mari, L.: Outline of a general model of measurement. Synthese **7**, 123–149 (2010)
3. Rossi, G.B.: Cross disciplinary concepts and terms in measurement. Measurement **42**, 1288–1296 (2009)
4. Morawski, R.Z.: Unified approach to measurand reconstruction. IEEE Trans. Instrum. Meas. **43**, 226–231 (1994)
5. Rossi, G.B.: A probabilistic model for measurement processes. Measurement **34**, 85–99 (2003)
6. Cox, M.G., Rossi, G.B., Harris, P.M., Forbes, A.: A probabilistic approach to the analysis of measurement processes. Metrologia **45**, 493–502 (2008)
7. Sommer, K.D.: Modelling of Measurements, System Theory, and Uncertainty Evaluation. In: Pavese, F., Forbes, A. (eds.) Data modeling for metrology and testing in measurement science, pp. 275–298. Birkhauser-Springer, Boston (2009)
8. Papoulis, A.: Probability, Random Variables and Stochastic Processes, 2nd edn. McGraw-Hill, Singapore (1984)
9. Thompson, J.R., Tapia, R.A.: Non parametric function estimation, modeling and simulation, SIAM, Philadelphia (1990)
10. Gentle, J.E.: Nonparametrical estimation of probability density functions. In: Gentle, J.E. (ed.) Computational Statistics. pp. 487–514. Springer, New York (2009)

Chapter 6
Inference in Measurement

6.1 How Can We Learn from Data?

In the previous chapter, we developed a general probabilistic model of the measurement process. We have now to learn how to use it in practice. This implies being able to apply correctly the principles and methods of statistical-probabilistic inference [1–3]. Basically a probabilistic inference is a way to learn from data. In the current practice of measurement, statistical methods are applied routinely, sometimes without a full understanding of their implications. This is not good practice. Measurement is itself both a science and a technology and must be applied with continuous awareness of the assumptions implied by the various methods. So from now onwards, we will try and develop some skill in data processing. This will be pursued in steps. In this chapter, we will first discuss probabilistic inferences in general, without going into details, and considering very simple illustrative examples only. Our goal here is to understand the rationale and the philosophy behind them. We will not be concerned with technicalities at all. Then, still in this chapter, we will begin to consider how to apply these ideas to measurement. We will discuss in particular the problem of systematic effects in measurement that, as we have seen in Chap. 2, has been, and somewhat still is, a highly controversial topic [4]. We will provide a solution to that problem, that has, possibly, a sounder foundation than other current approaches [5]. Then, we will continue this training in data processing in some chapters of the third part of the book, which is devoted to applications. We will discuss uncertainty evaluation (Chap. 9), today a major concern for metrologists and, in general, for people involved in measurement. Then, we will consider the application of measurement to decision-making (Chap. 10): here, it is particularly important to keep uncertainty under control and to carefully evaluate it, since it may cause wrong decisions. Decision-making also involves inferences, and it will provide us an opportunity for a deeper understanding of probabilistic inferences. Then, we will consider inter-comparisons (Chap. 11), a key tool for assuring the quality of measurements throughout the world, as we know from Chaps. 1 and 2. In this area, probabilistic inferences are also involved.

G. B. Rossi, *Measurement and Probability*, Springer Series in Measurement Science and Technology, DOI: 10.1007/978-94-017-8825-0_6,

Let us then start by reviewing the important difference between a model and an inference: these two issues are often related, but they are conceptually distinct and it is important not to confuse them.

6.2 Probabilistic Models and Inferences

In my experience, there are two main ways of working with probability in a research environment: *developing models* and *making inferences*. Although often related, these two processes are logically different and their difference is important.

A model, as we already know, may be understood as an abstract system aiming at representing, from some standpoint and up to some limitations, a real system, or a class of real systems. In particular, a probabilistic model is one that includes probabilistic variables, or probabilistic relations, or both of them.

A probabilistic inference, instead, concerns a real system: assumptions are made, data are acquired and, on the basis of both of them, probabilistic statements about the system under investigation are derived.

Let us illustrate these ideas with simple examples.

6.2.1 The Bernoullian Model

As a simple example of a probabilistic model, consider the case of coin tossing. Such an experiment is based on repeated independent executions of an elementary experiment, which has just two possible outcomes, heads, (A), or tails, $(\bar{A})$.

The model for the elementary experiment is simply $P(A) = p$, $P(\bar{A}) = q = 1 - p$. Then, if $\mathbf{y}$ is the N-dimensional vector of the outcomes of N trials—each elements of $\mathbf{y}$ equals either A or $\bar{A}$—the model for the overall experiment is

$$P(\mathbf{y}) = p^{n_A}(1-p)^{N-n_A}, \tag{6.1}$$

where n_A is the number of A-outcomes in $\mathbf{y}$. From a different standpoint, if we regard n_A as a probabilistic (or random[1]) variable, its probability distribution is given by

$$P(n_A) = \binom{N}{n_A} p^{n_A}(1-p)^{N-n_A}, \tag{6.2}$$

and is presented in Fig. 6.1, for $p = 0.5$.

Note that such a distribution depends upon the parameter p, i.e. the probability of A, heads. If we regard such a parameter as a probabilistic variable also, we can

[1] Here, the term "random variable" would be appropriate, since the Bernoullian model is intended to describe randomness.

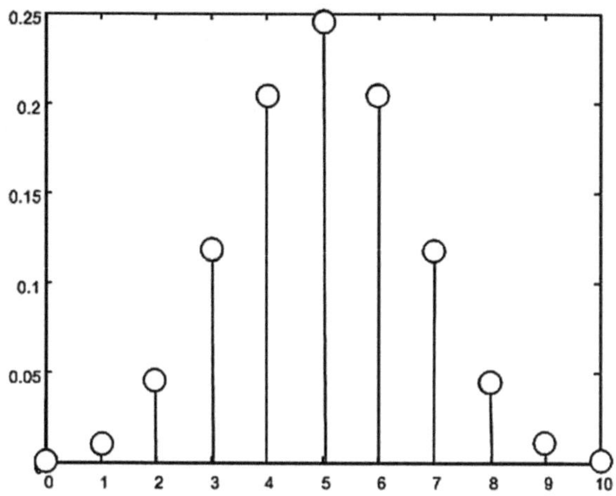

Fig. 6.1 Probability distribution for n_A, according to the Bernoullian model, for $p = 0.5$

consider the joint probability distribution of n_A and p, shown in Fig. 6.2. This distribution will be useful in the following.

As we can see from this simple example, a model is just a mathematical description of a class of phenomena and does not strictly imply performance of any specific experiment.

6.2.2 A Classification of Probabilistic Inferences

A statistical-probabilistic inference, instead, concerns a way of deducing consequences, to be stated in terms of probabilities, from a set of observations and a set of assumptions. Here, the experiment must be performed and the goal is to *learn from data*.

According to authoritative Refs. [2, 3], we consider here two main classes of such inferences, *(hypothetic-)deductive* and *inductive*. The latter may be further parsed into two groups: *hypothetic-inductive* and *predictive*, or purely inductive.

6.2.2.1 Hypothetic-Deductive Inferences and Their Epistemological Value

The term *hypothetic-deductive* is due to Costantini [3] and alludes to the general approach to the verification of scientific theories, as introduced by Galileo Galilei [6, 7] and successively adopted as a founding principle for modern science [8].

Modern science, according to Antiseri [8], may be characterised as a kind of knowledge which is *public*, *progressive* and *controllable*. Controllability is most important for us. The control of a scientific theory, in principle, may be done by performing an experiment whose result may be predicted by the theory and by checking

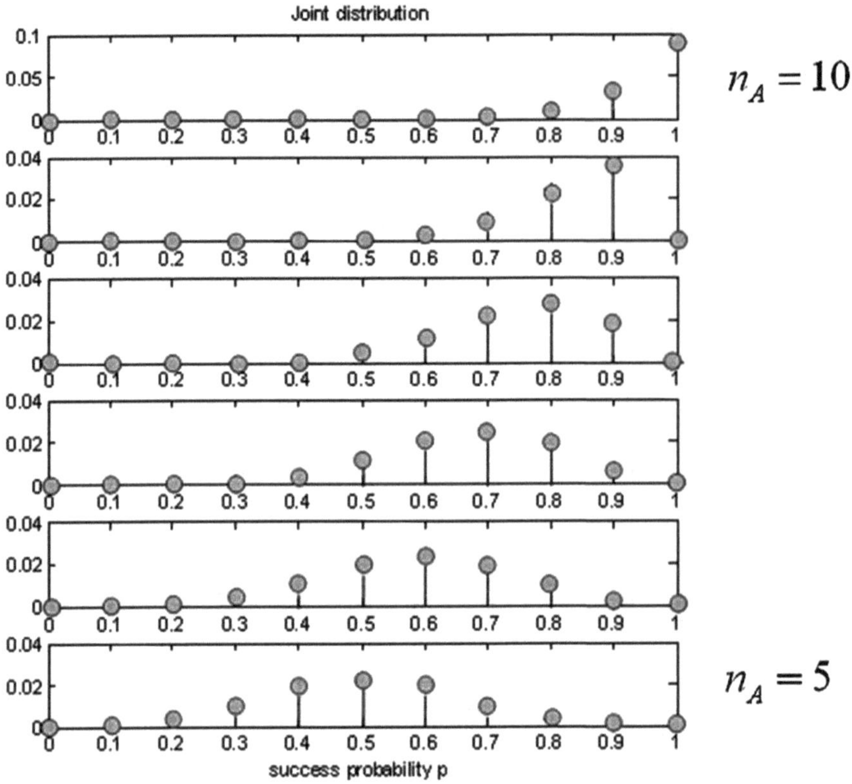

Fig. 6.2 Joint probability distribution for n_A and for p. For rending the figure more readable, the values for $n_A < 5$ are not displayed, since they are (skew) symmetric to the others

whether the actually observed result is in agreement with the predicted one. The theory to be checked constitutes the *hypothesis*, from which we *deduce* the expected result of the experiment, to be compared to the one actually observed. This is why this (general) verification approach may be called hypothetic-deductive. If the result is as expected, we say that the theory has been *corroborated* by the experiment; otherwise, we say that it has been *falsified*, and, in principle, it should be rejected.

The possibility for some theory of being controlled by experiments whose results may falsify it, that is its *falsifiability*, is, according to Popper, an essential requirement for asserting the *scientificity* of the theory, currently widely accepted in the scientific community [9].

I suggest that the same approach can be followed for assessing the scientific validity of models[2] or even of individual (scientific or technical) statements, like those expressing a measurement result [5].

[2] In my view, there is no substantial difference between models and theories: theories, in my view, are just very general models.

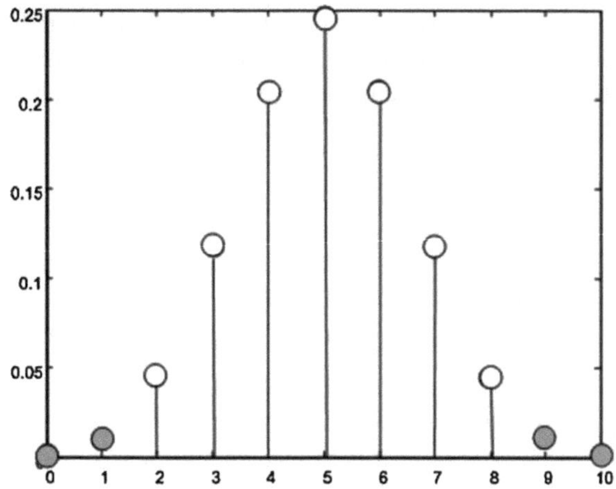

Fig. 6.3 Acceptance (*white*) and rejection (*grey*) regions for n_A

In this perspective, the control of a probabilistic model is a critical issue, since a probabilistic model is unable to predict exactly what will happen. This point has been raised to deny the scientificity of probabilistic theories and models. In this regard, I agree with Costantini [10], who claims that probabilistic theories and models may be falsified by hypothetic-deductive inferences. Let us see how do they work.

Hypothetic-deductive inferences are based on *significance tests* [11]. In these tests, a probabilistic model is assumed for describing the behaviour of a real system. Thanks to the model, the behaviour of the system in some given situation can be predicted, in probabilistic terms. Then, an experiment is performed reproducing such a situation and the real outcome of the system is observed and compared with the prediction. If there is *agreement* between prediction and observation, the model is maintained; otherwise, it is rejected. The point is how to assess that agreement: let us discuss this with a simple example.

Consider again the coin-tossing experiment, but suppose now that we visualise *one specific coin* and that, after carefully observing it, we assume the additional hypothesis of *symmetry*. Then, we can fix the needed probability $P(A) = p = p_0 = 1/2$ and the model, for the coin under investigation, is thus totally specified and may be expressed by a probability distribution for n_A, given by formula (6.2), with $p = 1/2$. For a number of trials $N = 10$, the possible values for n_A are the integer numbers from 0 to 10 that constitute the *space of the outcomes*. Then, we divide the space of outcomes into two subspace: R, where the deviations from the expected value are *not significant*, and $\bar{R}$, where they are *significant*. For example, we may choose $R = [2, 8]$, as shown in Fig. 6.3.

The rationale is that there is a high probability that the observed result falls in that region. In fact, such a probability can be calculated, thanks to the model, and it is $P(n_A \in R) = 0.98$. Obviously, there is some conventionality in the choice of such a "reasonably high" probability. In practical application, experience and consensus amongst experts can guide such choice. Despite this amount of conventionalism, the procedure is sound, well founded and widely applied [12]. Once the acceptance/rejection regions have be fixed, we have just to perform the experiment, with the real coin, and observe the result. If we obtain, e.g. $n_A = 9$, we reject our hypothesis, that is we conclude that

$$P(A) \neq 1/2. \tag{6.3}$$

Note that this is a probabilistic statement, although expressed by an inequality.

To sum up, let us outline the logical structure of a *hypothetic-deductive* probabilistic inference. We can summarise such a procedure in the following steps.

(a1) We hypothesise a probabilistic model,
(a2) on the basis of which we deduce the probability distribution for the observations in a given experiment, which allows us
(a3) to define an acceptance region R for the observation, that is a region in which the observation complies with the model;
(a4) then, we perform the experiment and acquire $\mathbf{y}$:
(a5) if $\mathbf{y}$, or some function of it—such as n_A in our example—falls into the acceptance region, the model is corroborated; otherwise, it is falsified by the observation and we may consider abandoning it.

6.2.2.2 Two Kinds of Inductive Inferences

In an inductive inference, instead, the probability distribution is not previously assumed, but is inferred—*induced*—from data. This inference, in turn, may be done in two main ways, either *by assuming some probabilistic data-producing mechanism*, or *by considering the way we learn from experience*. The former strategy is typical of Bayesian inference: since it hypothesises some kind of data producing mechanism, we will call it *hypothetic-inductive*; the latter instead is typical of a predictive approach and so we will call it a *predictive* inference [3].

For understanding how do they work, let us continue discussing the coin-tossing experiment. Suppose that we want to *estimate* the parameter p. We can do this by using the Bayes-Laplace rule, Bayes-Laplace rule that we have presented in Sects. 4.1.2, 4.1.3. We restate it as

$$P(C|E) = \frac{P(E|C)P(C)}{P(E)} \propto P(E|C)P(C), \tag{6.4}$$

where C stands for an (unknown) "cause", and E as an (observed) "effect".[3] The key point in this inference is to assume a proper model for describing the relation between the parameter p, the cause, and the observable effects, here the possible

[3] The interpretation of this rule as a way for assessing the probability of causes, after observing the effects, is traceable to Laplace himself [13].

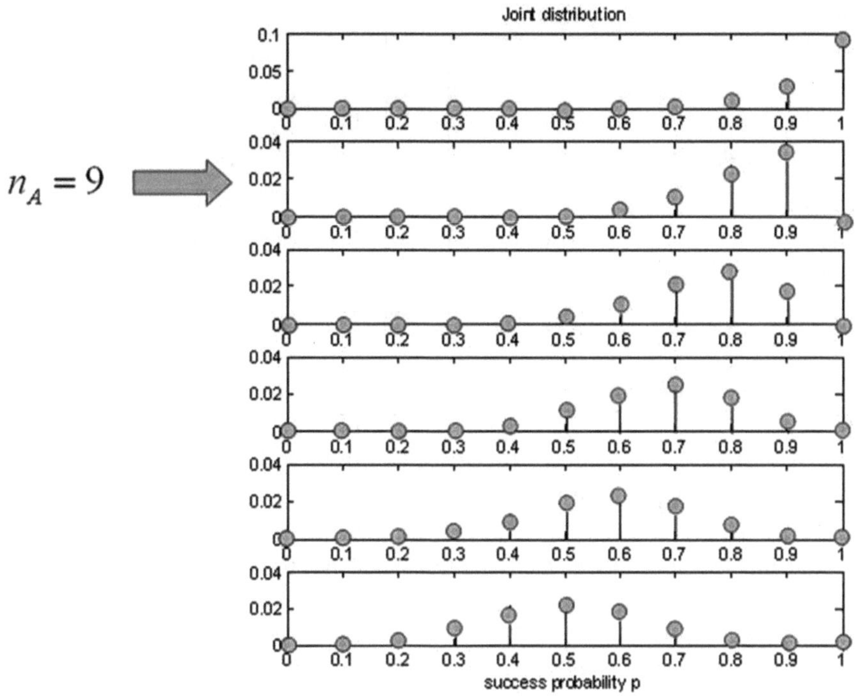

Fig. 6.4 Identifying the final distribution, for $n_A = 9$

values of n_A. One such a model is clearly provided by the previously established Bernoullian frame, formula (6.2). We rewrite it as

$$P(n_A|p) = \binom{N}{n_A} p^{n_A}(1-p)^{N-n_A}, \tag{6.5}$$

outlining the fact that such a model actually provides the sought link between p and n_A. Note that the right side of formula 6.5 is to be interpreted as a function of both n_A and p and corresponds to reading Fig. 6.2 by columns and re-scaling them in such a way that each column sums to one. By applying the Bayes–Laplace rule, we obtain

$$P(p|n_A) \propto p^{n_A}(1-p)^{N-n_A}, \tag{6.6}$$

where now the right side of the equation is a function of p, with n_A fixed and corresponding to the result of our experiment. This corresponds *to reading* Fig. 6.2 *by rows and again scaling such rows so that they sum to one*. For example, if we find, as above, $n_A = 9$, as shown in Fig, 6.4, the corresponding distribution for p is shown in Fig. 6.5, for discrete (a) and for continuous values (b), respectively.

The expected value of p is

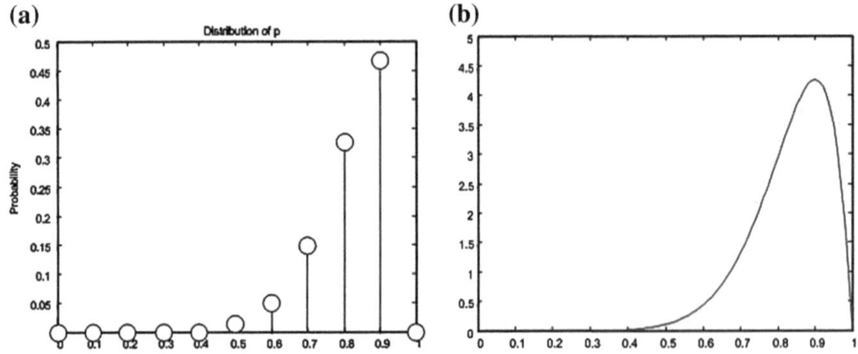

Fig. 6.5 The final probability distribution for n_A, discrete (**a**) and continuous (**b**)

$$\hat{p} = \frac{n_A + 1}{N + 2} = 0.83. \tag{6.7}$$

Note that the same model, the Bernoullian one, has been used in these two inferences, but in two different ways. So, the difference between a model and an inference should now be clear: a model is a general description of a class of phenomena, which may, or not, be used to support inferences concerning actual manifestations of such phenomena; an inference instead is a process in which we learn from data, using (or not, as we will show in the following) a model. Again it is important to elicit the logical structure of this inference, which may be expressed in the following steps.

(b1) We *hypothesise* a probabilistic relation, in a given experiment, linking the observation $\mathbf{y}$ to a parameter x, expressed by a conditional distribution $P(\mathbf{y}|x)$;
(b2) we perform the experiment and acquire the observation $\mathbf{y}$;
(b3) on the basis of the observation and of the hypothesised probabilistic relation, we assign (*induce*) a probability distribution to x.

Lastly, let us briefly touch on predictive inferences. Although often included in the general framework of Bayesian statistics, they follow a different logical pattern and merit being distinguished, as Costantini does [3]. They follow an approach traceable to Kemeny [14] and very well illustrated by Geisser [15]. These inferences aim at assigning a probability distribution to the possible future observations, on the basis of the past observations. In the coin-tossing experiment, e.g. we may look for $P(A|\mathbf{y})$ and $P(\bar{A}|\mathbf{y})$. A major difference with the hypothetic-inductive approach is that here, we do not assume any model for the process, rather we make assumptions *on the way we learn from observations*.

Without going into details, by assuming some *exchangeability* (the order of the observations is not relevant) and *invariance* (the roles of A and $\bar{A}$ are exchangeable) conditions, it is possible to obtain a so-called *lambda*-representation, which reads

$$P(A|\mathbf{y}) = \frac{\lambda}{\lambda + N} p_0 + \frac{N}{\lambda + N} \frac{n_A}{N}, \tag{6.8}$$

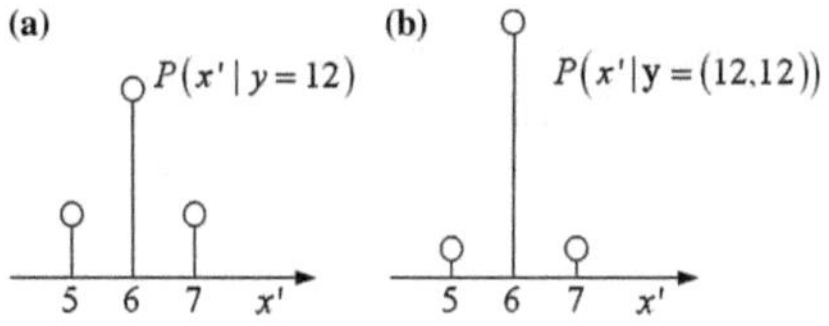

Fig. 6.6 The final probability distribution: **a** single observation; **b** observation repeated twice, with the same result obtained in the repetition

that is to say that the predictive probability is a weighted mean between the initial probability, p_0, and the relative frequency, n_A/N, where λ is the weight of the initial probability [10]. For instance, in our numerical example, if we assume $p_0 = 0.5$, $\lambda = 10$ and obtain $n_A = 9$, we conclude that

$$P(A|\mathbf{y}) = 0.7. \tag{6.9}$$

The logical sequence of steps for this kind of inference is thus as follows.

(c1) We assume some properties that characterise the way we learn from experience, such as exchangeability and invariance conditions,
(c2) we perform the experiment and acquire the observation $\mathbf{y}$,
(c3) on the basis of the observation, we assign a probability distribution to the possible future outcomes.

6.3 Measurement Evaluation

Let us now apply the principles we have so far discussed to measurement. *Measurement evaluation* is another way to denote restitution, when we put an emphasis on the evaluation of uncertainty. Let us now then reconsider restitution (or evaluation) starting from the numerical example firstly presented in Sect. 5.3, Fig. 5.6, and then reconsidered with inclusion of a systematic effect in Sect. 5.5, Figs. 5.11 and 5.12. At the end of Sect. 5.5, we invited the reader to think about the case of measurement based on repeated observation. We discuss this important issue now.

Consider first the basic case of Fig. 5.6. It is quite easy to recognise the pattern of hypothetic-inductive, or Bayesian inference. In fact, here we assume the probabilistic model of observation, expressed by the conditional distribution $P(y|x)$, we perform measurement, acquire instrument indication and, on the basis of that, we provide a distribution for the measurement value. This distribution, $P(x'|y = 12)$, is presented in Fig. 6.6a.

Let us probe further the potential of this kind of inference. Consider the case where we repeat observation more times, for example twice, obtaining a vector of indications, $\mathbf{y} = (y_1, y_2)$. For maximum simplicity, suppose also that $y_1 = 12$ and also $y_2 = 12$. The conditional distribution that describes observation is now $P(\mathbf{y}|x)$, that is the joint distribution of the observations, conditioned by the value of the measurand x. If we assume that the variations in the observations are independent, the joint distribution will be simply the product of the individual distributions, that is $P(\mathbf{y}|x) = P(y_1|x)P(y_2|x)$. Accordingly, restitution will be given by

$$P(x'|\mathbf{y}) = \left[\frac{P(\mathbf{y}|x)}{\sum_x P(\mathbf{y}|x)}\right]_{x=x'}. \tag{6.10}$$

It is easy to check that the resulting distribution is now

$$\begin{aligned}
P(x' = 5|\mathbf{y} = (12, 12)) &= 0.09,\\
P(x' = 6|\mathbf{y} = (12, 12)) &= 0.82,\\
P(x' = 7|\mathbf{y} = (12, 12)) &= 0.09,
\end{aligned}$$

presented in Fig 6.6b. It is apparent that the uncertainty has been reduced by the repetition of the measurement, since now the final distribution is more concentrated around the central value.

With this kind of inference, it is also possible to learn from observation in another way. In fact, until now we have assumed that the distribution that characterises observation is completely known: this may be not the case in a real situation, since for example the dispersion may vary in dependence on the operating conditions. This may be accounted for by assuming a distribution that depends upon an unknown dispersion parameter, to be estimated during the experiment. For a gaussian distribution that parameter is usually the standard deviation. To keep things simpler, in our example, we assume that the distribution has the following parametrical expression, for $i = 1, 2, \ldots 10$:

$$\begin{aligned}
P(y = 2i - 2|x = i) &= \frac{1-p}{2},\\
P(y = 2i|x = i) &= p,\\
P(y = 2i + 2|x = i) &= \frac{1-p}{2}.
\end{aligned}$$

Assume also, for maximum simplicity, that parameter p may assume just two values, $p_1 = 0.4$ or $p_2 = 0.6$, and start by assigning an equal probability to them. Since our goal is to learn about both x' and p, we have to make a joint inference. In the case of a single observation, this is accomplished by the rule

$$P(x', p|y) = \left[\frac{P(y|x, p)}{\sum_x P(y|x, p)}\right]_{x=x'}. \tag{6.11}$$

If we perform measurement just once and obtain $y = 12$, the resulting joint distribution is

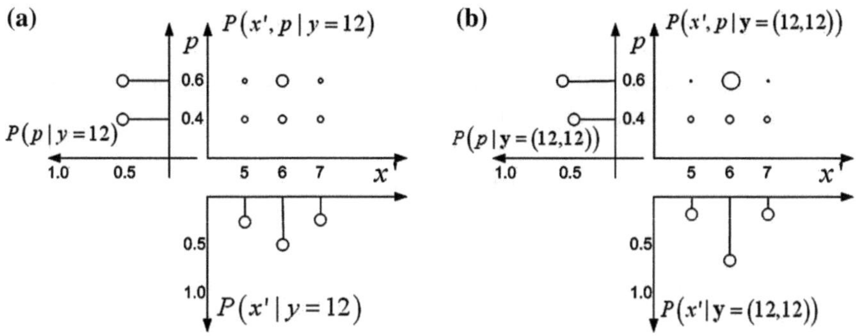

Fig. 6.7 The probability distributions for x' and p, joint and marginal, for (**a**) $y = 12$ and for (**b**) $\mathbf{y} = (12, 12)$

$$
\begin{aligned}
P(x' = 5, p = 0.4|y = 12) = 0.15, \\
P(x' = 6, p = 0.4|y = 12) = 0.20, \\
P(x' = 7, p = 0.4|y = 12) = 0.15, \\
P(x' = 5, p = 0.6|y = 12) = 0.10, \\
P(x' = 6, p = 0.6|y = 12) = 0.30, \\
P(x' = 7, p = 0.6|y = 12) = 0.10,
\end{aligned}
$$

shown in Fig. 6.7a.

From the joint distribution, we can obtain the marginal ones, $P(x'|y)$ and $P(p|y)$, also shown in the same part of the figure. Note that the final distribution for p is still uniform, since we cannot learn about p from a single observation.

On the other hand, if we repeat observation a few times and collect the vector $\mathbf{y}$, the inferential rule becomes

$$
P(x', p|\mathbf{y}) = \left[\frac{P(\mathbf{y}|x, p)}{\sum_x P(\mathbf{y}|x, p)} \right]_{x=x'}, \tag{6.12}
$$

which allows us to learn about p also. For example, if we repeat observation twice and obtain, as in the previous case, $y_1 = 12$ and $y_2 = 12$, the resulting joint distribution is

$$
\begin{aligned}
P(x' = 5, p = 0.4|\mathbf{y} = (12, 12)) = 0.12, \\
P(x' = 6, p = 0.4|\mathbf{y} = (12, 12)) = 0.20, \\
P(x' = 7, p = 0.4|\mathbf{y} = (12, 12)) = 0.12, \\
P(x' = 5, p = 0.6|\mathbf{y} = (12, 12)) = 0.05, \\
P(x' = 6, p = 0.6|\mathbf{y} = (12, 12)) = 0.46,
\end{aligned}
$$

$$P(x' = 7, p = 0.6|\mathbf{y} = (12, 12)) = 0.05,$$

shown in Figure 6.7b, with the related marginal distributions. Note that now the distribution for p has also changed and is

$$P(p = 0.4|\mathbf{y} = (12, 12)) = 0.44,$$
$$P(p = 0.6|\mathbf{y} = (12, 12)) = 0.56.$$

This change can be intuitively understood in this way: since we have obtained the same result twice, the experiment shows no dispersion, and consequently, the probability of the value of p that corresponds to a lower dispersion, that is $p = 0.6$, has increased.

To sum up, we see that through this hypothetic-inductive inference, we can learn from data in two ways, namely

- by improving our estimate of the measurand and
- by improving our knowledge of side parameters, p in this case.

At this point, we may be tempted to conclude that the restitution, or evaluation, process consists in a hypothetic-inductive inference. Yet this would be a wrong conclusion, since until now we have not yet considered the systematic effect. So we should now discuss the problem proposed at the end of Sect. 5.5.

Consider then again the case of Fig. 5.11, with θ still having a uniform distribution. We have now to use formulae 5.42 and 5.43, with y replaced by $\mathbf{y}$, which yields

$$P(x'|\mathbf{y}, \theta) = [\frac{P(\mathbf{y}|x, \theta)}{\sum_x P(\mathbf{y}|x, \theta)}]_{x=x'}, \tag{6.13}$$

$$P(x'|\mathbf{y}) = \sum_{\theta} P(x'|\mathbf{y}, \theta) P(\theta). \tag{6.14}$$

Suppose then we repeat observation twice and we obtain, as previously assumed, $\mathbf{y} = (12, 12)$. The result is illustrated in Fig. 6.8, to compared to that of Fig. 5.12.

Note that now the final distribution for the two repeated observations is closer to the uniform shape, than it was with a single observation. This can be explained in the following way: accounting for multiple observations is equivalent to make some averaging, which reduces the uncertainty related to random variations. So the contribution of the systematic effect tends to prevail. Since the distribution corresponding to the systematic effect is uniform, the final distribution tends to become uniform as well.[4]

Let us now further discuss the logic of this last restitution (or evaluation), which correspond to the general case, where both random variations and systematic effects are present. In order to implement formula 6.14, we have had to assign a probability distribution to θ, $P(\theta)$, and then to use it to obtain the final distribution. Yet our final

[4] We will consider a similar example in Sect. 9.1.3 to probe further this important topic.

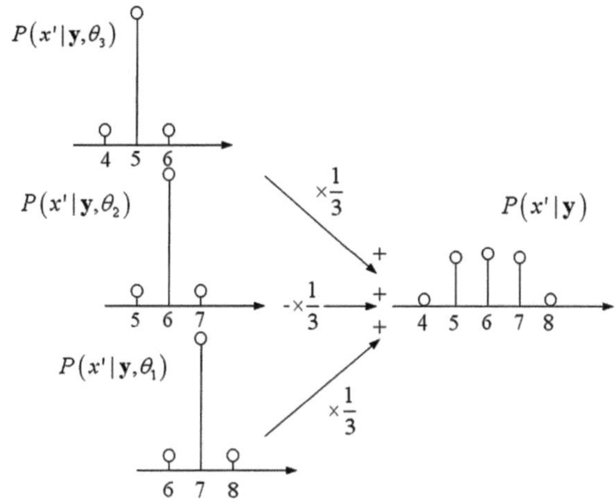

Fig. 6.8 The probability distribution $P(x'|\mathbf{y})$, when a systematic effect is present

knowledge about θ is equal to the initial one, i.e. $P(\theta|\mathbf{y}) = P(\theta)$, which implies that we *have not learnt from data about* θ. So θ has not undergone a hypothetic-inductive inference, rather it has been treated, necessarily, in a purely deductive way: a probability distribution for it has been assumed and consequences have been derived, without any interaction with the data. In fact, *this is always the case when we have a systematic effect*. Thus in general, the restitution, or evaluation process, requires the following logical steps:

(d1) to assume a probabilistic relation between the value of the measurand and the indications of the measuring system, parametrical with respect to some influence parameters: this relation is a model of the observation process;
(d2) to assume a probability measure over the space of the influence parameters;
(d3) to perform observation and acquire the indications of the measuring system;
(d4) to apply, in the restitution phase, the Bayes–Laplace rule and obtain a probability distribution for the measurand, still conditioned upon the influence parameters;
(d5) to decondition the probability distribution with respect to the influence parameters, which concludes the restitution phase and the overall measurement process.

If we analyse this procedure in the light of what we have so far exposed, we recognise in steps d1, d3 and d4 a Bayesian inference, so that we may say that the measurement process *embeds* a Bayesian inference.

On the other hand, we also note that steps d2 and d5 are not typical of a Bayesian inference. They include the assumption of a probability distribution for some parameters (step d2) and their use according to the rules of the calculus of probability (step d5). We say that these two steps form a *deductive process*: so we conclude that in general, *in a measurement process, we have the combination of a hypothetic-inductive inference and of a deductive process*.

The presence of a deductive process associated with measurement is undoubtedly a problem, since measurement is normally intended as the way for learning from experience *par excellence*. So, how can we accept that measurement includes a deductive process? How can we deal with this problem?

6.4 Measurement Verification

We consider a solution to the problem of systematic effects in measurement based on the practice of *measurement verification*. As we have just seen, systematic effects can be treated in probabilistic terms, but this implies that the restitution phase of the measurement process includes a deductive process, which is not controlled by the data acquired during the measurement process. How can we then ensure the *scientificity* of the overall measurement process? In this regard, recall what we have briefly discussed in Sect. 6.2.2.1, where we suggested that probabilistic theories and models can be falsified by hypothetic-deductive inferences. Here, we suggest to apply this general principle to provide a sound foundation to this part of measurement practice.

In this perspective, the solution of the problem of the systematic effect becomes simple and straightforward: *the validity of the measurement process, which includes a deductive treatment of systematic effects, may be controlled by a significance test, that is by a hypothetic-deductive inference.*

Let us briefly see how this inference can be stated. Consider a measurement process described by the conditional distribution $P(\hat{x}|x)$, related to the expected measurement value, $\hat{x}$. Remember that this distribution accounts for systematic effects too, if they are present. Suppose that we dispose of a *standard* whose value, x_0, is known with uncertainty negligible for our purpose. Then, the probabilistic model to be checked is the distribution $P(\hat{x}|x_0)$, which under these assumptions is fully specified and is a function of $\hat{x}$ only. Then, we can measure the standard through the measurement process under consideration and perform a significance test on the difference $\hat{x}_0 - x_0$, where $\hat{x}_0$ is the measurement value obtained by measuring the standard.

For a significance level α, the acceptance region will be $R = [a, b]$, such that

$$\sum_{\hat{x}=a}^{b} P(\hat{x} - x_0|x_0) = \alpha \tag{6.15}$$

and we have to check whether

$$\hat{x}_0 \in R \tag{6.16}$$

or not. As an example, consider again the measurement process of the previous section whose observation is characterised by the distribution of Fig. 5.11. The corresponding overall distribution $P(\hat{x}|x)$ is shown in Fig. 6.9a.

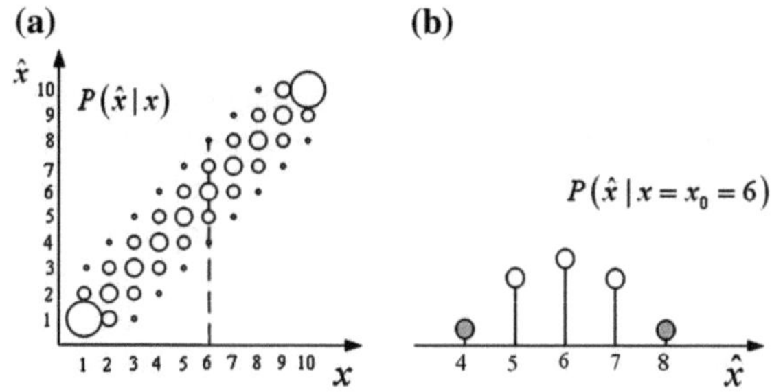

Fig. 6.9 Verification of the measurement process of the previous section: **a** probability distribution $P(\hat{x}|x)$; **b** distribution $P(\hat{x}|x_0 = 6)$ to be used for the significance test

Suppose that we have a standard object, whose value is $x_0 = 6$, with uncertainty negligible with respect to that of the measurement process. Then, the distribution to be used for the significance test is $P(\hat{x}|x_0 = 6)$, shown in Fig. 6.9b. We may fix the acceptance region as the interval [5, 7]: so if we obtain, e.g. $\hat{x}_0 = 7$, the verification is passed, whilst if we obtain $\hat{x}_0 = 8$, we can suspect that the systematic effect has been underestimated.

In practical application, the procedure just described may be implemented in different ways, that are part of a good practice in measurement. They include, e.g. the verification of the calibration of a measuring system or the control of a measurement process by check standards. Even inter-comparisons, including key comparisons, may be considered in this perspective, although they require a more complicated mathematical model that cannot be examined here.[5]

To sum up, we suggest that the probabilistic treatment of systematic deviations may be considered a sound scientific approach, since the hypotheses implied in it may be, in general, at least in principle, checked by an auxiliary experiment that we call *measurement verification* and that consists in a hypothetic-deductive inference [5, 16].

6.5 Summary

In this chapter, we have briefly reviewed some key ideas in probabilistic inferences, and we have considered three main kinds of such inferences. Then, we have applied these ideas to a discussion of the logic of the measurement process, based on the general probabilistic model that we have presented in the previous chapter. We have seen that the restitution process, which corresponds to what in the technical literature is often called measurement evaluation, may be interpreted as including a hypothetic-inductive (Bayesian) inference. This inference allows us to learn from data in a few ways,

- by locating the position of the measurement value along the reference scale,
- by reducing the uncertainty through the repetition of observations and

---

[5] They will be addressed in Chap. 11 and constitute a very important application of the ideas here presented.

– by directly estimating the dispersion due to random variations.

On the other hand, it is impossible to learn, during the measurement, about the systematic effects. To treat these in probabilistic terms also, as recommended by current international written standards, it is necessary to include some prior knowledge about them, which affects the final result in a purely deductive way. This poses a problem, from an epistemological standpoint, since this additional information is not subject, in the measurement process, to empirical validation. In order to overcome this problem, we have suggested to perform, whenever possible, some kind of measurement verification, through a proper significance test. Such a test has been discussed and illustrated by a simple example.

With this chapter, we have completed most of the part of this book devoted to the theory. We have just to briefly consider multidimensional measurement, in the next chapter. Yet this topic will be just briefly addressed, the main scope of this book remaining one-dimensional measurement. Note that in this chapter, as in general in this part of the book, we have used only very simple illustrative examples, since the aim is here to illustrate the main ideas. Such examples are not directly representative of any real situation. On the other hand, real case studies will be presented in the third part of the book, devoted to applications.

References

1. Estler, W.T.: Measurement as inference: fundamental ideas. Ann. CIRP **48**, 1–22 (1999)
2. Hacking, I.: An Introduction to Probability and Inductive Logic. Cambridge Press, Cambridge (2001). (talian edition: Il Saggiatore, Milano 2005)
3. Costantini, D.: I fondamenti storico-filosofici delle discipline statistico probabilistiche. Bollati Boringhieri, Torino (2004)
4. Wöger, W.: IEEE Trans. Instrument. Measure. **36**, 655 (1987)
5. Rossi, G.B.: Probabilistic inferences related to the measurement process. J. Phys. **238**, 012015 (2010). (Conference Series)
6. Geymonat, L.: Galileo Galilei. Einaudi, Torino (1963)
7. Rossi, P.: La nascita della scienza moderna in Europa. Laterza, Roma (1997)
8. Reale, G., Antiseri, D.: Storia della filosofia. Bompiani, Milano (2008)
9. Popper, K. R.: La logica della scoperta scientifica. Einaudi, Torino (1998)
10. Costantini, D.: Verso una visione probabilistica del mondo. GEM, Padova (2011)
11. Fisher, R.A.: Statistical Methods and Scientific Inference. Oliver and Boyd, Edinburgh (1956)
12. Pavese, F., Forbes, A. (eds.): Data modeling for metrology and testing in measurement science. Birkhauser-Springer, Boston (2009)
13. Laplace, P.S.: Mémoire sur la probabilité des causes par les évenemens. Mem Acad R Sci **6**, 621–656 (1774)
14. Kemeny, J.G.: Carnap theory of probability and induction. In: Carnao, R., Schilpp, P.A. (eds.) The Philosophy of Rudolf Carnap, p. 711. Cambridge University Press, Cambridge (1963)
15. Geisser, S.: Bayesian analysis. In: Zellner, A. (ed.) Econometrics and Statistics. North Holland, Amsterdam (1980)
16. Rossi, G.B.: Probability in metrology. In: Pavese, F., Forbes, A. (eds.) Data Modeling for Metrology and Testing in Measurement Science. Birkhauser-Springer, Boston (2009)

Chapter 7
Multidimensional Measurement

7.1 What Happens when Moving from One to Two Dimensions

Multidimensional measurement concerns properties that depend on more than one attribute [1, 2]. If there are p such attributes, the state of an object a, with respect to the property under consideration x, may be thought of as a point in a p-dimensional state space. Thus, measurement aims at representing such a state in a q-dimensional numerical space, with $q \leqslant p$, in such a way as to map the empirical relations into corresponding numerical ones. A thorough treatment of multidimensional measurement is beyond the scope of this book. We just present some basic ideas in order to suggest that a probabilistic development similar to that of the one-dimensional case is possible, at least in principle, and to encourage readers to develop their own ideas in this fascinating area.

In order to introduce some of the problems of multidimensional measurement, let us discuss a simple example. Consider coordinate measurement, that is, measurement of length-related features, by *coordinate measuring machines*, for testing the dimensional conformance of workpieces to their design features [3]. The workpiece in Fig. 7.1, for example, constitutes a plate with two holes.

Let such holes be our "objects", a and b, respectively. Basically, a coordinate measuring machine drives a touch probe in a three-dimensional measuring space and measures the coordinates of some touch points, as programmed by the operator. Then, the required features—part lengths, positions, diameters and angles—can be indirectly measured as functions of the coordinates of the selected touch points. Such functions are based on analytical geometry results or on more sophisticated optimisation procedures. In this example, each hole in the measurement plane can be characterised by measuring three non-coincident points on its circumference. So the *state* of each "object" (hole) can be thought of as a point in the six-dimensional space of the coordinates of the three points,

$$(P_1, P_2, P_3) = ((x_1, y_1), (x_2, y_2), (x_3, y_3)).$$

G. B. Rossi, *Measurement and Probability*, Springer Series in Measurement Science and Technology, DOI: 10.1007/978-94-017-8825-0_7,

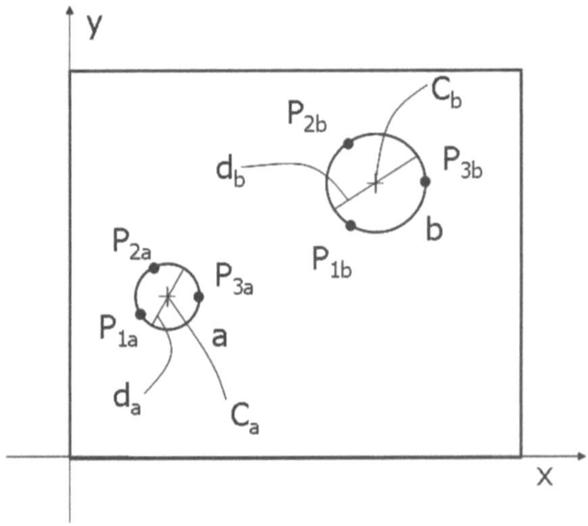

Fig. 7.1 Measurement of the features of two *holes* in a plate

Yet this is not the final result of measurement: typically, we will be interested in such parameters as the *diameter* (d) or the *centre position* ($C = (x_C, y_C)$) of the hole. From simple geometry, we know how they are related to the directly measured attributes, the coordinates of the three selected points. It is now interesting to note the differences between these two properties, diameter and centre position, that are both outcomes from the same initial information.

In the case of the diameter, we have a *one-dimensional order property*, for which we can write

$$a \succsim_d b \Leftrightarrow d(a) \geqslant d(b),$$

where $\succsim_d$ is the weak order relation related to the property "diameter" and d is the measure of the diameter. In the case of Fig. 7.1, $a \succ_d b$ holds, that is, a is greater than b, as far as the diameter is concerned.

For the centre position points, instead, we do not have any order; rather, we can consider whether they are coincident or not, and we have

$$a \sim_C b \Leftrightarrow C(a) = C(b),$$

or we may consider their *distance*, $\delta(a, b)$. If we have four objects, a, b, c, d, we can compare their distances, $\delta(a, b)$ and $\delta(c, d)$.[1]

To sum up, one class of problems in multidimensional measurement consists in finding a proper q-dimensional representation for each property of an object characterised by a p-dimensional state, with $q \leqslant p$. As a special, very important, case, we can sometimes attain a one-dimensional representation: this usually occurs if it is possible to identify an order relation for the property under investigation, as in the case of the diameter. If this is the case, multidimensional measurement reduces to

[1] We will discuss the important notion of distance in a moment.

derived measurement, as pointed out, e.g. by Muravyov and Savolainen [4]. Alternatively, we have to remain in a multidimensional space. In this case, *distance* is usually the most important empirical relation to be considered.

So far, we have considered coordinates of points as input data. Yet there is another, very important, possibility, where the input consists of (perceived) empirical distances amongst objects. This is what we have to discuss now.

7.2 Distances and Metrics

In mathematical analysis, a *distance* is a biargumental, symmetric, non-negative function, $d(x, y)$, that satisfies the triangular inequality: $d(x, y) + d(y, z) \geq d(x, z)$. In measurement, we have also to regard "distance" as an empirical property, often called (dis)similarity, that can be directly perceived or otherwise experimented with. For example, we can consider the degree of similarity of a set of colour, of sounds, of emotional states and of sensory—acoustic, visual and olfactory—environments. We will denote empirical distance by the symbol δ and its numerical representation by d.

To understand how distance information can yield a representation in a proper numerical space, suppose that we have three objects, $A = \{a, b, c\}$, and that, in respect of some property of them, we have the following information about distances:

$$\begin{aligned} d(a, b) &= 1, \\ d(b, c) &= 2, \\ d(a, c) &= 3. \end{aligned}$$

Here, it is apparent that the three objects, in respect of the property under consideration, can be represented on a *straight line*, that is, on a one-dimensional space, as shown in Fig. 7.2.

Note now that the same evidence could have been expressed, in terms of empirical distances, without numbers, by the following relations:

$$\begin{aligned} &\delta(a, c) \succ \delta(b, c) \succ \delta(a, b), \\ &\delta(a, c) \sim \delta(a, b) \circ \delta(b, c), \end{aligned}$$

where, as usually, $\circ$ is a concatenation operator. The representation in Fig. 7.2 is still appropriate, as would be any other one differing just by a linear transformation.

Thus, to sum up, in this example, we have seen how it is possible, starting from a set of empirical *distance relations*, to find a proper space, a straight line in this case, and a numerical representation of each object, expressed through coordinates in that space, that complies with empirical relations.

Consider now another example. Again, we start by expressing the initial information on distance in a numerical fashion, which is easier to understand, and then, we will restate it in terms of empirical relations. Suppose the input data are now

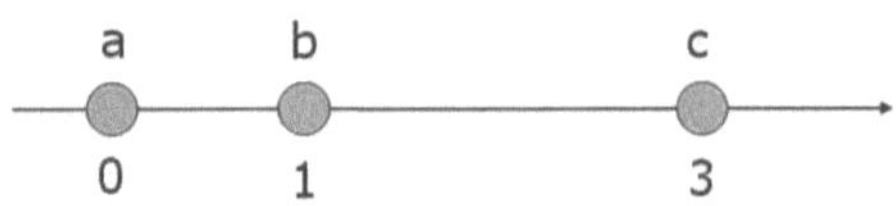

Fig. 7.2 Representation of the objects of A on a *straight line*

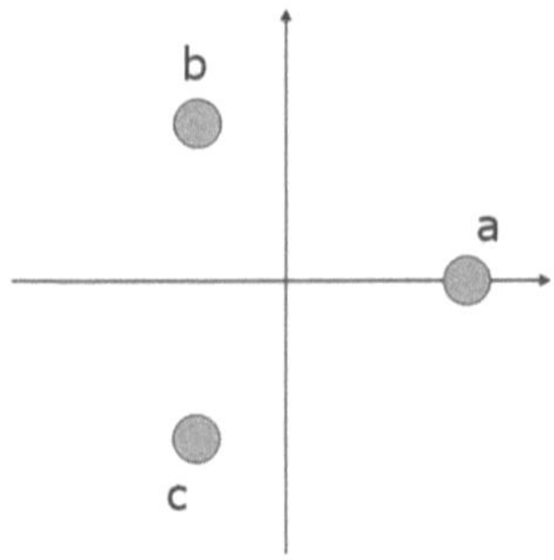

Fig. 7.3 Representation of the objects of A on a plane

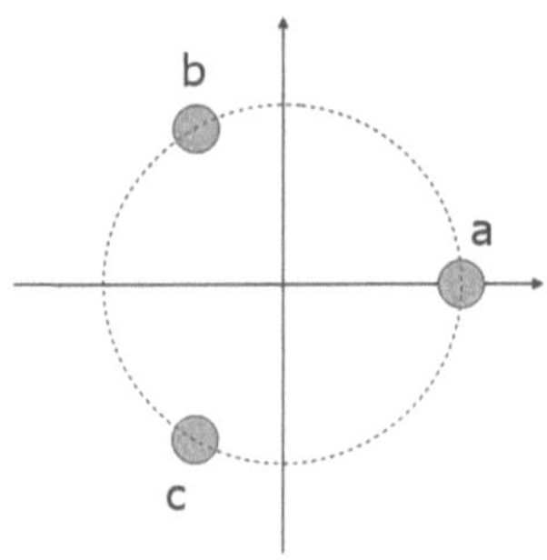

Fig. 7.4 Representation of the objects of A on a *circle*

$$d(a,b) = 1,$$
$$d(b,c) = 1,$$
$$d(a,c) = 1.$$

It is apparent that now, we can no longer represent the objects on a straight line; rather, we can do that on a plane, as shown in Fig. 7.3.

Again, the initial information can be stated in terms of empirical relations, as

$$\delta(a,c) \sim \delta(b,c) \sim \delta(a,c).$$

Note now that although the representation on the plane works, it is possible to find a more efficient one by noting that all the objects can be represented by points on a circumference, as demonstrated in Fig. 7.4.

Thus, a one-dimensional representation is again possible, by fixing a starting point, a positive rotation convention and the unitary arc. For example, if we take the position of a as the starting point, we assume the counterclockwise rotation as positive and we set the overall length of the circle to 2π, we obtain for the coordinate of the objects in this "circular" space

$$x_a = 0, x_b = 2\pi/3 \text{ and } x_c = 4\pi/3,$$

respectively. But this cannot be achieved by a simple linear transformation; rather, a *change in the metrics* is required, since now the distance between two generic objects is no longer expressed by the length of the segment that connects them but by the length of the arc that connects them. A good example of this situation is provided by the distance amongst towns on earth.

To sum up, these very simple examples suggest a general framework for scale construction in multidimensional measurement, which includes the following steps:

1. to obtain empirical data, expressed either as relations amongst the coordinates of points that represent the objects in a p-dimensional manifold or as relations amongst the distances of the objects; initial data can be expressed either on a nominal (categorical) or on an ordinal or a linear scale;
2. to search for an appropriate metrics;
3. to search for a minimum dimension representation;
4. to obtain the final scale, either as one-dimensional—as happens for derived measurement—or still as multidimensional; in the one-dimensional case, the key relation is order, and in the multidimensional case, it is distance.

Starting from relations amongst point coordinates is called in psychophysics the *content model* approach, to distinguish it from the *distance model* [5]. We will also briefly discuss the latter in the following. We have thus to understand in greater depth the notion of distance. For achieving this in the simplest way, we go back to one-dimensional measurement and we consider both *nominal* and *distance* structures.

7.3 Nominal and Distance Structures

7.3.1 Nominal Structures

A nominal structure is one in which only an equivalence relation, "$\sim$", is defined, that is, a relation that is reflexive, symmetric and transitive.

Definition 7.1 Let $(A, \sim)$ be a structure where, for each $a, b, c \in A$:

(i) $a \sim a$ (reflexivity);
(ii) $a \sim b \Rightarrow b \sim a$ (symmetry);
(iii) $a \sim b$ and $b \sim c \Rightarrow a \sim c$ (transitivity).

Then, $(A, \sim)$ is a nominal structure.

Theorem 7.2 (Representation for nominal structures) *Let A be a finite (not empty) set of objects carrying the property x, and let $(A, \sim)$ be a nominal structure. Then, there exists a measure function $m : A \to \mathbb{R}$ such that, for each $a, b \in A$,*

$$a \sim b \Leftrightarrow m(a) = m(b).$$

Proof Since $\sim$ is an equivalence relation and A is finite, with, say, N elements, it is possible to partition it in $n \leq N$ equivalence classes, $A_1, A_2, \ldots, A_n$. Then, we can form a series of standards, $S = s_1, s_2, \ldots, s_n$, by simply picking one element from each equivalence class. Then, for each $a \in A$, $1 \leq i \leq n$, we define the measure function m by

$$m(a) = i \Leftrightarrow a \sim s_i.$$

It is easy to check that such a function satisfies the representation. □

Let us now consider uniqueness.

Theorem 7.3 (Uniqueness for nominal structures) *Let A be a finite (not empty) set of objects carrying the property x and let* $(A, \sim)$ *be a nominal structure and* $m : A \to \mathbb{R}$ *a measure function for it. Then, any other measure function* m' *is such that, for each* $a \in A$,

$$m'(a) = \phi(m(a)),$$

where ϕ *is a one-to-one function, that is,*

$$\phi(u) = \phi(v) \Leftrightarrow u = v.$$

Proof We have first to prove that if m is a proper measure function, $m' = \phi(m)$, with ϕ being one-to-one function, also is. Indeed, for each $a, b \in A$,

$$a \sim b \Leftrightarrow m(a) = m(b) \Leftrightarrow \phi(m(a)) = \phi(m(b)) \Leftrightarrow m'(a) = m'(b).$$

Then, we have to show that if both m and m' are valid measure functions, then their exists a one-to-one function ϕ, such that $m' = \phi(m)$. In fact, if both m and m' are valid measure functions,

$$m(a) = m(b) \Leftrightarrow a \sim b \Leftrightarrow m'(a) = m'(b).$$

Then, the function ϕ, defined, for each $a \in A$, by $m'(a) = \phi(m(a))$, is one-to-one. □

7.3.2 Distance Structures

Consider now the notion of *(empirical) distance* in one dimension.[2] It is close to the already discussed notion of *difference*, but with two main differences:

[2] The theory developed in this subsection is related to the theory of absolute difference structures ([6] pp. 170–177) and also accounts for the theory of proximity measurement ([7], Chap. 14, pp. 159–174).

- it is positive or null and
- it does not imply an order amongst the objects.

The first aspect may be considered by substituting the sign reversal axiom, $\Delta_{ab} \succsim \Delta_{cd}$ implies $\Delta_{dc} \succsim \Delta_{ba}$, which holds true for differences, by requiring that, for $a \neq b$, $\delta_{ab} \sim \delta_{ba} \succ \delta_{aa} \sim \delta_{bb}$. The point of order is more critical, since for developing a representation, we still have to establish an order, but such an order may be purely conventional and may be reversed safely: in other words, if we invert the sign to the numbers, the representation is still valid. Thus, in contrast to (algebraic) difference structures, we cannot claim that order is an empirical property of the objects, since here it just constitutes an artificial and conventional step for developing the representation. What plays the role of order is now *betweenness*: we say that b is between a and c, and we write $a|b|c$, if $\delta_{ac} \succsim \delta_{ab}, \delta_{bc}$. We then have to assume that betweenness is well behaved: this basically implies that if $a|b|c$ and $b|c|d$, then also both $a|b|d$ and $a|c|d$.

These are the most relevant changes from difference structures; the other required properties concern monotonicity and solvability and are very similar to those assumed for difference structures.

Let us now formalise some of these ideas. We start by defining betweenness.

Definition 7.4 (*Betweenness*) Let $(A, \succsim_\delta)$ be a structure where A is a finite set and

(i) $\succsim_\delta$ is a weak order on $A \times A$ and
(ii) for $a \neq b$, $\delta_{ab} \sim_\delta \delta_{ba} \succ_\delta \delta_{aa} \sim_\delta \delta_{bb}$.

We say that b is between a and c, and we write $a|b|c$, if and only if $\delta_{ac} \succsim_\delta \delta_{ab}, \delta_{bc}$.

Note that betweenness is symmetric in the first and third argument: $a|b|c \Leftrightarrow c|b|c$. Furthermore, for any a, b *and* c, at least one of $a|b|c$, $a|c|b$ or $b|a|c$ must hold.

We now define a *distance structure*.

Definition 7.5 (*Distance structure*) Let A be a finite set. Then, $(A, \succsim_\delta)$ is a (finite) distance structure if and only if $\succsim_\delta$ is a binary relation on $A \times A$, and for all $a, b, c, a', b', c' \in A$, the following five axioms hold:

(i) $\succsim_\delta$ is a weak order;
(ii) for $a \neq b$, $\delta_{ab} \sim_\delta \delta_{ba} \succ_\delta \delta_{aa} \sim_\delta \delta_{bb}$;
(iii) • if $b \neq c$, $a|b|c$ and $b|c|d$, then both $a|b|d$ and $a|c|d$,
  • if $a|b|c$ and $a|c|d$, then $a|b|d$;
(iv) if $a|b|c$, $a'|b'|c'$ and $\delta_{ab} \sim_\delta \delta_{a'b'}$, then $\delta_{bc} \succsim_\delta \delta_{b'c'}$ if and only if $\delta_{ac} \succsim_\delta \delta_{a'c'}$;
(v) if $\delta_{ab} \succsim_\delta \delta_{cd}$, then there exists d', with $a|d'|b$ and $\delta_{ad'} \sim_\delta \delta_{cd}$.

We also define a metric and a metric space.

Definition 7.6 (*Metric space*) Let X be a set and d a real-valued function on $X \times X$, such that

(i) $d(x, y) \geq 0$ and $d(x, x) = 0$,
(ii) $d(x, y) = d(x, y)$,

(iii) $d(x, y) + d(y, z) \geq d(x, z)$.

Then, we say that d is a metric and (X, d) is a metric space.

We may now formulate a representation theorem for distance structures yielding to what we will call a *metric scale*.

Theorem 7.7 (Representation for distance structures) *Let A be a finite (not empty) set of objects carrying the property x, and let $(A, \succsim_\delta)$ be a distance structure. Then, there exists a function $m : A \to \mathbb{R}$ and a function $d : A \times A \to \mathbb{R}$ such that, for each $a, b, c, d \in A$,*

$$\delta_{ab} \succsim_\delta \delta_{cd} \Leftrightarrow |m(a) - m(b)| \geq |m(c) - m(d)| \Leftrightarrow d(a, b) \geq d(c, d).$$

Furthermore, d is a metric for A.

Proof (outline) The proof closely resembles that for difference structures, so we simply informally sketch the main idea under it. In the case of difference structures, it was possible *to deduce an order amongst the objects from the given order amongst the intervals*. That was possible since differences are *antisymmetric*, that is $\Delta_{ab} \succsim_d \Delta_{cd}$ implies $\Delta_{dc} \succsim_d \Delta_{ba}$. In contrast, distances are *symmetric*, that is $\delta_{ab} \sim_\delta \delta_{ab}$. The idea is thus *to establish a conventional order amongst the objects which is compatible with the given order amongst the intervals* and then to proceed in much the same way as for difference structures, that is, to construct an equally spaced reference scale and to define a measure function referred to that scale.

For establishing such a conventional order, it is possible to proceed as follows.

Take a pair $x, y \in A$, and conventionally set $x \succsim y$. Then, for each pair $a, b \in A$, set $a \succsim b$ whenever either

- $a|x|y$, $b|x|y$ and $a|b|x$, or
- $a|x|y$ and $x|b|y$, or
- $a|x|y$ and $x|y|b$, or
- not$(a|x|y)$, not$(b|x|y)$ and $x|a|b$.

Then, it is possible to check that the relation $\succsim$ just defined is a weak order on A. Thus, it is possible to identify, as usually, a series of standards, $S = \{s_0, s_1, \ldots, s_n\}$, and to define, for each $a \in A$, a measure function through the rule

$$m(a) = i \Leftrightarrow a \sim s_i.$$

Proceeding in much a similar way as we have done for difference structures, it is possible to prove that the series of standards S is equally spaced for distances, that is, for each $i \in \{0, 1, \ldots, n-1\}$, $\delta_{s_{i+1}s_i} \sim_\delta \delta_{s_1 s_0}$. Then, the representation follows.

$$\delta_{ab} \succsim_\delta \delta_{cd} \Leftrightarrow |m(a) - m(b)| \geq |m(c) - m(d)|$$

Let us now introduce the distance $d : A \times A \to \mathbb{R}$. For each $a, b \in A$, let

$$d(a, b) = |m(a) - m(b)|.$$

Then, the second part of the representation trivially holds. Since $\succsim_\delta$ is a weak order and A is finite, there is a maximum interval, call it ab. We have just to check that d is a metric for A. From its definition, it immediately follows that

$$d(a, b) \geq 0, d(a, a) = 0 \, and \, d(a, b) = d(b, a).$$

We have thus to check that for each $a, b, c \in A$, $d(a, b) + d(b, c) \geq d(a, c)$. Indeed,

$$\begin{aligned} d(a, c) &= |m(a) - m(c)| = |m(a) - m(b) + m(b) - m(c)| \\ &\geq |m(a) - m(b)| + |m(b) - m(c)| = d(a, b) + d(b, c) \end{aligned}$$

which completes the proof. □

Let us consider now uniqueness.

Theorem 7.8 (Uniqueness for distance structures)
Let A be a finite (not empty) set of objects carrying the property x and let $(A, \succsim_\delta)$ *be a difference structure and* $m : A \to \mathbb{R}$ *a measure function for it. Then, any other measure function* m' *is such that, for each* $a \in A$,

$$m'(a) = \alpha m(a) + \beta,$$

with $\alpha \neq 0$.

Proof We have first to prove that if m is a proper measure function, also $m'(a) = \alpha m(a) + \beta$, with $\alpha > 0$, is appropriate. In fact, if $|m(a) - m(b)| \geq |m(c) - m(d)|$, then also $|[\alpha m(a) + \beta] - [\alpha m(b) + \beta]| \geq |[\alpha m(c) + \beta] - [\alpha m(d) + \beta]|$ holds true.

Then, we have to prove that if both m and m' satisfy the representation, then they must be related by $m' = \alpha m + \beta$. Let $S = (s_0, \ldots, s_n)$ be a series of standards for the structure under consideration[3] and m any proper measure function for it. Since S is equally spaced for distances, the distance $|m(s_i) - m(s_{i-1})|$ must be the same for each i, $1 \leq i < n$, call it $d_0 > 0$. Then, for each i, $m(s_i) = m(s_0) + id_0$.

Consider now another valid measure, m'. In contrast to the difference-structures case, here we have two possibilities: m' has been obtained either by assuming *the same conventional order* as m or by assuming the *opposite one*.

In the former case, we proceed as we have done with difference structures: we note that we also have $m'(s_i) = m'(s_0) + id'$, where $d' = |m'(s_i) - m'(s_{i-1})|$. Then, for each $a \in A$, there will be a standard $s_i \in S$ such that $a \sim s_i$. Then, $m(a) = m(s_i) = m(s_0) + id$ and also $m'(a) = m'(s_i) = m'(s_0) + id'$. Thus,

[3] Note that we write here $S = (s_0, \ldots, s_n)$, rather than $S = \{s_0, \ldots, s_n\}$ as we usually do, since we want to regard S as an ordered set.

Fig. 7.5 A hierarchy for one-dimensional structures and related scales

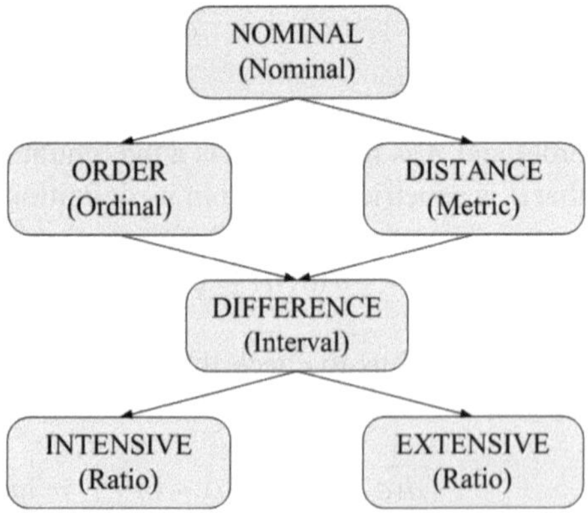

$$m'(a) = \frac{d'}{d}m(a) + m'(s_0) - \frac{d'}{d}m(s_0) = \alpha m(a) + \beta.$$

In the latter case, we will have a "reversed" series of standard, that is, $S' = (s_n, \ldots, s_0)$, and the related measure m' would satisfy $m'(s_i) = m'(s_0) - id'$, with $d' = |m'(s_i) - m'(s_{i-1})|$. Thus, for each $a \in A$, there will be a standard $s_i \in S$ such that $a \sim s_i$. Then, $m(a) = m(s_i) = m(s_0) + id$ and also $m'(a) = m'(s_i) = m'(s_0) - id'$. Thus,

$$m'(a) = -\frac{d'}{d}m(a) + m'(s_0) + \frac{d'}{d}m(s_0) = \alpha m(a) + \beta,$$

with $\alpha < 0$. □

On the basis of these new results, one-dimensional structures and related scales, a hierarchy for them can be established, as shown in Fig. 7.5. Proceeding from the weakest to the strongest, we have nominal scales first. Then, the next structural property may be either an order on the objects or a distance. We have seen that these two properties are independent and so ordinal and metric scales follow and are at the same level in the hierarchy. Then, a scale which is both ordinal and metric is an interval one. Lastly, the strongest scale is the ratio one, which can be achieved by adding either an empirical ratio (intensive structure) or an empirical addition (extensive structure) property.

7.4 Probabilistic Representation for Nominal and Metric Structures

Let us now consider the probabilistic formulation of the above representations. Consider first a formal definition of a probabilistic nominal structure.

Definition 7.9 (*Probabilistic nominal structure*) Let A be a finite (not empty) set of objects manifesting the property x. Let $\mathcal{A} = (A, \sim)$ denote a generic (empirical) nominal structure on A, and let $\mathcal{E}$ be a finite collection of distinct nominal structures on A. Then, a(n) (empirical) probabilistic nominal structure is a probability space $\mathcal{S}_{\mathcal{E}} = (\Omega, \mathcal{F}, P)$, where the elements of Ω are in a one-to-one correspondence with the elements of $\mathcal{E}$, $\mathcal{F}$ is an algebra on Ω and $P : \mathcal{F} \to [0, 1]$ is a probability function.

The representation theorem runs as follows.

Theorem 7.10 (Probabilistic representation of nominal structures) *Let A be a finite (not empty) set of objects manifesting the property x, and let* $\mathcal{S}_{\mathcal{E}} = (\Omega, \mathcal{F}, P)$ *be a probabilistic nominal structure on A. Then, there is a probabilistic function* $m = \{m_\omega : A \to \mathbb{N}, \omega \in \Omega\}$ *and a vector probabilistic variable* $\mathbf{x} = (x_a | x_a : \Omega \to \mathbb{N}, a \in A)$, *such that, for each* $a, b \in A$,

$$P(a \sim b) = P(m(a) = m(b)) = P(x_a = x_b).$$

Proof For each $\omega \in \Omega$, there is one and only one structure, $\mathcal{A}_\omega = (A, \sim_\omega) \in \mathcal{E}$, that corresponds to it. Let N be the number of elements in A, $n_\omega \leq N$ the number of equivalence classes in $\mathcal{A}_\omega$, $n = max\{n_\omega | \omega \in \Omega\}$ and $X = \{1, 2, \ldots, n\}$.

Let $m_\omega : A \to X \subset \mathbb{N}$ be a measure function that satisfies the representation of the equivalence relation $\sim_\omega$, associated with $\mathcal{A}_\omega$. We now define the probabilistic function m as the set of all such functions,

$$m = \{\{m_\omega | m_\omega : A \to X, \omega \in \Omega\}\},$$

with their associated probabilities:

$$P(m_\omega) = P(\omega).$$

Similarly, we introduce the vector probabilistic variable

$$\mathbf{x} = (x_a | x_a : \Omega \to \mathbb{N}, a \in A),$$

where each component is defined by

$$x_a(\omega) = m_\omega(a),$$

with

$$P(\mathbf{x}_\omega) = P(\omega).$$

In order for $\mathbf{x}$ to be well defined on X^N, we assign null probability to the points of X^N not included in the previous assignment.

We thus obtain

$$P(a \sim b) = P\{\omega | a \sim_\omega b\} = P\{\omega | m_\omega(a) = m_\omega(b)\} = P(m(a) = m(b)).$$

On the other hand, we also obtain

$$P(a \sim b) = P\{\omega | a \sim_\omega b\} = P\{\omega | x_a(\omega) = x_b(\omega)\} = P(x_a = x_b),$$

which completes the proof. □

The treatment of distance structures develops along the same lines.

Definition 7.11 (*Probabilistic distance structure*) Let A be a finite (not empty) set of objects manifesting the property x. Let $\mathcal{A} = (A, \succsim_\delta)$ denote a generic (empirical) distance structure on A, and let $\mathcal{E}$ be a finite collection of distinct distance structures on A. Then, a(n) (empirical) probabilistic distance structure is a probability space $\mathcal{S}_\mathcal{E} = (\Omega, \mathcal{F}, P)$, where the elements of Ω are in a one-to-one correspondence with the elements of $\mathcal{E}$, $\mathcal{F}$ is an algebra on Ω and $P : \mathcal{F} \to [0, 1]$ is a probability function.

Theorem 7.12 (Probabilistic representation of distance structures) *Let A be a finite (not empty) set of objects manifesting the property x, and let* $\mathcal{S}_\mathcal{E} = (\Omega, \mathcal{F}, P)$ *be a probabilistic distance structure on A. Then, there is a probabilistic function* $m = \{m_\omega : A \to \mathbb{N}, \omega \in \Omega\}$ *and vector probabilistic variable* $\mathbf{x} = (x_a | x_a : \Omega \to \mathbb{N}, a \in A)$, *such that, for each* $a, b, c, d \in A$,

$$P(\delta_{ab} \succsim_\delta \delta_{cd}) = P(|m(a) - m(b)| \geq |m(c) - m(d)|) = P(|x_a - x_b| \geq |x_c - x_d|).$$

Proof follows the same pattern of Theorem 7.10, and thus, it is omitted.

7.5 Additional Notes on Multidimensional Measurement

In general, objects are *identified* by a q-tuples $a = (a_1, \ldots, a_q)$ of attributes, that is, by elements of the Cartesian product $A = A_1 \times A_2 \times \cdots A_q$; the term "product structures" is thus also used [4] [1]. In the most general case, the components of a may be on a nominal scale (at the empirical level, they basically serve to *identify* the object a). From a foundational standpoint, we may distinguish three main cases.

(a) Objects are clustered on the basis of the properties that they share: this leads to an essentially nominal scale.
(b) Objects are mapped into points in a numerical metric space, in such a way that the relations that hold amongst their empirical distances are mapped into corresponding numerical distances.
(c) Objects are mapped into an oriented straight line, on the basis of some order holding amongst them: in this case , multidimensional measurement degenerates into *conjoint measurement*. Derived measurement may be seen as a special

[4] Here we assume, for the sake of simplicity, that the state space and the corresponding numerical space have the same dimension, that is p=q.

case of conjoint measurement, occurring when the mapping is founded on some accepted (natural, physical) law linking the attributes under consideration.

Here, we only briefly touch on case (b), by mentioning the conditions that must hold for the representation to exist, closely following Ref. [7], Chap. 14, to which the reader is referred for a complete treatment of the subject.

Suppose we look for a metric of the kind

$$d(a, b) = [\Sigma_{i=1}^{q} |m(a_i) - m(b_i)|^r]^{1/r}.$$

This representation implies four fundamental assumptions ([7], p. 175).

1. Decomposability: the distance between objects is a function of component-wise contributions.
2. Intra-dimensional subtractivity: each component-wise contribution is the absolute value of an appropriate scale difference.
3. Inter-dimensional additivity: the distance is a function of the sum of component-wise contributions.
4. Homogeneity: straight lines are additive segments.

Under these assumptions, it is possible to obtain a fundamental representational result [7]. Let $A = A_1 \times A_2 \times \cdots A_q$ be a set of multiattribute objects and $\succcurlyeq$ a weak order on $a \times A$. If the previous conditions are satisfied, it is possible to find a unique $r \geq 1$ and a real-valued functions m_i defined on A_i, such that, for all $a, b, c, d \in A$,

$$\delta_{ab} \succcurlyeq \delta_{cd} \Leftrightarrow d(a, b) \geq d(c, d),$$

where

$$d(a, b) = [\Sigma_{i=1}^{q} |m(a_i) - m(b_i)|^r]^{1/r}.$$

Under proper conditions, these metric representations can be turned into probabilistic ones, to some extent, which give rise to the notion of *probabilistic metric space* [8]. A key reference in that regard remains the book by Schweizer and Sklar, published in 1983 [9], to which the reader is referred for probing further this important and wide subject.

References

1. Roberts, F.S.: Measurement Theory, with Applications to Decision-Making, Utility and the Social Sciences. Addison-Wesley, Reading (1979). Digital Reprinting. Cambridge University Press, Cambridge (2009)
2. Luce, R.D., Krantz, D.H., Suppes, P., Tversky, A.: Foundations of Measurement. Academic Press, New York (1990)
3. Bosch, J.A. (ed.): Coordinate Measuring Machines and Systems. Marcel Dekker, New York (1995)

4. Muravyov, S., Savolainen, V.: Special interpretation of formal measurement scales for the case of multiple heterogeneous properties. Measurement **29**, 209–224 (2001)
5. Ekman, G.: Comparative studies on multidimensional scaling and related techniques. Reports from the Psychological Laboratories—Supplement Series, vol. 3 (1970)
6. Krantz, D.H., Luce, R.D., Suppes, P., Tversky, A.: Foundations of Measurement. Academic Press, New York (1971)
7. Suppes, P., Krantz, D.H., Luce, R.D., Tversky, A.: Foundations of Measurement. Academic Press, New York (1989)
8. Regenwetter, M., Marley, A.A.J.: Random relations, random utilities, and random functions. J. Math. Psychol. **45**, 864–912 (2001)
9. Schweizer, B., Sklar, A.: Probabilistic Metric Spaces. North Holland, New York (1983). Reprinted by Dover (2005)

Part III
Applications

Chapter 8
Perceptual Measurement

8.1 Measuring the Impossible

Perceptual measurement, that is, measurement of quantities related to human perception and interpretation, has been historically a controversial issue, as we have seen in Chap. 1. The Report of the Committee of the British Association for the Advancement of Science (1939) sanctioned a division between the two scientific communities of physicists/engineers on one side and psychologists on the other [1]. Since then, progress was achieved on both sides, but with lack of communication and of constructive interaction [2]. We briefly mention some achievements, to give a feeling of the current state of the art.

In psychology, a new school of measurement emerged, properly known as psychometrics [3]. Whilst in psychophysics, the properties (perceptions) of an object are traditionally investigated, as perceived by a *standard observer*, and in psychometrics, the focus changes to the properties (traits or states) of individuals, measured as responses to *standard test items*. In psychophysics, the methods proposed by Stevens were developed and extensive experience attested to their applicability and performance results [4]. Further major contributions were made by scientists in the psychology area, in the development of the representational theory of measurement [5].

As regards physical measurement, it is very important to recall that in 1960, luminous intensity was included amongst the base quantities of the International System of Units (SI) [6]. Since this quantity measures the physiological response to a physical stimulus, we can say that the idea of explicitly accounting for people's responses was officially accepted. Generally speaking, interest in so-called physiological quantities grew and this concept is now considered a key topic in metrology [7]. The interest in truly perceptual measurement also grew recently: in 2003, a Report of the National Physical Laboratory (NPL) addressed *soft metrology* amongst the "new directions" to be taken. Soft metrology includes "measurement techniques and models which enable the objective quantification of properties which are determined by human perception", where "the human response may be in any of the five senses: sight, smell, sound, taste and touch" [8].

G. B. Rossi, *Measurement and Probability*, Springer Series in Measurement Science and Technology, DOI: 10.1007/978-94-017-8825-0_8,

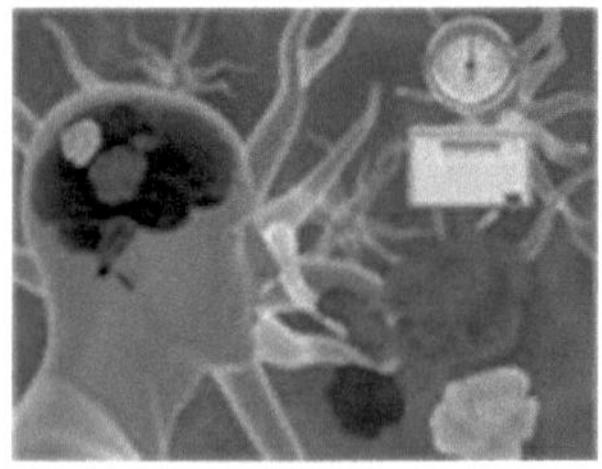

Fig. 8.1 The logo of the MINET project, suggesting that regularity in perception ensures the measurability of perception-related quantities

In 2004, a real turning point occurred with the European Research Call on "Measuring the Impossible" [9]. This was a part (technically a "pathfinder") of a more general session on New and Emerging Science and Technology (NEST), which underlines the new and visionary atmosphere that backgrounds such matters. In this Call, an explicit mention was made of many reasons that push towards more research effort in this area. They include *scientific arguments*, "many phenomena of significant interest to contemporary science are intrinsically multidimensional and multidisciplinary, with strong crossover between physical, biological and social sciences", *economic aspects*, "products and services appeal to consumers according to parameters of quality, beauty, comfort, which are mediated by human perception" and *social reasons*, "public authorities, and quasi public bodies such as hospitals, provide citizens with support and services whose performance is measured according to parameters of life quality, security or well-being". Several projects were developed on this topic, and a coordination action, MINET —Measuring the Impossible Network (2007–2010), was undertaken [10], Fig. 8.1.

It coordinated fourteen projects and promoted discussion, cooperation and synergy amongst researchers operating in this field. Its main results included the implementation of an interactive website, the creation of a database on available expertise, the organisation of workshops, conference sessions, think tanks and an extensive study-visit programme.

In 2008, an international training course on perceptual measurement was held in Genova, Italy, and in 2012, a book on "Measurement with Persons" was published, probably the first that tackled this subject in a truly inter-disciplinary fashion [11]. An expert group report was also issued at the end of the MINET project (2010), to address future research needs in this area [12].

It is hopeful that such an important experience does not remain isolated and that future research is promoted and funded in this important area. This would greatly contribute to the reconciliation of the above-mentioned division and to future positive collaboration of the involved scientific communities.

After this general background presentation, let us now discuss, in some detail, a special but important example of perceptual measurement: the case of perceived sound intensity or *loudness*, demonstrating the application of the general approach to measurability pursued in this book and its consequences.

8.2 Measuring the Intensity of a Sensation

8.2.1 Premise: Some Acoustic Quantities

Before discussing loudness measurement, let us briefly recall some basic quantities in physical acoustics [13]. Readers already familiar with this topic may skip this section.

Sound is a perceived acoustic phenomenon, so it has a dual nature, physical and perceptual. In physical terms, acoustic phenomena are pressure waves whose propagation in space is best described by *(average) acoustic intensity*, I, which is the energy, associated with the acoustic phenomenon, that crosses a unitary surface, in a given direction, in some observation time, T. Under typical propagation conditions, I is proportional to the mean square value of acoustic pressure variations, p, associated with the phenomenon:

$$I \propto \frac{1}{T}\int_T p(t)^2 \mathrm{d}t = p_{\mathrm{rms}}^2. \tag{8.1}$$

Acoustic pressure is the quantity actually measured by accurate acoustic sensors, *condenser microphones*. A typical *phonometer*, the standard professional instrument for acoustic measurement, includes a condenser microphone and a digital data acquisition and processing system, based on a dedicated microprocessor. It acquires pressure records and performs the calculation for obtaining sound intensity.

Since the range of acoustic pressure values is very large, roughly from 20 μPa to 20 Pa, a logarithmic scale is used instead, by introducing the *sound pressure level*, L_p, and the *sound intensity level*, L_I, defined as

$$L_\mathrm{p} = 10\log\left(\frac{p_{\mathrm{rms}}^2}{p_{\mathrm{ref}}^2}\right) = 20\log\left(\frac{p_{\mathrm{rms}}}{p_{\mathrm{ref}}}\right) = 10\log\left(\frac{I}{I_{\mathrm{ref}}}\right) = L_\mathrm{I} \tag{8.2}$$

and expressed in *decibels*, where $p_{\mathrm{ref}} = 20\ \mu\mathrm{Pa}$ and $I_0 = 10^{-12}\ \mathrm{Wm}^{-2}$ are conventional reference values, chosen in such a way that approximately

$$L_\mathrm{p} = L_\mathrm{I}. \tag{8.3}$$

So we can take L_p, which results directly from microphone-based measurements, as a measure of the physical intensity of the phenomenon. When needed, the intensity value can be explicated by

$$\frac{I}{I_{\mathrm{ref}}} = 10^{\frac{L_\mathrm{p}}{10}}. \tag{8.4}$$

Sound pressure level provides, with a single number, an overall description of the phenomenon. When a greater detail is required, time and/or spectral patterns come into play.

The appropriate spectral measure for acoustic phenomena is *sound intensity density*, $i(f)$, which is a function of frequency that describes how the overall intensity is distributed in the frequency domain. Thus the sound intensity in the band, (f_1, f_2), having bandwidth $\Delta f = f_2 - f_1$, is

$$I_{f_1 f_2} = \int_{f_1}^{f_2} i(f)\mathrm{d}f. \tag{8.5}$$

Sound intensity density can be obtained by spectrum analysis of pressure records, which can be performed by phonometers or by other signal processing devices. Since the frequency range for acoustic phenomena is also large, it is common practice to parse it in bands whose width increases alongside frequency and to calculate the intensity in each such band. This yields an *octave- or one-third octave-band analysis*. This is equivalent to having a series of band-pass filters and to calculating the sound intensity, or, equivalently, the sound pressure level, at the output of each filter. The upper cut-off frequency, f_2, and the lower cut-off frequency, f_1, of each such filter are related by

$$f_2 = 2^{\alpha} f_1, \tag{8.6}$$

where $\alpha = 1$ for octave-band and $\alpha = 1/3$ for one-third octave-band filters, respectively. The corresponding centre frequency, f_0, is defined as

$$f_0 = \sqrt{f_1 f_2}. \tag{8.7}$$

The standardised series of centre and cut-off frequency for one-third octave-band filters is shown in Table 8.1 [14]. An example of one-third octave-band analysis is displayed, later in this chapter, in Fig. 8.10.

We have now enough physical background for discussing loudness measurement, starting from very simple pure tones, up to real-world sounds.

8.2.2 Loudness of Pure Tones

In the first chapter of this book, we have suggested that in order to measure a property, the following steps must be successfully completed:

1. define the class of objects manifesting the property,
2. identify the empirical relations that define the property,
3. construct a reference measurement scale and
4. devise at least one measuring system based on that reference scale.

Let us then apply these ideas to loudness measurement, which constitutes a good example of more general and fundamental problem of measuring the intensity of a sensation. The first step consists in selecting a class of "objects", which, in our

Table 8.1 Normalised frequencies for third-octave band analysis

Centre frequency	Cut-off frequencies	Centre frequency	Cut-off frequencies	Centre frequency	Cut-off frequencies
	14.1		178.0		2,818
16.0		200.0		3,150	
	17.8		224.0		3,548
20.0		315.0		4,000	
	22.4		355.0		4,467
25.0		400.0		5,000	
	28.2		447.0		5,623
31.5		500.0		6,300	
	35.5		562.0		7,079
40.0		630.0		8,000	
	44.7		708.0		8,913
50.0		800.0		10,000	
	56.2		891.0		11,220
63.0		1,000		12,500	
	70.8		1,122		14,130
80.0		1250		16,000	
	89.1		1413		17,780
100.0		1,600		20,000	
	112.0		1,778		22,390
125.0		2,000			
	141.0		2,239		
160.0		2,500			
	178.0		2,818		

case, are sounds. We start by considering the class of *pure tones*, that is, sounds characterised by a pressure record

$$\{p(\tau) = p_0(\cos(2\pi f \tau)|0 \le \tau \le T\} \quad (8.8)$$

where

- τ is the time;
- T is the observation time that must be large enough to ensure that the sound is perceived as *stationary*; experimental studies by Zwicker [15] showed that the *minimum* value is $T = 500$ ms; in our laboratory, for example, we often take $T = 5$ s;
- p_0 (Pa) is the amplitude (modulus) of the pressure signal and
- f (Hz) is the frequency.

The corresponding sound pressure level, in dB, is

$$L_p = 20 \log \left(\frac{p_0}{\sqrt{2} p_{\text{ref}}} \right). \quad (8.9)$$

This choice is a good starting point for this investigation, since pure tones are easily reproducible and allow exploring the entire audible range, by letting f vary between, roughly, 20 Hz and 20 kHz and the sound pressure level L_{p} between 0 and 100 dB. Then, each "object", i.e. each sound, may be represented by a *point in the plane* (f, L_{p}). The choice of this class, A, of objects completes Step 1 of the measurement assessment procedure.

Step 2 requires the identification of empirical properties that give rise to a measurement scale. Let us consider *order* first. The question is, is it possible to order pure tones according to their loudness? The response is yes, provided that the influence of frequency on loudness perception is properly accounted for. Such response comes from an experimentation performed first by Fletcher and Munson in the first half of the twentieth century and consists in auditory tests with representative groups of people. They were able to draw on the (f, L_{p}) plane *isophonic curves*, such that the points on each curve represent sounds that are perceived as *equally loud*. The original curves were then improved by additional experimentation, and at present, a standardised version is available and reported in ISO standard 226 [16] and they appear as in Fig. 8.2. The level associated with each contour increases when moving from bottom to top. Consider, for example, sounds a and b in the Fig. 8.2: although the former has a higher sound pressure level, that is, a higher *physical* intensity, it has a lower loudness level, that is, a lower *perceived* intensity, since it lies on a "lower" contour. Thus,

$$a \prec b. \tag{8.10}$$

Defining a procedure, based on a reliable experimental basis, for assessing empirical order, as we have just done, constitutes Step 2 of the procedure.

Step 3 concerns the definition of the scale. An ordinal scale called *loudness level* (LL) and expressed in *phon* has been defined in the following way. Take conventionally tones at 1 kHz as *reference sounds*. Then, for each sound a, consider the 1 kHz sound, a', that lies on the same contour, and take as loudness level measure for a the sound pressure level (in dB) of a', that is,

$$LL(a) = L_{\mathrm{p}}(a') \Leftrightarrow a \sim a'. \tag{8.11}$$

It is easy to check that a scale so defined correctly represents perceived order. In our example, we obtain

$$LL(a) = LL(a') = 40 \text{ phon},$$
$$LL(b) = LL(b') = 50 \text{ phon},$$

and, consequently,

$$LL(a) < LL(b), \tag{8.12}$$

as expected.

Lastly, Step 4 requires that, at least, one measurement procedure is devised, based on this scale. The following, indirect, procedure can be readily implemented. A record of the sound to be measured must first be acquired by professional and

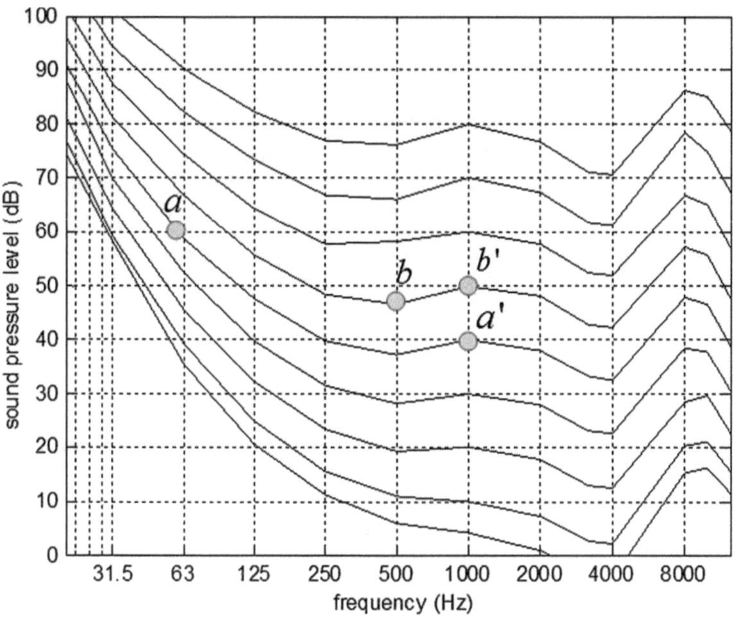

Fig. 8.2 Equal loudness level curves, according to ISO standard 226 (1987)

metrologically certified instrumentation, such as a phonometer or, more generally, a measuring chain that includes a condenser microphone, a preamplifier, an amplifier, an analogue-to-digital converter and a dedicated computer, including a data acquisition interface, a data storage device, professional data-processing software and a display. The recorded signal must then be processed to measure the corresponding sound pressure level, L_{p}, and frequency, f. Then, using the appropriate formulae provided in the ISO 226 standard, which provide the same result as the procedure illustrated in Fig. 8.2, it is possible to identify the LL value to be assigned to the measurand sound.

Since this procedure can be successfully applied, according to the assumed measurability criterion, we can conclude that *loudness of pure tones can be actually measured, at least on an ordinal scale*.

Anyway, note that since this is an ordinal scale, we cannot make any meaningful statement concerning difference or ratio. For example, although $LL(b)/LL(a) = 50/40 = 1.25$, we cannot conclude that b is 25 % louder than a, since ratio and, consequently, percentage are not defined on such a scale.

Is it then possible to measure loudness of pure tones on a ratio scale? Yes, it is, provided that *additional empirical evidence* is accounted for. In Chap. 1, we mentioned Stevens's power law that links the physical intensity of a stimulus to the corresponding perceived intensity. In the case of loudness, L, we can write, for a generic sound a,

$$L(a) = \alpha \left(\frac{I(a)}{I_{ref}} \right)^{\beta}, \tag{8.13}$$

where I is the acoustic intensity (remember formula 8.4). Select now, as *reference sounds, the class of* 1 kHz *pure tones*. Experimental results show that the power law is verified for this class of sounds, with $\beta = 0.3$. The second parameter, α, can be obtained by conventionally fixing the measurement unit for loudness. Then, the *sone* scale was defined by taking as unitary sound a 1 kHz tone with $L_{\rm p} = 40$ dB. Therefore, for a generic 1 kHz sound, a', we obtain

$$L(a') = \frac{1}{16} \left(\frac{I(a')}{I_{\rm ref}} \right)^{0.3}. \tag{8.14}$$

An equivalent expression can be obtained using sound pressure level, $L_{\rm p}$, instead of acoustic intensity:

$$L(a') = \frac{1}{16} 2^{L_p(a')/10}. \tag{8.15}$$

Note now that for any 1 kHz sound, a',

$$L(a') = LL(a'). \tag{8.16}$$

Thus, we can rewrite the above equation as

$$L(a') = \frac{1}{16} 2^{LL(a')/10}. \tag{8.17}$$

The corresponding curve is shown in Fig. 8.3.

Lastly, for any pure tone a, there will a 1 kHz reference sound, a', having equal loudness and, consequently, equal loudness level:

$$a \sim a' \Leftrightarrow L(a) = L(a') \Leftrightarrow LL(a) = LL(a'). \tag{8.18}$$

By substituting in the above formula, we obtain

$$L(a) = \frac{1}{16} 2^{LL(a)/10}, \tag{8.19}$$

that defines *loudness*, in *sone*, for pure tones (at any audible frequency).[1] For example, in the case of our two above sounds, we obtain

$L(a) = 1$ sone,

[1] In reality, this formula holds true only for sounds having $L_{\rm p} \geq 40$ dB. Yet a proper expression can be found also for sounds that do not satisfy this condition, and the scale can thus be extended over the entire audibility range.

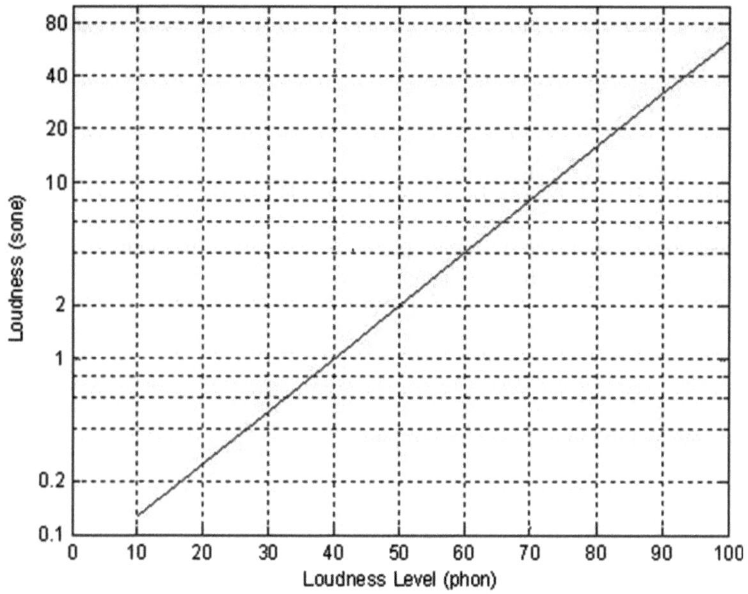

Fig. 8.3 Conversion of loudness level (phon) into loudness (sone)

$$L(b) = 2 \text{ sone},$$

and we can thus correctly conclude that *b is twice louder than a*.

Similarly to what we have done for the ordinal scale, we can now briefly summarise that four-step procedure that allows us to measure the loudness of pure tones on a ratio scale.

1. The class of objects includes pure audible tones.
2. Empirical properties are summarised by equal-intensity contours, Fig. 8.2, plus Stevens's law, Fig. 8.3.
3. The reference measurement scale is constituted by 1 kHz tones, to which a loudness value is assigned according to formula (8.17).
4. Measurement can be taken for any tone a, by measuring $LL(a)$ first as previously described and then converting it into $L(a)$ by formula (8.19) or by Fig. 8.3.

Note that, from another standpoint, the above consideration also allows us to produce standard reference pure tones of any desired loudness.

We are now able to treat pure tones. Yet although such sounds are very useful for providing a first approach to the study of acoustic perception and for producing synthetic sounds for laboratory experimentation, our final goal is to measure real-world sounds and noise.

We thus move now to *pink noises*, another very useful class of synthetic sounds, and, lastly, to stationary sounds, for which we will briefly discuss two main measurement approaches: *direct*, where persons act as measuring instruments, and *indirect*, where the measurement procedure is based on post-processing recorded acoustic signals.

8.2.3 Loudness of Pink Noise

Pink noise, in acoustics, is a stationary phenomenon, whereby sound intensity density is inversely proportional to frequency, that is

$$i(f) = cf^{-1}, \tag{8.20}$$

where c is a constant. So the energy in each one-third octave band is constant. In fact, for any such band (f_1, f_2), we obtain

$$I_{f_1 f_2} = \int_{f_1}^{f_2} \frac{\mathrm{d}f}{f} = c \ln\left(\frac{f_2}{f_1}\right) = c \ln\left(\sqrt[3]{2}\right) \triangleq I_0. \tag{8.21}$$

If I is the overall intensity and n the number of bands, we have

$$I = nI_0. \tag{8.22}$$

Let p_0 be root mean square value of the acoustic pressure corresponding to I_0. Then,

$$L_{p_0} = 10 \log\left(\frac{I_0}{I_{\mathrm{ref}}}\right), \tag{8.23}$$

and

$$L_\mathrm{p} = 10\left[\log\left(10^{L_{p0}/10}\right) + \log n\right] = L_{p_0} + 10 \log n. \tag{8.24}$$

For one-third octave analysis, $n = 32$, and we simply obtain

$$L_{p_0} = L_\mathrm{p} - 15, \tag{8.25}$$

which allows us to generate pink noise of any required sound pressure level.

In some regard, pink noise is even a simpler sound than pure tones, since it is uniquely defined by its sound pressure level alone. On the other hand, unfortunately, we do not have a simple law for expressing its loudness. At the current state of the art, we can only calculate it by some model, as the one we will present in Sect. 8.2.6. In this way, we obtain the results in Table 8.2 [17].

Table 8.2 Loudness of pink noise, as a function of sound pressure level

L_p	45	50	55	60	65	70	75	80	85
L	3.96	5.93	8.60	12.16	16.95	23.38	31.95	43.43	58.81

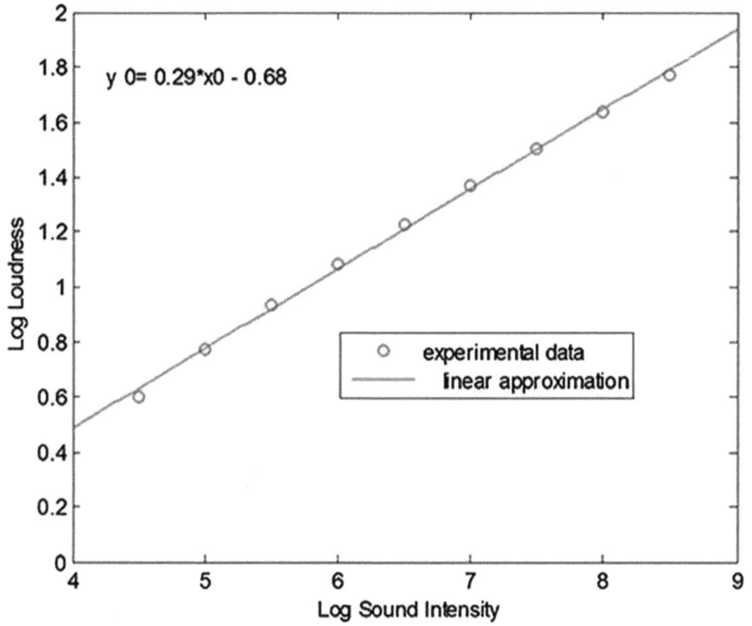

Fig. 8.4 Linear fitting for the loudness of pink noise, as a function of sound pressure level

It is practical to approximate such a functional relation by a power law:

$$L = \alpha_0 (I/I_{\text{ref}})^{\beta_0}. \tag{8.26}$$

Taking the logarithm of both sides, we obtain

$$\log L = \beta_0 \log(I/I_{\text{ref}}) + \log \alpha_0, \tag{8.27}$$

that is,

$$y_0 = k_0 x + h_0, \tag{8.28}$$

where $y_0 = \log L$, $k_0 = \beta_0$, $x = \log(I/I_{\text{ref}})$ and $h_0 = \log \alpha_0$. We can estimate the parameters k_0 and h_0 by linear regression and then obtain the parameters of the (approximate) power law. This is illustrated in Fig. 8.4.

Regression yields

$$k_0 = +0.29,$$
$$h_0 = -0.68,$$

and thus, we obtain

$$\alpha_0 = 0.21,$$
$$\beta_0 = 0.29.$$

This result is very useful for the measurement procedure we are now going to study.

8.2.4 Direct Measurement of Loudness: Master Scaling

The *master scaling* approach was conceived by Berglund [18] as a noteworthy development of the original *magnitude estimation* method by Stevens [19], with the aim of ensuring a proper quality control of perceptual measurement, especially when performed in the field and by a small group of people (or even by a single person!).

Suppose that a single person is asked to "measure" the loudness of two distinct working environments on two different days. If we want to compare such results, the comparison will be affected by uncertainty due to intra-individual variability. "Characteristically, sensory systems are dynamic and adaptable", as Berglund notes [20], and "each observer's magnitude estimates consist of context-dependent relative values rather than absolute values or invariable". Alternatively, we can consider the case of two persons, measuring contemporarily the noise in two different areas. Here, comparison will be affected by inter-individual variability. In both cases, master scaling helps in reducing uncertainty.

The idea, simple and very effective, is as follows: let the person listen to a fixed set of reference sounds (the master) and to rate them, together with the noise to be measured. Then, the rating provided by the person can be *referred* to the *invariant* reference set, and the results will then be comparable with those obtained in a different moment or by a different person.

Let us now discuss the method in greater detail.

Suppose we ask a person (or a group of persons) to listen to $n + 1$ sounds, $\{s_1, s_2, \ldots, s_n, a\}$, the first n being reference sounds, such as pink noises, and a being the measurand. The physical intensities of the reference sounds, $\{I_1, I_2, \ldots, I_n\}$, are known, usually expressed as sound pressure levels, $\{L_{\mathrm{p}1}, L_{\mathrm{p}2}, \ldots, L_{\mathrm{p}n}\}$. Let $\{y_{\mathrm{L}1}, y_{\mathrm{L}2}, \ldots, y_{\mathrm{L}n}, y_{\mathrm{L}a}\}$ be the corresponding estimated loudness values provided by the person.

We assume that *the person responds according to the power law* (13) *where α and β are here regarded as unknown parameters dependent on the context and the* *individual*. Let us then see how such data can be processed to provide the final measurement result, $\hat{L}_a$, according to the master scale method.

Applying Eq. (8.13) to the subject's responses to the reference sounds and taking the logarithm, we obtain, for $i = 1, 2, \ldots, n$,

$$\log y_{\mathrm{L}_i} = \beta \log(I_i/I_{\mathrm{ref}}) + \log \alpha. \tag{8.29}$$

Let us now define $y_i = \log y_{\mathrm{L}_i}$, $x_i = \log(I_i/I_{\mathrm{ref}})$, $k = \beta$ and $h = \log \alpha$. Note that $x_i = L_{\mathrm{p}i}/10$. Then, the response of the person to the reference stimuli can be described by the probabilistic model

$$y_i = kx_i + h + v_i, \tag{8.30}$$

where v_i are independent realisations of a random variable v that account for random deviations from the power law. Since both x_i and y_i are known, the unknown, individual and context-dependent, parameters k and h can be obtained by solving a linear regression, least-squares, problem [21]. Let $\hat{k}$ and $\hat{h}$ be such estimates, then

$$y = \hat{k}x + \hat{h}, \tag{8.31}$$

is a *calibration equation*[2] for the person acting as a measuring instrument, and $\hat{\beta} = \hat{k}$ and $\hat{\alpha} = 10^{\hat{h}}$.

Let now $y_a = \log y_{\mathrm{L}_a}$ be the loudness value assigned by the person to the measurand, a. We obtain

$$\hat{x} = \frac{y_a - \hat{h}}{\hat{k}}, \tag{8.32}$$

where $\hat{x}$ represents *the value of a pink noise which is perceived by the person as equally loud as the measurand*.

On the other hand, the loudness associated with the master sounds can be approximately evaluated, as we have discussed in the previous section, by a power law,

$$L = \alpha_0 \left(I/I_{\mathrm{ref}}\right)^{\beta_0}, \tag{8.33}$$

where $\alpha_0 = 0.21$ and $\beta_0 = 0.29$. Defining, as we have done in the previous section, $y_0 = \log L$, $k_0 = \beta_0$, $x = \log(I/I_{\mathrm{ref}})$ and $h_0 = \log \alpha_0$, we can express the same law as

$$y_0 = k_0 x + h_0. \tag{8.34}$$

We can now apply this formula to $\hat{x}$, obtaining

$$y_0 = k_0\hat{x} + h_0 = \frac{k_0}{\hat{k}}(y_a - \hat{h}) - h_0, \tag{8.35}$$

---

[2] Calibration will be discussed in some detail in Chap. 10.

Table 8.3 Results of direct loudness measurement, by the master scaling method

s_i	s_1	s_2	s_3	s_4	s_5	s_6	s_7	a
y_{L_i}	3	5	7	9	10	12	15	12
Lp_i	56.4	58.9	61.2	63.3	65.9	68.7	71.0	74.7

and going back to the exponential formulation, we finally obtain

$$\hat{L}_a = \alpha_0 \left(\frac{y_{L_a}}{\hat{\alpha}} \right)^{\frac{\beta_0}{\hat{\beta}}}, \tag{8.36}$$

which allows assigning the measurement value $\hat{L}_a$ to the measurand [20].

Let us now illustrate all this with a numerical example. In an experiment we have done in our laboratory, we wanted to measure the intensity of some samples of environmental noise in a port area [22].[3] We used different methods, including the master scaling. We involved a group of some twenty people, but here, to simplify things, we will consider just the result for a single person, concerning $n = 7$ reference sounds and one measured sound, a, as reported in Table 8.3.

In Table 8.3, the sound pressure level, L_{p}, is also reported for each reference sound, as well as for the measured sound. We can estimate k and h by linear regression, as shown in Fig. 8.5.

We thus obtain

$$\hat{k} = +0.44,$$
$$\hat{h} = -1.90,$$

from which we calculate

$$\hat{\alpha} = 0.01,$$
$$\hat{\beta} = 0.44,$$
$$\hat{x} = 6.68.$$

Then, applying transformation (8.36), we finally obtain

$$L_a = 19 \text{ sone}.$$

[3] In the following, we will use some results from an experimental activity carried out in our laboratory, whose results have been published in Refs. [22, 23] mainly. Interested readers may found a full account of that experiment in those references.

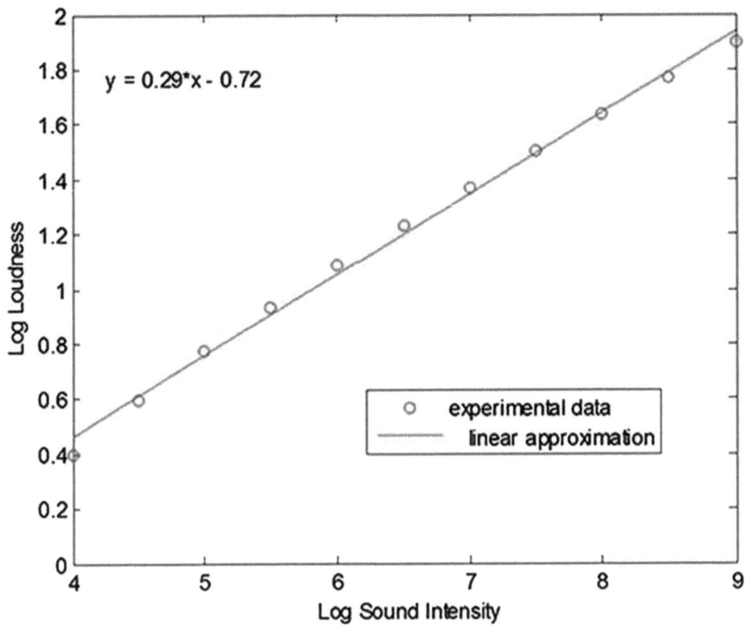

Fig. 8.5 Regression of loudness estimates for pink noise by a single person [22]

8.2.5 Direct Measurement of Loudness: Robust Magnitude Estimation

Another approach to the direct measurement of the intensity of a sensation has been recently developed in our laboratory [22, 23].[4] It is called *robust magnitude estimation*, and it is based on the theory of intensive structures that we have encountered in Chap. 3.

As we know, if a structure has an empirical difference order, $\succsim_d$, it undergoes an interval *difference* representation, and if m and m' are two appropriate measures, the following uniqueness condition holds true:

$$m(a) = \alpha_1(m'(a) + \beta). \tag{8.37}$$

On the other hand, if a structure has an empirical ratio order, $\succsim_r$, it may also be given an interval *ratio* representation, where any two measures, m and m'', satisfy

$$m(a) = \alpha_2 m''(a)^\gamma. \tag{8.38}$$

An intensive structure may be given *both* representations [24].

[4] See footnote 3.

Table 8.4 Results of direct loudness measurement, by robust magnitude estimation

s_i	s_1	s_2	s_3	s_4	s_5	s_6	s_7	a
$y_{\mathrm{L}d_i}$	2.1	5.1	7.9	10.9	12.7	15.6	17.0	11.7
$y_{\mathrm{L}r_i}$	3	5	7	9	10	12	15	12
$y'_{\mathrm{L}d_i}$	5.6	8.2	10.6	13.2	14.8	17.3	18.6	14.0
$y'_{\mathrm{L}r_i}$	5.5	8.3	10.8	13.2	14.4	16.6	19.8	16.6
y_{L_i}	5.5	8.2	10.7	13.2	14.6	17.0	19.2	15.3

Let us then consider an experiment similar to that in the previous section, except that now the person is asked to rate sounds in terms of *both loudness differences and loudness ratios*, and let $\{y_{\mathrm{L}d_1}, y_{\mathrm{L}d_2}, \ldots, y_{\mathrm{L}d_n}, y_{\mathrm{L}d_a}\}$ and $\{y_{\mathrm{L}r_1}, y_{\mathrm{L}r_2}, \ldots, y_{\mathrm{L}r_n}, y_{\mathrm{L}r_a}\}$ be the corresponding results. Furthermore, it is possible to fit data in such a way that, for $i = 1, 2, \ldots, n+1$,

$$y'_{\mathrm{L}d_i} \triangleq \alpha_1(y_{\mathrm{L}d_i} + \beta) \cong y_{\mathrm{L}d_i}, \tag{8.39}$$

$$y'_{\mathrm{L}r_i} \triangleq \alpha_2 y^{\gamma}_{\mathrm{L}r_i} \cong y_{\mathrm{L}r_i}, \tag{8.40}$$

and it results that

$$y'_{\mathrm{L}d_i} \cong y'_{\mathrm{L}r_i}, \tag{8.41}$$

then by a singular value decomposition in the plane $(y'_{\mathrm{L}d}, y'_{\mathrm{L}r})$ [21], it is possible to obtain y_{L_i} such that

$$y_{\mathrm{L}_i} \cong y'_{\mathrm{L}d_i} \cong y'_{\mathrm{L}r_i}, \tag{8.42}$$

which provides the required scaling of the series of sounds on a ratio scale, as produced by the person that performs the test. A more accurate result can be obtained by averaging over a group of persons [22].

To illustrate this procedure, consider again the same experiment of the previous section, where two sets of responses, $y_{\mathrm{L}d_i}$ and $y_{\mathrm{L}r_i}$, are now obtained, reported in the second and third lines of Table 8.4, respectively, and visualised in Fig. 8.6.

From Eqs. (8.37) and (8.38), taking the logarithms, we obtain

$$\log \frac{\alpha_1}{\alpha_2} + \log(y_{\mathrm{L}d_i} + \beta) - \gamma \log y_{\mathrm{L}r_i} \cong 0, \tag{8.43}$$

that can be solved, by a total least-squares procedure, for the parameters α_1/α_2, β and γ [21]. Note that the parameters α_1 and α_2 appear through their ratio only. In order to identify them individually, it is necessary to fix the measurement unit. To that goal, assume that for one of the reference signals, s, we have previously determined, with some accepted standardised method, the corresponding loudness value, L_{s}. Then, we can impose the additional condition:

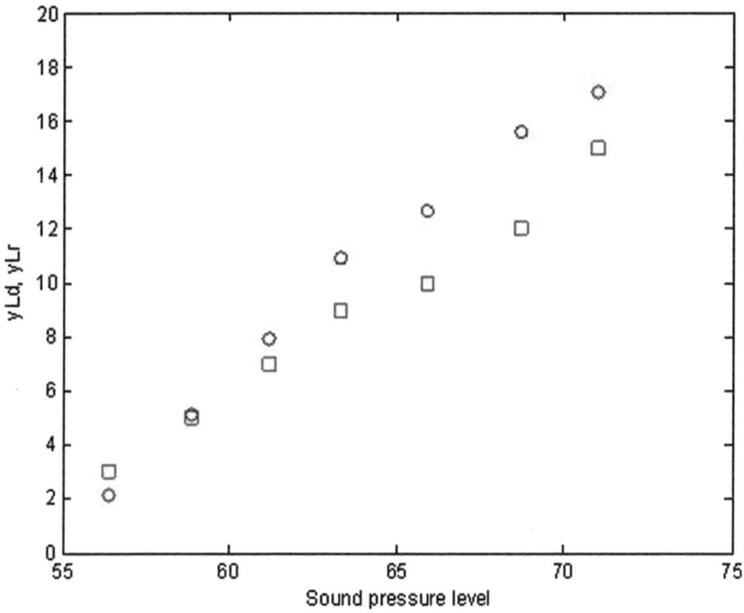

Fig. 8.6 Results from interval estimation and ratio estimation, for a single person: $y_{\mathrm{L}d_i}$ are represented by circles, $y_{\mathrm{L}r_i}$ by squares [23]

$$L_s = \alpha_2 y_{\mathrm{L_s}}^{\gamma}, \tag{8.44}$$

that is,

$$\alpha_2 = \frac{L_\mathrm{s}}{y_{\mathrm{L_s}}^{\gamma}}. \tag{8.45}$$

Applying this procedure to the experimental data in Table 8.4, we obtain

$$\begin{aligned} \alpha_1 &= 0.88, \\ \alpha_2 &= 2.32, \\ \beta &= 4.20, \\ \gamma &= 0.79. \end{aligned}$$

Through this re-scaling, we obtain the $\{y'_{\mathrm{L}d_i}\}$ and $\{y'_{\mathrm{L}r_i}\}$ values, reported in lines 4–5 in the Table 8.4, and compared in Fig. 8.7.

In Fig. 8.7, each sound is represented by a point in the plane $(y'_{\mathrm{L}d}, y'_{\mathrm{L}r})$: if the results were coincident, such points should lay midway between the two axes: we note that this is almost the case, and thus, the two results are indeed quite similar. The final result of this scaling phase, that is, the calculation of the $\{y_{\mathrm{L}_i}\}$ values for the reference sounds, can be achieved by a singular value decomposition in the plane

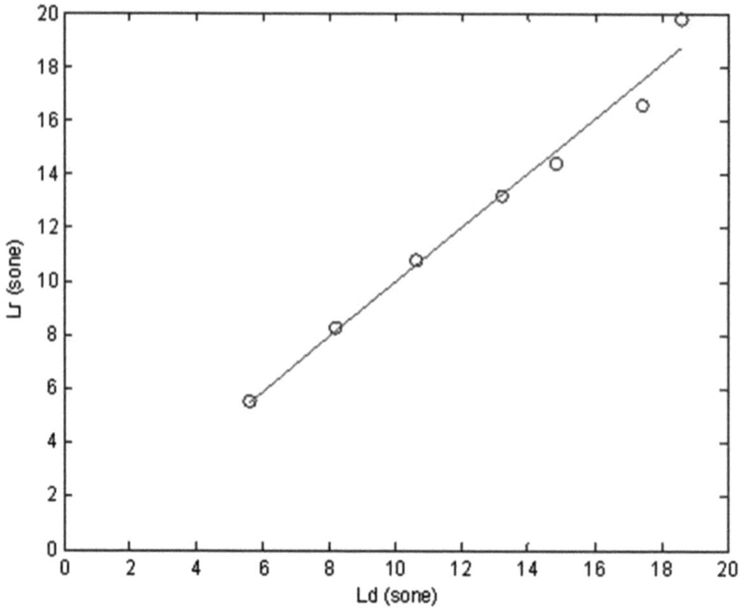

Fig. 8.7 Compatibility assessment for results from interval estimation and ratio estimation, after application of admissible transformations [23]

(y'_{Ld}, y'_{Lr}), that is, the plane of Fig. 8.7. Yet since the two sets of values are quite close to each other, a similar result may be simply obtained by taking the mean, that is,

$$y_{L_i} = \frac{1}{2}(y_{Ld_i} + y_{Lr_i}). \tag{8.46}$$

Such results are shown in the sixth line of Table 8.4 and in Fig. 8.8, as a function of the sound pressure level.

Interestingly enough, from these data, we can also obtain an estimate of the uncertainty of this reference scale: in fact, the distances of the points (y'_{Lr_i}, y'_{Ld_i}) from the identity straight line $y'_{Lr_i} = y'_{Lr_i}$, that is,

$$\mathrm{d}(i) = \frac{|y'_{Ld_i} - y'_{Lr_i}|}{\sqrt{2}}, \tag{8.47}$$

provide an estimate of the "errors", and thus, it makes sense to take their mean square values as an uncertainty figure for the scale:

$$u_0 = \sqrt{\frac{1}{n}\sum_{i=1}^{n} \mathrm{d}(i)^2}. \tag{8.48}$$

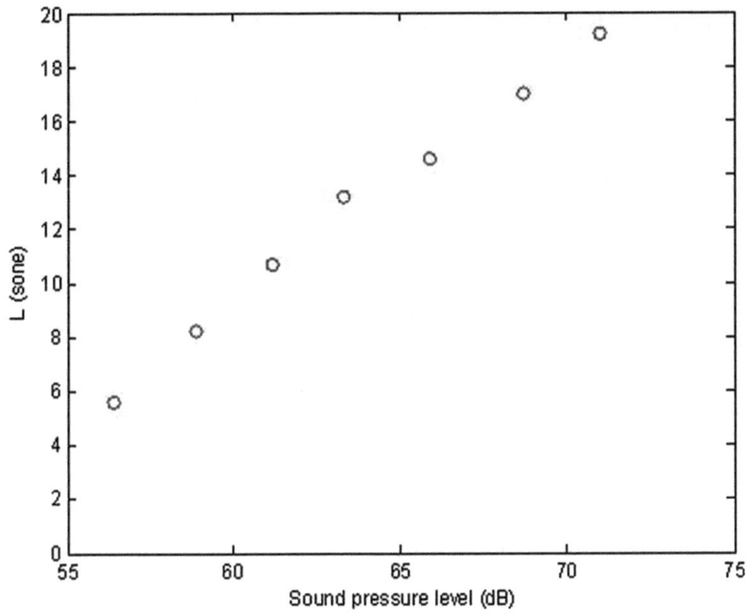

Fig. 8.8 Measured loudness versus sound pressure level [23]

We obtain

$$u_0 = 0.4 \text{ sone},$$

that is, an uncertainty of 2 % of full range. Note that this is the uncertainty of the reference scale. The uncertainty of measurement results based on that scale is obviously greater and can be evaluated when a set of results obtained from a group of persons is available. This has been done for the experiment under consideration [22], and the result was

$$u_0 = 2.5 \text{ sone},$$

that is, an uncertainty of 12 % of full range.

For example, the final measurement result, for the same sound examined in the previous section, can be stated as

$$\hat{L}_a = 15.3 \text{ sone, with a standard uncertainty of 2.5 sone.}$$

Note that in the current practice, in this kind of measurements, uncertainty is not explicitly expressed. Thus, this is somewhat a more advanced approach.

8.2.6 Indirect Measurement: Loudness Model

To complete this concise review of methods for measuring the loudness of stationary sounds, we briefly mention the "indirect" approach that consists in acquiring a sound record and in processing it to obtain a measure of the loudness associated with that sound. This requires a model of the way the sound signal is processed in the auditory system, so that the numeric processing of the signal mimics the physiological and perceptual process. Two main models have been developed so far, due to Zwicker's [15] and to Moore's [25] studies, respectively. They are different in some regards, but the basic idea is the same and so, and for the purpose of this qualitative introduction, we will treat them together.

Let us then briefly review the main phenomenological points that are accounted for in these models. As we have already mentioned, the auditory sensitivity depends on frequency, being maximum in the range of, say, one to five kilohertz and decreasing when moving towards both extremes of the audition range. Furthermore, the sensitivity depends on the intensity of the sound, according to the power law. These properties, the dependence on frequency and on intensity, were sufficient for developing a model for *pure tones*, as we have already seen (formulae 8.11 and 8.19). The point now is that in a stationary sound, there is no longer a single frequency but instead a continuous spectrum of frequencies. So, how to combine the responses to the various frequencies that are contemporarily present in the sound? A simple idea could be to make a weighted mean of such responses, using a proper frequency weighting function in the frequency domain. Such weighing should give maximum emphasis to spectral components that fall in the area of maximum sensitivity, $1 \div 5$ kHz, and put decreasing weight on components out of this band. In fact, this does make sense as a first approximation, and it is actually used in practice. The result of such a processing is called a *weighted sound pressure level* [14]. Unfortunately, this procedure does not completely account for what really happens in the auditory system. In fact, two additional phenomena need considering.

Firstly, the *perception resolution in the spectral domain*, that is, the capability of distinguishing two adjacent frequencies, is not constant in the frequency domain; rather, it decreases as long as frequency increases. This can be partially accounted for by performing a constant-percentage spectrum analysis, as we mentioned earlier (Sect. 8.2.1). In reality, a one-third octave-band analysis, although providing a first approximation, does not exactly correspond to human response. Other more sophisticated ways of partitioning the spectral domain, often called "critical band analysis", have been proposed, although this is still a quite controversial point. In fact, a major difference between Zwicker's and Moore's models consists in a different approach to this problem. In any case, the spectral resolution of the auditory system needs to be accounted for.

The second important point that needs considering is the phenomenon of *(spectral) masking*. In simple words, this means that frequency components that are close to each other interact in a nonlinear way, in that, for example, a "strong" component tends to mask, that is, to make inaudible, a "weaker" one. So, to sum up, a proper model must account for all these phenomena.

Qualitatively, a processing algorithm based on such a model includes the following steps:

1. To estimate the one-third octave power spectrum of the signal.
2. To re-scale such a spectrum accounting for the frequency resolution of the auditory system.[5]
3. To account for the masking phenomenon, by a kind of spectral "convolution" with a masking "window".
4. To scale the amplitude of spectral component according to the power law and to the perception threshold, which also depends on frequency; the result of these transformations is expressed by the so-called specific loudness, which is a function of normalised frequency that expresses how the (total) loudness is distributed along the normalised frequency.
5. Lastly, the integral of the specific loudness provides the desired loudness value, in *sone*.

To illustrate the result of such a procedure in a practical case, let us consider the sound a that we have encountered in the previous sections.[6] The corresponding signal is reported in Fig. 8.9, the one-third octave spectrum in Fig. 8.10 and the specific loudness in Fig. 8.11.

The final loudness value obtained with this method is

$$\hat{L}_a = 18.0 \text{ sone}.$$

If we compare this result with those obtained in the previous sections, we may note that in this particular case, the master scaling methods provide a higher value, whilst the robust magnitude estimation provide a lower one. Of course, this is just a single example and no general conclusion can be drawn from it. We can also note that the difference between this estimate and the one produced by the robust magnitude estimation methods is within two times the quoted standard uncertainty for that method.

8.3 State of the Art, Perspective and Challenges

The case of loudness is emblematic of the state of the art of perceptual measurement. As we have seen, there are different available methods for measuring loudness, both instrumental and based on persons, whose validity has been amply investigated in the

---

[5] As we mentioned, the way this re-scaling is actually done constitutes a major difference between the two models.

[6] See again footnote 3.

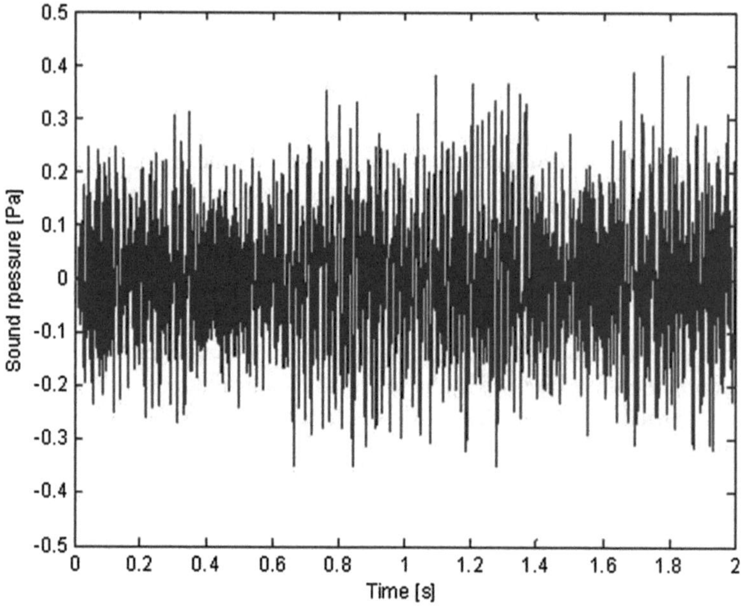

Fig. 8.9 Time representation of sound *a*

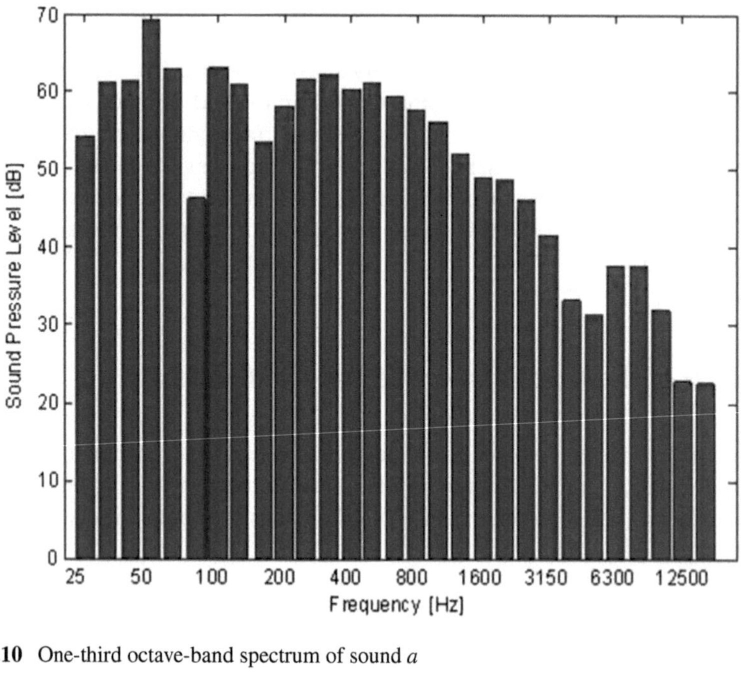

Fig. 8.10 One-third octave-band spectrum of sound *a*

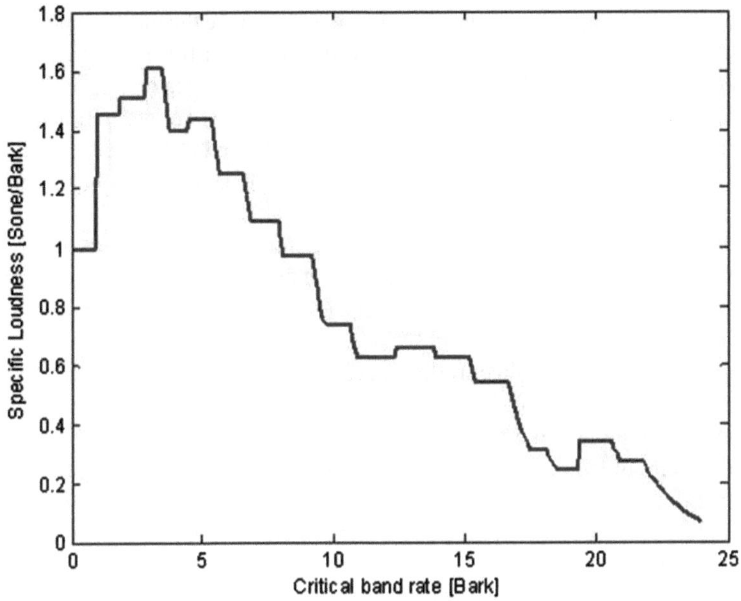

Fig. 8.11 Specific loudness of sound *a*

scientific literature. Still, its condition is quite different from that of typical physical measurements, included in the International SI.

For the sake of comparison, consider length as an example. Length measurement takes advantage of the existence of an internationally agreed primary method, based on laser interferometry. Some laboratories in the world, associated with national metrological institutes, are accredited for implementing such a primary method. Then, there are accepted procedures for calibrating length-measuring devices and laboratories accredited for doing that, so that properly trained instrument users can make reliable length measurements and provide results with an associated uncertainty statement.

Consider now loudness. In fact, some kind of standardisation exists: for pure tones, there is the ISO 226 standard [16]; for stationary sounds, there are ISO 532 (1975) and DIN 45631 (1991) [26], based on Zwicker's model, and ANSI s3.4 (2007) [27], based on Moore's model. For non-stationary sounds, we can mention DIN 45631/A1 [28]. Yet there is not at present any internationally agreed primary method, nor accepted procedures for calibrating instruments in respect of it, for accrediting laboratories and for providing results with associated uncertainty statements. The consequence of this situation is that although such measurements are used and provide precious results in research environments, their application in daily life suffers for this lack of standardisation. For example, in the case of measurement necessary for assessing the exposure of workers to noise in a working environment, the quantity actually used is weighted sound pressure level. As we have seen in the previous section, this does not provide an appropriate measure of loudness, since the masking phenomenon and other perception features are not properly accounted for. Examples have been provided in the literature of evident discrepancies between the results of this method and measurements with persons [15], yet since there is not an internationally agreed way of measuring loudness, this rough method is still used in practice. This is a good example of the negative consequences of not having yet an international organisation for perceptual measurements.

On the other hand, their practical—not just scientific—importance is very high and concerns at least the areas of perceived quality of products and service, environment, ergonomics, safety, security and clinics [2].

Perceived quality of products and services has been in the past a major motivation for supporting research in this product development area. In the future, it will still play a role, since the shortage of energy sources and the concern for pollution may increase the demand for durable, high-quality goods.

Outdoor and indoor environments will be of major concern in the years to come. Research projects concerned with the characterisation of landscapes and soundscapes (a combination of sounds deriving from an immersive environment) may be mentioned, including measurement campaigns for reducing loudness or the odour intensity of fertilisers near industrial plants [29]. This study area, known as "environmental psychophysics", faces the challenges of characterising multisensory exposures that vary over time and are often obscured by background conditions and that require carefully designed and controlled measurement procedures. The indoor environment is also of great importance, because people spend about 90 % of their time indoors, either at work or at home. The quality of the indoor environment depends on the quality of its subsystems, i.e. air quality, soundscapes, visual–tactual surfaces and their integration. To make progress in this area, perceptual studies and measurements must be combined with the sophisticated modelling of complex systems.

Ergonomics may be defined as "The scientific discipline concerned with the understanding of the interactions among human and other elements of a system, and the profession that applies theory, principles data and methods to design in order to optimize human well-being and overall system performance". The relationship between human beings and their environment is experienced through the senses, and their perceptual measurements are key ways for obtaining valuable scientific and professional data. A typical ergonomic concern is the measurement of comfort. In transportation systems, discomfort is often associated with noise and vibration exposures in which case perception plays a central role [30]. Ergonomics sets out to ensure, on the one hand, a good quality of life for operators, and on the other hand, the best performance of the system in question. Consider the case of a driver: ensuring that he/she is working in optimal conditions favours the safety of people; in the case of a watchman, performance affects security. Security is another important application area. Face recognition for suspect identification is a good example. So far, several approaches have been made with a view to automating this task and approaches related to the psychology of face recognition look promising [31, 32].

Clinical applications are also important. The measurement of the perceptual intensities of touch, warmth and cold in pain-affected skin areas of the human body may help to optimise treatment [29]. Changes in sensorial sensitivity can be used in diagnostics or to monitor rehabilitation processes. Humanoid robotics sets out to develop machines that, to some extent, resemble some aspect of human behaviour. They must be fitted with sophisticated sensor interfaces that mimic some aspect of human perception and may be used in rehabilitation and special assistance programmes.

To guarantee such a progress, the major challenge is the rapprochement of the scientific communities involved. Three major steps may be envisaged in such a path.

The first step requires changes in the attitudes of both parties. Physicists and engineers should be more open to accept that measurement can also be taken with persons acting as measuring instruments, in properly designed and conducted experiments. Psychologists and behavioural scientists should perhaps develop greater sensitivity to the benefit of an international organisation supporting the measurement of some key perceptual quantities, such as loudness and annoyance.

Another major step would be to make a joint effort to develop and apply a common understanding and theory of measurement. To this goal, I have tried to contribute by developing this very book.

Lastly, it is essential that common projects are developed, as those related to the already mentioned European Call on "Measuring the Impossible". Similar initiatives should continue in the future [12].

The future will certainly present important challenges: it will be necessary to find new and creative ways to support human activities and take environmental care in managing our energy resources. This will require increasing collaboration between science and technology and amongst all sciences.

References

1. Ferguson, A., Myers, C.S., Bartlett, R.J.: Quantitative estimates of sensory events. Final Rep. Br. Assoc. Adv. Sci. **2**, 331–349 (1940)
2. Rossi, G.B., Berglund, B.: Measurement of quantities involving human perception and interpretation. Measurement **44**, 815–822 (2011)
3. Lord, F.M., Novick, M.R.: Statistical Theory of Mental Test Scores. Addison Wesley, Reading (1968)
4. Baird, J.C., Noma, E.: Fundamentals of Scaling and Psychophysics. Wiley, New York (1978)
5. Narens, L., Luce, R.D.: Measurement: the theory of numerical assignment. Psychol. Bull. **99**, 166–180 (1986)
6. BIPM: Principles governing photometry. Imprimerie Durand, Luisant (1983)
7. Nelson, R.A., Ruby, L.: Physiological units in the SI. Metrologia **30**, 55–60 (1993)
8. Pointer, M.R.: New directions—soft metrology requirements for support from mathematics statistics and software. NPL report CMSC 20/03 (2003)
9. European Commission: Measuring the impossible. EUR 22424, European Communities ISBN 92-79-03854-0 (2007)
10. Pendrill, L.R., et al.: Measurement with persons: a European network. Measure **5**, 4254 (2010)
11. Berglund, B., Rossi, G.B., Townsend, J., Pendrill, L. (eds.): Measurement with Persons. Taylor and Francis, New York (2012)
12. Galanter, E., et al.: Measuring the Impossible—Report of the MINET High-Level Expert Group. EU NEST, Bruxelles (2010)

13. Pierce, A.D.: Acoustics—An Introduction to Its Physical Principles and Application. Acoustical Society of America, USA (1989)
14. Yang, S.J., Ellison, A.J.: Machinery Noise Measurement. Clarendon Press, Oxford (1985)
15. Zwicker, E., Fastl, H.: Psycho-Acoustics. Springer, New York (1999)
16. ISO: ISO standard 226: acoustics—normal equal loudness levels (1987)
17. Schlittenlacher, J., et al.: Loudness of pink noise and stationary technical sounds. Paper presented at Inter-Noise, Osaka, Japan. 4–7 Sept 2011
18. Berglund, B.: Quality assurance in environmental psychophysics. In: Bolanowski, S.J., Gescheider, G.A. (eds.) Ratio Scaling of Psychological Magnitudes. Erlbaum, Hillsdale (1991)
19. Stevens, S.S.: Measurement, psychophysics and utility. In: Churchman, C.W., Ratoosh, P. (eds.) Basic Concepts of Measurements, pp. 1–49. Cambridge University Press, Cambridge (1959)
20. Berglund, B.: Measurement in psychology. In: Berglund, B., Rossi, G.B., Townsend, J., Pendrill, L. (eds.) Measurement with Persons, pp. 27–50. Taylor and Francis, New York (2012)
21. Forbes, A.B.: Parameter estimation based on the least-squares method. In: Pavese, F., Forbes, A. (eds.) Data Modeling for Metrology and Testing in Measurement Science, pp. 147–176. Birkhauser-Springer, Boston (2009)
22. Crenna, F., Rossi, G.B., Bovio, L.: Loudness measurement by robust magnitude estimation. Paper presented at the 14th joint int. IMEKO TC1+TC7+TC13 symposium, Jena. 31 Aug 2 Sept 2011
23. Rossi, G.B., Crenna, F.: On ratio scales. Measurement **46**, 29–36 (2013). doi:10.1016/j.measurement.2013.04.042
24. Miyamoto, J.M.: An axiomatization of the ratio difference representation. J. Math. Psychol. **27**, 439–455 (1983)
25. Moore, B.C.J.: Psychology of Hearing. Academic Press/Elsevier, San Diego (2003)
26. DIN: DIN 45631: procedure for calculating loudness level and loudness (1991)
27. ANSI: ANSI S3.4-2007: procedure for the computation of loudness of steady sounds (2007)
28. DIN: DIN 45631/A1: calculation of loudness level and loudness from the sound spectrum—Zwicker method—amendment 1: calculation of the loudness of time-variant sound (2008)
29. Berglund, B., Harju, E.: Master scaling of perceived intensity of touch, cold and warmth. Eur. J. Pain **7**, 323–334 (2003)
30. Crenna, F., Belotti, V., Rossi, G.B.: Experimental set-up for the measurement of the perceived intensity of vibrations. Paper presented at the XX IMEKO world congress metrology for green growth, Busan, Republic of Korea. 9–14 Sept 2012
31. Townsend, J.T., Burns, D., Pei, L.: The prospects for measurement in infinite-dimensional psychological spaces. In: Berglund, B., Rossi, G.B., Townsend, J., Pendrill, L. (eds.) Measurement with Persons, pp. 143–174. Taylor and Francis, New York (2012)
32. Crenna, F., Rossi, G.B., Bovio, L.: Measurement of the perceived similarity in face recognition. Paper presented at the XX IMEKO world congress metrology for green growth, Busan, Republic of Korea. 9–14 Sept 2012

Chapter 9
The Evaluation of Measurement Uncertainty

9.1 How to Develop a Mathematical Model of the Measurement Process

9.1.1 Statement of the Problem

The evaluation of measurement uncertainty is a main task in measurement, as we have amply discussed in Chap. 2, and one of the main aims of this book is to provide a sound foundation for that. We have already taken some steps in this direction: in Chap. 5, we have presented a general probabilistic model of the measurement process which provides the basic theoretical framework for the evaluation. In Chap. 6, we have started to consider the application of that framework, discussing the logical steps involved, by means of simple illustrative examples. We are now ready to demonstrate the practical application of this approach and to provide guidelines for dealing with real-world cases.

According to Cox and Harris [1], uncertainty evaluation comprises two phases: formulation and calculation. The former, which is under the responsibility of a measurement expert, consists in developing a model of the specific measurement process under consideration. The latter, that in some cases may be delegated to a numerical analyst, consists in carrying out the necessary uncertainty calculations based on that formulation.

Here, we will mainly provide support to the formulation phase, by showing how to systematically develop a model of the measurement process based on the general framework that we have presented in this book [2]. Then, we will briefly touch on the calculation phase, by presenting some prototype measurement software, developed in our Measurement Laboratory in Genova [3], and finally, we present a case study [4].

As we have seen in Chap. 5, the basic formulae for uncertainty evaluation are, for a scalar measurand, in terms of continuous variables,

G. B. Rossi, *Measurement and Probability*, Springer Series in Measurement Science and Technology, DOI: 10.1007/978-94-017-8825-0_9,

$$p(x'|\mathbf{y}) = \int_{\Theta} \left[\frac{p(\mathbf{y}|x, \boldsymbol{\theta})}{\int_X p(\mathbf{y}|x, \boldsymbol{\theta})\mathrm{d}x} \right]_{x=x'} p(\boldsymbol{\theta})\mathrm{d}\boldsymbol{\theta} \tag{9.1}$$

and, for a vector measurand,

$$p(\mathbf{x}'|\mathbf{y}) = \int_{\boldsymbol{\Theta}} \left[\frac{p(\mathbf{y}|\mathbf{x}, \boldsymbol{\theta})}{\int_{\mathbf{X}} p(\mathbf{y}|\mathbf{x}, \boldsymbol{\theta})\mathrm{d}\mathbf{x}} \right]_{\mathbf{x}=\mathbf{x}'} p(\boldsymbol{\theta})\mathrm{d}\boldsymbol{\theta}. \tag{9.2}$$

Basically what follows may be seen as a guide to the practical application of these formulae.

9.1.2 Linear Models

Linear models are very important since in many cases, a measuring device may be assumed to have a linear behaviour.[1] In fact instrument manufacturers usually do their best for developing linear devices, that are appreciated by practitioners for their simplicity. Furthermore, even in the case of nonlinear systems, it is often appropriate to consider for uncertainty evaluation a linearised model in the neighbourhood of the operating point, since variations around that point produced by uncertainty sources are often small as compared to the operating range.

Consider then the simple model

$$y = kx + w, \tag{9.3}$$

where $k > 0$ is the sensitivity of the measuring system and w is a probabilistic (or random) variable, describing an additive random effect such as measurement noise. Consider first the case where a single indication, y, is acquired and there is no systematic effect. Then, formula (9.1) simplifies into

$$p(x'|y) = \left[\frac{p(y|x)}{\int_X p(y|x)\mathrm{d}x} \right]_{x=x'} \propto [p(y|x)]_{x=x'}. \tag{9.4}$$

In practice, it is important to note that the distribution at the left side of this expression is proportional to the one at the right side, regarded as a function of x, with y fixed and equal to the actually observed value, with x then replaced by x'. The proportionality holds true since the division may be interpreted as a scaling operation that ensures that the distribution on the left side has a unitary integral, as required by probability distributions.

[1] This and the following two sections are amply based on Ref. [2], to which the reader is referred for additional details.

Let us now calculate the distribution $p(y|x)$. To that purpose, note that $p(y|x)$ is the distribution of y, for x fixed at a specific value. If we fix the value of x in formula (9.3), we see that the probabilistic variable y differs from w only in the term kx, which is an additive constant. The distribution of y, for x fixed, is thus the distribution of w, which we denote by $p_w(\cdot)$, calculated for the argument y, translated by the term kx. We thus obtain

$$p(y|x) = p_w(y - kx). \tag{9.5}$$

For obtaining $p(x|y)$, remember that it is proportional to $p(y|x)$, regarded as a function of x, and properly scaled. Since in the expression for $p(y|x)$, just found, regarded as a function of x, the dependent variable, x, is multiplied by k, in order to ensure unit area we have to multiply by k.[2] Then, after replacing x by x', we finally obtain

$$p(x'|y) = kp_w(y - kx'). \tag{9.6}$$

9.1.3 Systematic Effects and Random Variations

In the case of an additive systematic effect θ, the model becomes

$$y = kx + w + \theta. \tag{9.7}$$

For a single measurement, restitution yields simply

$$p(x'|y) = \int_{\Theta} kp_w(y - kx' - \theta)p(\theta)\mathrm{d}\theta. \tag{9.8}$$

The integration may be done once a proper distribution for θ has been assumed. Often this is a uniform distribution, but the formula holds true in the general case.

In order to appreciate the structural difference between a systematic and a random effect, it is interesting to study the case of measurement based on repeated observations. Now, the appropriate model is

$$y_i = kx + w_i + \theta \tag{9.9}$$

---

[2] For understanding this result without performing analytical calculation, consider the following geometrical argument. Multiplying the argument of a function by a constant factor k is equivalent to scaling the function along the abscissa by the same factor. For example, if $k > 1$, the result is a contraction of the graph of the function. The integration is equivalent to calculating the area under the graph, which, after contraction, is reduced by the factor k. To restore a unit area, it is thus necessary to further multiply by k. A similar argument holds true for $k < 1$, which corresponds to a dilation.

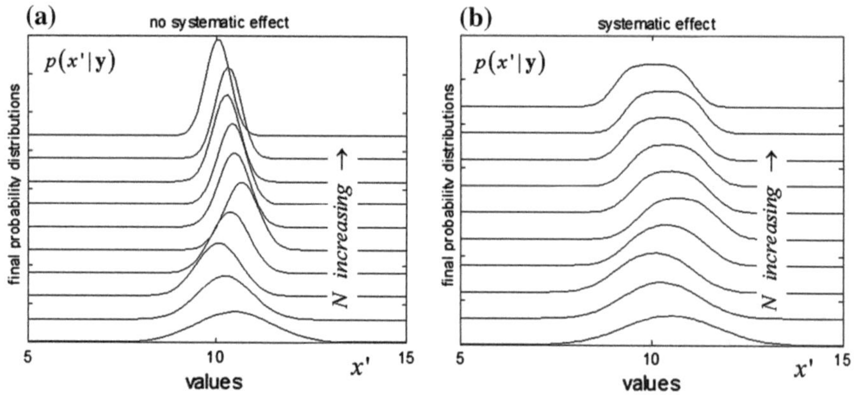

Fig. 9.1 Convergence of final distribution in measurement based on repeated observations, as the number of observations increases: **a** no systematic effect; **b** systematic effect described by a uniform distribution

where i is a discrete index and w_i a series of independent probabilistic variables, each with the same distribution of w. Note that both θ and x remain constant through the repeated observations and thus they are not indexed by i. Let us collect the variables describing possible indications into the vector $\mathbf{y} = (y_1, \ldots, y_N)$. Observation is described by

$$p(\mathbf{y}|x) = \prod_i p_w(y_i - kx - \theta) \tag{9.10}$$

and restitution by

$$p(x'|\mathbf{y}) = \int_{\Theta} \frac{\prod_i kp_w(y_i - kx' - \theta)}{\int_{X'} \prod_i kp_w(y_i - kx' - \theta)\mathrm{d}x'} p(\theta)\mathrm{d}\theta. \tag{9.11}$$

Consider now how the final distribution changes, in a given experiment, as N increases. To simulate such an experiment, let us assign a Gaussian distribution to w and let us consider two cases: (a) no systematic effect; (b) a systematic effect θ with a uniform distribution. Consider the following numerical data: $x = 10$, $\sigma = 1$, $k = 1$, θ either null (a) or having a uniform distribution on the interval $[-1, +1]$ (b),[3] and N increasing from 1 to 10. The result of the simulation is reported in Fig. 9.1.

In Fig. 9.1a, the final distribution converges to the value of the measurand as N increases. In Fig. 9.1b, instead, since θ is described by a uniform distribution, the process converges to such a distribution, centred on the value of the measurand, and this is just what may be expected, since the number of indications does not affect the uncertainty contribution due to the systematic effect.

[3] Numerical values are assumed to be expressed in arbitrary consistent units.

9.1.4 Observability

So far, we have treated influence parameters that give rise to a systematic deviation. We can now consider parameters concerning the probability distributions involved, typically dispersion parameters. Remember what we noted in Chap. 6: for some parameters, it is possible to learn from data, whilst for others it is not. Here, we can learn about dispersion or correlation parameters, whilst we cannot for parameters that account for systematic effects. Actually, this is another way to approach systematic effects: we may indeed distinguish between *observable* and *unobservable* parameters, where "observable" here means "such that information on it can be obtained through observation data".

In our example (formula 9.9), for instance, we may assume that the distribution of the random variable is Gaussian and its standard deviation, σ is unknown. If we repeat observation N times, we can obtain some information on it: it is common practice [5] to estimate σ by

$$s = \frac{1}{N-1}\sqrt{\sum_t (y_i - \bar{y})^2} \qquad (9.12)$$

where

$$\bar{y} = \frac{1}{N}\sum_i y_i. \qquad (9.13)$$

On the other hand, we cannot obtain any information on θ; otherwise, it would not be a *systematic* effect. So, let us see how we can treat this case in our probabilistic framework. The model is still provided by formula (9.9), but now in describing restitution, we have to account for the dependence upon σ as well. Introducing the standard Gaussian distribution, $\varphi(\xi) = (2\pi)^{-1/2}\exp(-\xi^2/2)$, the distribution of w is

$$p_w(\xi) = \sigma^{-1}\varphi(\sigma^{-1}\xi). \qquad (9.14)$$

Observation is then described by

$$p(\mathbf{y}|x,\sigma,\theta) = \prod_i \sigma^{-1}\varphi[\sigma^{-1}(y_i - \theta - kx)], \qquad (9.15)$$

where the dependence upon σ is apparent.

For restitution, we can use formula (9.2), considering σ as an additional measurand. We thus introduce the measurand vector $\mathbf{x} = (x, \sigma)$ and obtain

$$p((x',\sigma')|\mathbf{y}) = \int_\Theta \left[\frac{p(\mathbf{y}|x,\sigma,\theta)p(\sigma)}{\int_{X,\Sigma} p(\mathbf{y}|x,\sigma,\theta)p(\sigma)\mathrm{d}\sigma \mathrm{d}x}\right]_{(x,\sigma)=(x',\sigma')} p(\theta)\mathrm{d}\theta. \qquad (9.16)$$

This is the joint final distribution of x' and σ', from which we obtain the marginal distributions

$$p(x'|\mathbf{y}) = \int_{\Sigma} p(x', \sigma'|\mathbf{y})\mathrm{d}\sigma'], \tag{9.17}$$

and

$$p(\sigma'|\mathbf{y}) = \int_{X} p(x', \sigma'|\mathbf{y})\mathrm{d}x']. \tag{9.18}$$

If we assume for σ, the non-informative prior $p(\sigma) \propto \sigma^{-1}$ [6, 7], we obtain

$$p(\sigma'|\mathbf{y}) \propto \sigma^{-N} \exp\left[-\frac{1}{2}\frac{(N-1)s^2}{\sigma^2}\right], \tag{9.19}$$

where s is defined as above by formula (9.12) [8]. This confirms that we have gained information on σ through the measurement, whilst the distribution for θ remains unchanged [2]. The distribution for x' depends upon the form of the distribution $p(\theta)$; we will provide a numerical example in the next section.

9.1.5 Low-Resolution Measurement

Low-resolution measurements have practical and theoretical import. They occur often in practice. In traditional instruments, it was common to design the device in such a way that the resolution was comparable to the uncertainty, so that in providing the final result, an implicit approximate evaluation of the uncertainty was communicated through the least significant digit of the number. This yielded small observable variations in repeated observations. In modern instrumentation, it is often possible to select the measuring range and the overall gain: it is still common practice to do that in such a way that the above condition is fulfilled. Low resolution thus does not imply poor measurement conditions. In perceptual measurement, it is also common to use limited resolution scales, often with some ten values, in order to reduce intra- and inter-subjects variability.

From the theoretical side, low-resolution measurement is also very interesting, since it implies a quantisation transformation which is highly nonlinear and thus constitutes a good trail for checking uncertainty evaluation procedures: some literature has been produced in this regard [3, 9–11].

To study this effect, we modify the model in formula (9.9) by introducing a quantisation operator $Q(\cdot)$, yielding

$$y_i = Q(kx + w_i + \theta). \tag{9.20}$$

Let us now characterise the quantisation operator, by studying firstly the transformation

$$v = Q(u), \tag{9.21}$$

where u is a *continuous* probabilistic variable and v a *discrete* one. Let q be the quantisation interval, that is the distance between two successive values in the discrete representation. Then, the values of v can be expressed as lq, where l is an integer number and the quantisation transformation is defined by

$$v = Q(u) = lq \Leftrightarrow lq - \frac{q}{2} < u \leq lq + \frac{q}{2}. \tag{9.22}$$

The (discrete) distribution of v is then

$$P(v = lq) = \int_{lq-\frac{q}{2}}^{lq+\frac{q}{2}} p_u(\xi)\mathrm{d}\xi \tag{9.23}$$

or, in a shorthand notation,

$$P(v) = \int_{v-\frac{q}{2}}^{v+\frac{q}{2}} p_u(\xi)\mathrm{d}\xi = \int_{-\frac{q}{2}}^{+\frac{q}{2}} p_u(v+\xi)\mathrm{d}\xi. \tag{9.24}$$

We can now return to our model. For a single observation y, the conditional distribution is now

$$P(y|x,\sigma,\theta) = \int_{-\frac{q}{2}}^{+\frac{q}{2}} \sigma^{-1}\varphi[\sigma^{-1}(y - kx - \theta + \xi)]\mathrm{d}\xi \tag{9.25}$$

and for N observations

$$P(\mathbf{y}|x,\sigma,\theta) = \prod_i \int_{-\frac{q}{2}}^{+\frac{q}{2}} \sigma^{-1}\varphi[\sigma^{-1}(y_i - kx - \theta + \xi)]\mathrm{d}\xi. \tag{9.26}$$

It is important to note that this formulation is "exact": our probabilistic approach allows treating this case without any approximation! Restitution may be obtained by including this formula in formulae (9.16–9.18) above. The calculation requires some software, as the one that will be described in the next section.

Table 9.1 Results of low-resolution length measurements [11]

y_1	y_2	y_3	y_4	y_5	y_6	y_7	y_8	y_9	y_{10}
7.5	7.5	7.4	7.5	7.5	7.4	7.5	7.5	7.5	7.4

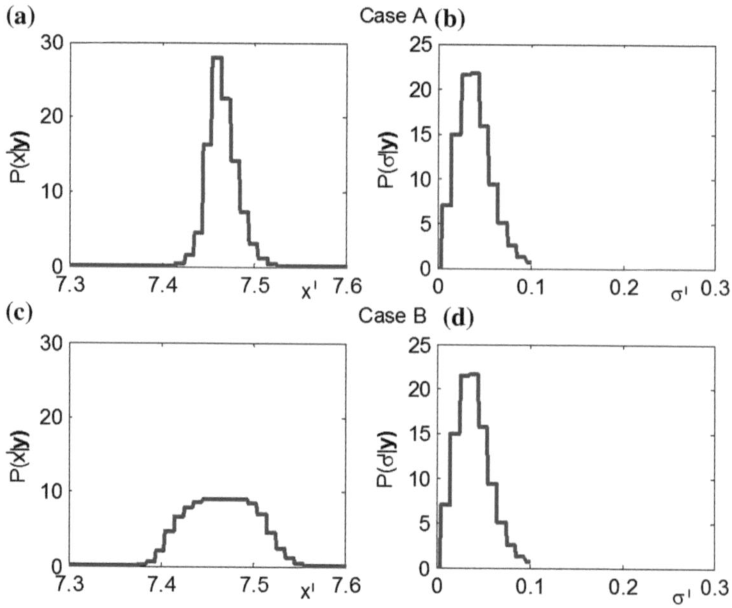

Fig. 9.2 Results for low-resolution length measurements. *Case A* Final distribution for x (**a**) and for σ (**b**). *Case B* Final distribution for x (**c**) and for σ (**d**)

As an illustrative numerical example, consider a length measurement, based on 10 repetitions, performed by a device with resolution $q = 0.1$ mm [11]. Consider two cases,

(a) no systematic effect,
(b) additive systematic effect, bounded within $\pm\Delta\theta = \pm 0.05$ mm.

Suppose that, in both cases, the numerical results are as in the Table 9.1.

The result can be obtained, in both cases, by applying formula (9.26). As mentioned, this requires a numerical approach [3]. Basically, the variables involved can be turned into discrete ones, by assuming a proper discretisation interval for all the involved variable, here denoted by Δx . This is usually fixed as a fraction of the quantisation interval q. In this case, since $q = 0.1$ mm, $\Delta x = q/10 = 0.01$ mm is appropriate. The results are presented in Fig. 9.2. Note that the systematic effect θ affects the distribution for the measurand, $p(x'|\mathbf{y})$, whilst it does not influence the standard deviation. Thus, its final distribution, $p(\sigma'|\mathbf{y})$, is the same in both cases.

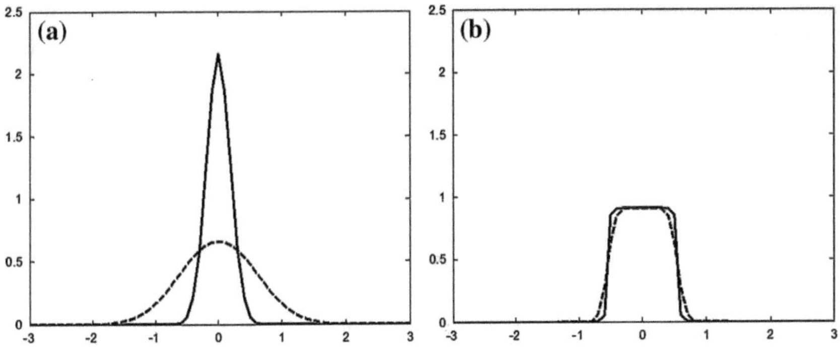

Fig. 9.3 Effect of quantisation on measurement based on repeated observations: final distribution for a single observation (*dotted line*) and for 10 repeated observations, for different values of the ratio σ/q: **a** $\sigma/q = 0.5$; **b** $\sigma/q = 0.1$

9.1.6 Practical Guidelines

The probabilistic restitution procedure allows us to assign the final distribution to the measurand and to calculate all required parameters in practical measurement, i.e. the measurement value, $\hat{x}$, and the standard, u, and expanded, U, uncertainty. In practical cases, however, it may desirable to avoid the explicit calculation of the final distribution and to obtain directly the final parameters. Even in these cases, the probabilistic approach helps, since it is possible to obtain general criteria on how to deal with low-resolution measurements.

The resolution effect is a nonlinear one. Basically, if the random effect, depending upon the standard deviation σ of the additive random component is important, then the quantisation effect can be treated as an additional random component. On the other hand, if the quantisation effect prevails, it must be treated as a systematic effect. This is illustrated in Fig. 9.3, where the final distribution for a measurement based on 10 observations is compared with that of a single observation.

It appears that the behaviour of the system depends upon the ratio σ/q: if this is "high", the effect of random variations prevails and quantisation can be neglected, and viceversa. For practical application, it is important to establish at what value of such ratio does the transition between the two operating conditions take place. The answer can be obtained through the probabilistic approach [10] and is shown in Fig. 9.4.

In the figure, the ratio of the uncertainty for measurement based on N repeated observations, u_N, to that based on a single observation, u_1, computed through the probabilistic model, is plotted as a function of the ratio σ/q, for $N = 10$ and for $N = 30$. If the quantisation effect is equivalent to an additional random component, the ratio u_N/u_1 must be equal to $1/\sqrt{N}$, that is to 0.3162, for $N = 10$ and to 0.1826, for $N = 30$. Looking at the figure, we note that the transition occurs, in both cases, for $\sigma/q = 0.5$.

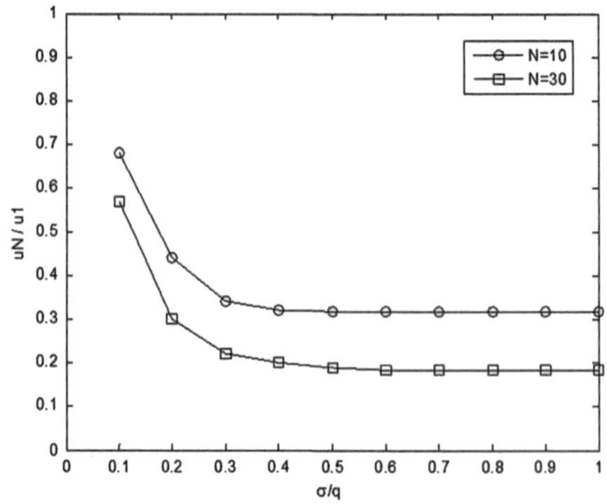

Fig. 9.4 Study of the effect of quantisation as a function of the ratio σ/q

Thus, we obtain the following practical rule: in the case of measurement based on repeated observations,

- if $\sigma/q > 0.5$, the quantisation effect is equivalent to an additional random effect; thus, if the standard deviation is estimated, as usually, through formula (9.12), it includes the effect of quantisation;
- otherwise, if $\sigma/q \leq 0.5$, quantisation produces mainly a systematic effect and thus a corresponding term, $u_q = q/2\sqrt{3}$, should be included in the uncertainty budget.

So the probabilistic approach can be used both for providing the complete result, expressed by a probability distribution, for sophisticated measurements, or for deriving simple practical rules, for usual daily measurements.

9.1.7 *Hysteresis Phenomena*

Quantisation is a nonlinear effect, since, as we have seen, the behaviour of the system *substantially* (and not just *proportionally*) changes in dependence of the values of involved parameters. Another highly nonlinear phenomenon is hysteresis, which also occurs in some measuring systems, especially in the case of measurement of properties of fluids, such as pressure or flow rate [5, 12, 13].[4]

The hysteresis phenomenon can be modelled by considering two (or more) different behaviours, depending on some (usually unknown) internal-state condition.

[4] Hysteresis occurs when the behaviour of a system depends on its past environment. This happens because the system can be in more than one internal state. Prediction of its future development would require knowledge either of its internal state or of its history. In typical measurement conditions, such a prediction is impossible and thus hysteresis constitutes a source of uncertainty.

Table 9.2 Measurement with hysteresis phenomena

Test case	σ	y_1	y_2	y_3	y_4	y_5	y_6	y_7	y_8	y_9	y_{10}
A	0.54	0.20	1.13	0.47	1.71	0.60	1.26	1.11	0.54	−0.09	0.97
B	0.98	1.00	0.84	1.55	0.06	−0.79	−0.55	1.37	1.29	−0.98	1.34

For example, a variation of the model expressed by formula (9.3), with hysteresis, may be

$$y = kx + w + h, \quad \text{for condition } \alpha,$$
$$y = kx + w - h, \quad \text{for condition } \beta,$$

where h is a constant value. Typical conditions α and β may be, e.g. "for ascending inputs" and "for descending inputs", respectively. Hysteresis is difficult to manage in a deterministic context, since this model constitutes a polydromic non-invertible function. In a probabilistic approach instead, the conditional distribution for the observation process can be stated as follows:

$$p(y|x) = p_\alpha p(y - kx - h) + p_\beta p(y - kx + h), \tag{9.27}$$

where p_α and p_β denote the probabilities of conditions α and β, respectively. When no additional information is available, it is reasonable to assume $p_\alpha = p_\beta = 0.5$. Then, restitution is provided by

$$p(x'|y) = p_\alpha k p_w(y - kx' - h) + p_\beta k p_w(y - kx' + h). \tag{9.28}$$

The extension to measurement based on repeated observations can be done in a similar way as discussed in the previous sections. The main problem with hysteresis is that we neither know in which condition the instrument is when we perform measurement nor whether it keeps the same condition throughout the eventual repeated observations or it commutes between the two.

In our laboratory, we wanted to study the behaviour of the probabilistic restitution procedure in this regard, and we used the simulated data in Table 9.2 [3].

In case A, we assumed that the instrument always behaves according to one curve, whilst in case B, we assumed it commutes randomly from one to the other. Thus, in case A, we have a lesser dispersion and the instrument behaves apparently better. In reality, data from case B are more informative since they are not biased, even if they have a greater dispersion. Interestingly enough, the restitution algorithm recognises that, in this behaving, in an "intelligent" way. The results are presented in Fig. 9.5. Noteworthy, in case A, the final distribution is still bimodal, since it is impossible to understand from data in which condition we are, whilst in case B it is unimodal and the final uncertainty is smaller, even if the apparent dispersion is greater!

Such a result would be impossible to achieve with the usual processing methods.

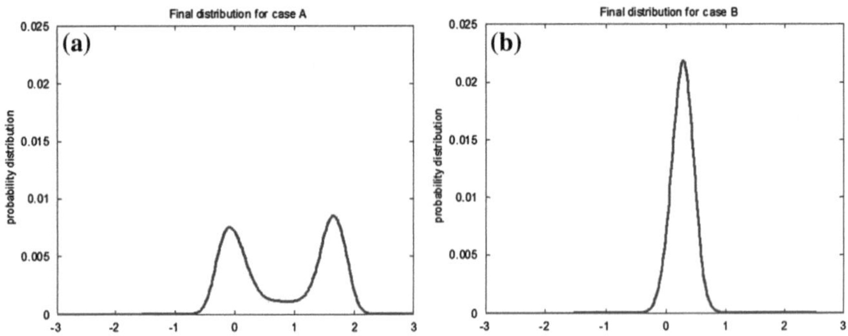

Fig. 9.5 Processing of data with an hysteresis phenomenon, in cases A (**a**) and B (**b**), as reported in Table 9.2

9.1.8 Indirect Measurement

Our probabilistic approach allows also treating indirect (or derived) measurement, where the value of the measurand is obtained by measuring other quantities functionally related to the measurand.[5] Such measurement can be dealt with by simply interpreting the vector $\mathbf{y}$ in formulae (9.1) and (9.2) not as describing indications from the same measurement process, as we have done in the previous sections, but as describing indications from different measurement channels, where each channel concerns one of the quantities functionally related to the measurand. For example, if the quantity of interest is density and we measure it through a mass and a volume measurement, using the formula $\rho = m/V$, the vector $\mathbf{y}$ would be in this case $\mathbf{y} = (y_m, y_V)$, where y_m and y_V are indications from a mass and a volume meter, respectively.

The point now is how to assign a probability distribution to ρ, based on the indications y_m and y_V. According to what we have so far discussed, we will be able to assign a distribution to m and V, $p(m'|y_m)$ and $p(V'|y_V)$, respectively. Then, we can simply define the measurement value as the probabilistic variable

$$\rho' = \frac{m'}{V'} \tag{9.29}$$

and consequently assign the distribution

$$p(\rho'|y) = \int \delta\left(\rho' - \frac{m'}{V'}\right) p(m'|y_m)p(V'|y_V)\mathrm{d}m'\mathrm{d}V'. \tag{9.30}$$

[5] Indirect or derived measurement has been previously treated in Sects. 3.7, 4.4 and 5.7. Note that we prefer to use the term "derived" when dealing with scales and the term "indirect" when we consider measurement, but the two terms are essentially equivalent, since the underlying idea is the same.

This is a straightforward procedure, which is in agreement with international recommendations, in particular to Supplement 1 to the GUM [1, 14]. The problem at this point is solely computational, which leads us to the software issue, to be discussed now, in general terms.

9.2 Measurement Software

To fully benefit from the probabilistic framework so far developed, it is necessary to have proper software for calculation of the involved probability distributions. The issue of software in measurement and in metrology has been debated in the last, say, 10 years in view of its criticality. In particular, it has been treated in the European Network SoftTools MetroNet, where a regular series of Euro Conference were organised, and a book was produced, edited by Pavese and Forbes [15], which constitutes a comprehensive compendium of mathematical and numerical methods in measurement. It also includes some free software. In particular, Chap. 12 of that book provides an up to date summary of the software issue [16].

Software development in measurement has to face some critical issues. As an example, in legal metrology,[6] it is necessary to ensure that the software embedded in a device is protected from fraudulent manipulation. In fact in modern digital measuring systems, the software has become a part of the system and it affects the overall performance. Since measuring devices are expected to be highly *reliable*, it is essential that such confidence is extended to the software, which must be developed in conformance to strict quality assurance procedures. For example, laboratories wishing to comply with the requirements of the ISO/IEC Standard 17025 [17] have to take special care in software development, and international guidelines have been produced in this regard [18]. Here, we do not treat this topic in detail, since in this book, we are interested in principles rather than in technicalities. Yet we think it may be useful to briefly present, as an example, a prototype software package that has been developed in our laboratory [3], just to give a feeling of the subject.

This package, called UNCERT,[7] allows treating the model

$$y_i = Q(kx + w_i + s + h), \tag{9.31}$$

$$s = \mathbf{a} \cdot \mathbf{v}^T, \tag{9.32}$$

where s accounts for the effect of a linear combination of influence quantities, collected in a vector $\mathbf{v} = (v_1, \ldots, v_n)$, with sensitivities forming the vector $\mathbf{a} = (a_1, \ldots, a_n)$. The first m $(0 \leq m \leq n)$ variables in $\mathbf{v}$ are assumed to be

[6] Legal metrology will be briefly addressed in Chap. 11.

[7] Here, we just mention the main features of the code and some of its applications. For additional details, the reader can consult Refs. [3] and [4] especially.

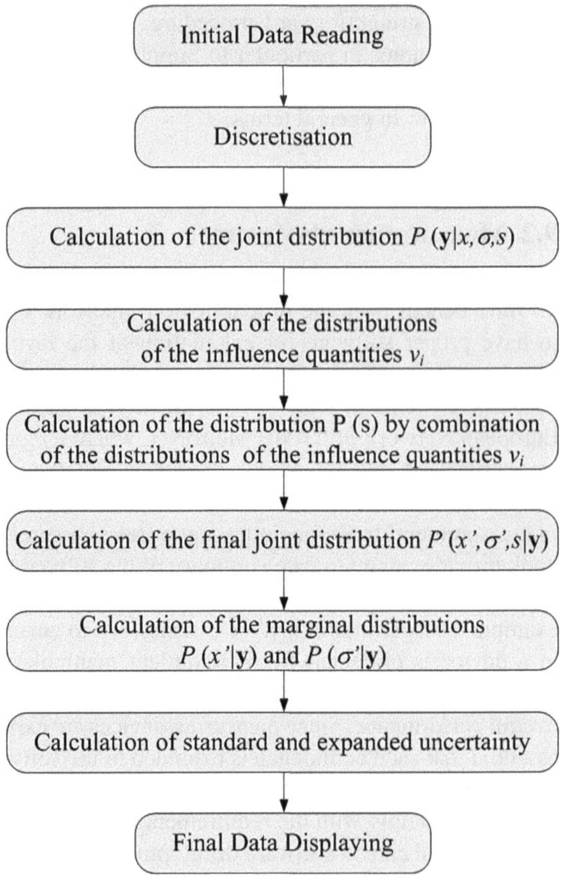

Fig. 9.6 Flowchart of the code UNCERT

correlated whilst the remaining $n - m$ are not, and the other terms are defined as in the previous sections.[8]

This model is indeed quite general and allows treating a wide class of problems: the examples on low-resolution measurement and on the hysteresis phenomenon have been developed with this program, as it will be the test case in the next section. The calculation is based on treating all quantities as *discrete probabilistic variables, with a careful choice of the quantisation interval*. Related quantisation and truncation effects have been carefully studied and kept under control. A flowchart of the code is presented in Fig. 9.6, where the treatment of hysteresis has been omitted for the sake of simplicity.

The program calculates separately the term $P(\mathbf{y}|x, \sigma, s)$ and the term $P(s)$, which results from a combination of the influence quantities collected in the vector $\mathbf{v}$.

---

[8] The term $\mathbf{a} \cdot \mathbf{v}^T$ is the scalar product of $\mathbf{a}$ times $\mathbf{v}$ and the operator T superscript denotes transposition.

For the subset of correlated variables a Gaussian approximation is used; the remaining ones are combined through convolution. Then, the term $P(\mathbf{y}|x, \sigma, s)P(s)$ is formed, which allows calculation of the joint final distribution $P(x', \sigma', s|\mathbf{y})$. Lastly, the marginal distributions for x' and σ' are computed, as well as any required parameter, such as the standard uncertainty u or the expanded uncertainty U.

The package is organised in modules that perform specific tasks, such as the calculation of the individual distributions of the v_i variables, their combination by convolution, the Gaussian approximation for the subset of mutually correlated variables and so forth. The modular architecture proved to be particularly convenient for validation, since the modules could be firstly tested individually and then in their assembly. The development process was managed according to the general principles of quality assurance, since our laboratory has an ISO 9001 certification, for "experimentation, research and education in measurement".

The procedure for the development of each module included the definition of its features, the preparation of the code, its testing and eventual correction and improvement, its final approval, documentation and inclusion in the official database of the laboratory. The overall package was also validated as a whole, considering the following items:

1. since the code deals with a combination of (probabilistic) *discrete* variables, the opposite cases of distributions with a small and with a large number of points were tested, as well as the case of a large number of variables;
2. back-to-back comparison with programs that provide a direct implementation of the GUM-Mainstream method was performed;
3. performance in the treatment of reference data, taken, e.g. from the GUM or from other standards, was evaluated.

The validation process took advantage of collaborative work that took place in the already mentioned European Network SoftTools MetroNet.

9.3 A Working Example

To conclude this brief survey of uncertainty evaluation, we present now a test case that has been utilised in the GUM [5], as Example H.1. We studied this case during a collaboration with the National Physical Laboratory, as a part of the activities of the above-mentioned EU Network: here, we just mention a few highlights, whilst for a full report, readers can consult Ref. [4].

Consider the measurement of the length of a gauge block, where the measurement process consists in comparing it with a standard of the same nominal length $x_0 = 50$ mm and taking $N = 5$ repeated observations. Assume the following notation:

- y_i is the ith indication of the comparator;
- w_i is the ith random variation of the indication;
- δy is a possible systematic deviation of the comparator;

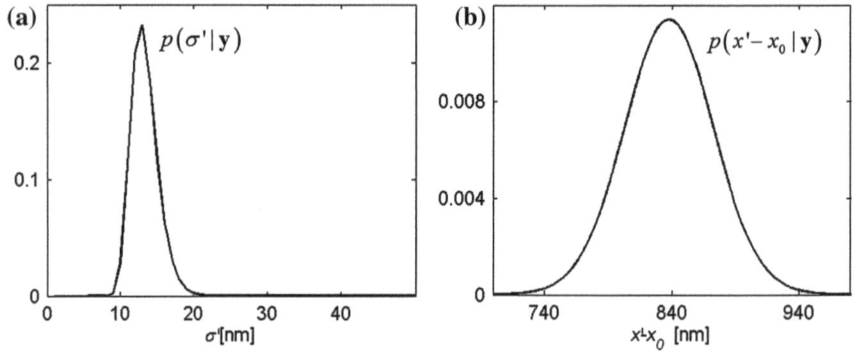

Fig. 9.7 Calculation results for the gauge block example [4]

- $l_s = l_0 + \delta_l$ is the length of the standard, where $l_0 = 50,000,623$ nm;
- $\Delta\theta = \theta - \theta_s$ is the difference in temperature between the gauge blocks;
- $\Delta\alpha = \alpha - \alpha_s$ is the difference in their coefficients of thermal expansion;
- $\theta = \theta_0 + \delta\theta$ is temperature of the measurand block, which undergoes cyclic variations from the reference temperature $\theta_0 = 20\,^\circ\text{C}$.

Then, after neglecting some higher-order terms, observation can be expressed by

$$y_i = x + w_i + \delta y - (l_0 + \delta l)(1 - \Delta\alpha(\theta_0 + \delta\theta) - \alpha\Delta\theta). \tag{9.33}$$

After collecting all the variables giving rise to a systematic deviation in a vector $\mathbf{v} = (\delta y, \delta l, \Delta\alpha, \delta\theta, \Delta\theta)$, we can express observation as

$$y_i = x + w_i + s(\mathbf{v}), \tag{9.34}$$

with

$$s(\mathbf{v}) = \delta y - (l_0 + \delta l)(1 - \Delta\alpha(\theta_0 + \delta\theta) - \alpha\Delta\theta), \tag{9.35}$$

which are related to formulae (9.31) and (9.32), apart from s now being a *nonlinear* function of $\mathbf{v}$. The linearised version can be treated with the package UNCERT; the nonlinear version requires some additional software that allows dealing with nonlinear functions of probabilistic variables such as the Monte Carlo method [1, 19, 20].

Simulating the random variations with a standard deviation $\sigma_0 = 13$ nm and assuming all the other numerical values for the parameters as in the GUM, we obtained the final distributions for the standard deviation (a) and for the measurand (b) as in Fig. 9.7.

References

1. Cox, MG., Harris, P.M.: SSfM best practice guide No. 6, uncertainty evaluation. Technical Report DEM-ES-011, National Physical Laboratory, Teddington, Middlesex, UK, (2006)
2. Cox, M.G., Rossi, G.B., Harris, P.M., Forbes, A.: A probabilistic approach to the analysis of measurement processes. Metrologia **45**, 493–502 (2008)
3. Rossi, G.B., Crenna, F., Codda, M.: Metrology software for the expression of measurement results by direct calculation of probability distributions. In: Ciarlini, P., Cox, M.G., Pavese, F., Richter, D., Rossi, G.B. (eds.) Advanced Mathematical Tools in Metrology VI. World Scientific, Singapore (2004)
4. Rossi, G.B., Crenna, F., Cox, M.G., Harris, P.M.: Combining direct calculation and the Monte Carlo Method for the probabilistic expression of measurement results. In: Ciarlini, P., Filipe, E., Forbes, A.B., Pavese, F., Richter, D. (eds.) Advanced Mathematical and Computational Tools in Metrology VII. World Scientific, Singapore (2006)
5. BIPM, IEC, IFCC, ISO, IUPAC, IUPAP, OIML.: Guide to the expression of uncertainty in measurement. ISO, Geneva, Switzerland. Corrected and reprinted 1995, (1993). ISBN 92-67-10188-9
6. Press, S.J.: Bayesian statistics. Wiley, New York (1989)
7. Gill, J.: Bayesian methods. Chapman and Hall/CRC, Boca Raton (2002)
8. Lira, I.: Bayesian assessment of uncertainty in metrology. Metrologia **47**, R1–R14 (2010)
9. Wong, P.W.: Quantization noise, fixed-point multiplicative round off noise, and dithering. IEEE Trans. Acoust. Speech Sig. Proc. **38**, 286–300 (1990)
10. Michelini RC, Rossi GB (1996) Assessing measurement uncertainty in quality engineering. In: Proceedings of the IMTC/96-IMEKO TC 7: Instrumentation and Measurement Technology Conference, Brussels, 4–6, 1996, p 1217–1221
11. Lira, I.H.: The evaluation of standard uncertainty in the presence of limited resolution of indicating devices. Meas. Sci. Technol. **8**, 441–443 (1997)
12. Bentley, J.P.: Principles of Measurement Systems, 4th edn. Pearson Education Ltd., Harlow (2005)
13. Morris, A.S., Langari, R.: Measurement and Instrumentation. Academic Press/Elsevier, Waltham (2012)
14. BIPM, IEC, IFCC, ISO, IUPAC, IUPAP, OIML: Guide to the expression of uncertainty in measurement (GUM)-Supplement 1: Propagation of distributions using a Monte Carlo method. International Organization for Standardization, Geneva (2006)
15. Pavese, F., Forbes, A. (eds.): Data Modeling for Metrology and Testing in Measurement Science. Birkhauser-Springer, Boston (2009)
16. Greif, N., Richter, D.: Software validation and preventive software quality assurance. In: Pavese, F., Forbes, A. (eds.) Data Modeling for Metrology and Testing in Measurement Science, pp. 371–412. Birkhauser-Springer, Boston (2009)
17. ISO: ISO/IEC 17025: General requirements for the competence of testing and calibration laboratories. ISO, Geneve (1999)
18. Wichmann B, Parkin G, Barker R (2007) Validation of software in measurement systems, Software support for metrology, Best practice guide 1, NPL Report DEM-ES 014, January 2007
19. Steele, A.G., Douglas, R.J.: Monte Carlo Modeling of Randomness. In: Pavese, F., Forbes, A. (eds.) Data Modeling for Metrology and Testing in Measurement Science, pp. 329–370. Birkhauser-Springer, Boston (2009)
20. Gentle, J.E.: Computational statistics. Springer, New York (2009)

Chapter 10
Inter-Comparisons and Calibration

10.1 A Worldwide Quality Assurance System for Measurement

How can we guarantee the quality of measurement, on a worldwide basis? This is possible, at least in physics, chemistry and engineering, thanks to the international system of metrology, that we have briefly introduced in Sect. 3.7.4. Basically, such a system operates at a national and international level. At a national level, it ensures the possibility of calibrating instruments by a network of accredited calibration laboratories and by one (or more) National Metrology Institute (NMI). Typically, the system works in this (simplified) way: instrument users, in any field and in each country, send their instruments for calibration to laboratories that use calibration devices or standards. These in turn may be calibrated by primary standards that are realised and maintained by the NMI. But how is it possible to guarantee the quality of these primary standards? There are no superior devices, at the national level, to which they may be compared. The only possibility is to act at the international level.

At present, this is managed through the Mutual Recognition Agreement (CIPM MRA) [1], which specifies the organisational and technical requirements for the mutual recognition of measurements performed by NMIs. A major tool for such a recognition are *key comparisons* [2]. They can be performed either directly against an international reference facility at the BIPM or through a stable travelling standard, which is circulated amongst several NMIs, which are asked to provide a measurement value for it, accompanied by an uncertainty statement. An international committee of NMI experts in the field evaluates the resulting data and provides practical information on the degree of comparability of the individual results. Similar exercises, called inter-comparisons, are performed amongst laboratories at lower levels of the metrological structure, and they are very effective for guaranteeing the performance of the overall system of metrology [3]. Their use could perhaps be extended to measurement in behavioural sciences also [4].

In this chapter, we provide a brief description of key comparisons and then, we also briefly address calibration.

G. B. Rossi, *Measurement and Probability*, Springer Series in Measurement Science and Technology, DOI: 10.1007/978-94-017-8825-0_10,

10.2 A Probabilistic Framework for Comparisons

10.2.1 How Key Comparisons Work

Let us now discuss how the data produced in a typical key comparison can be evaluated in order to produce the required information [5, 6]. Suppose that N NMIs participate in such a comparison and they measure, at the best of their capability, the same travelling standard, for example a stable gage block, whose length, x, is of interest. They will provide a measurement value, $\hat{x}_i$, and a standard uncertainty, u_i, for it, so that the resulting data set will be $\{(\hat{x}_i, u_i), i = 1, \ldots, N\}$. Typically, the goal of such an evaluation is to obtain a reference value for the standard, $\hat{x}$, with an associated uncertainty, u_x. Then, the "compatibility" of the result provided by each NMI with this reference value can be quantified by the difference, $d_i = \hat{x}_i - \hat{x}$, between the value provided and the reference value. A few procedures have been proposed, and the topic is still under discussion [6–9]. A thorough examination of them is beyond the scope of this book: we simply suggest a possible, conceptually simple, probabilistic approach, based on the notion of measurement scale, encouraging, as usually, the interested readers to consult the literature and to develop their own view.

10.2.2 Checking the Individual Results

Based on what we have discussed in Chap. 6, it is of utmost importance to note that in the evaluation of key comparisons, two distinct and different inferential processes are involved:

(a) the assessment of the reliability of the results of each participating NMI and
(b) the assignment of a probability distribution to the unknown value of the travelling standard.

In our lexicon, the former implies a *hypothetic-deductive* inference, the latter an *inductive* one. Let us then discuss separately both of them.

Each result provided by each NMI may be regarded as a *probabilistic model* for the travelling standard, expressed by a probability distribution, $p_i(x)$, over its possible values. In particular, if the data actually provided is, as usual, the pair $(\hat{x}_i, u_i)$, we can assume a Gaussian distribution,

$$p_i(x) = u_i^{-1}\varphi[u_i^{-1}(x - \hat{x}_i)]. \tag{10.1}$$

In order to check the validity of this model, we have to perform a *significance test*. To do that, ideally we should know the value of the standard, say x_0. In this case, we should select a conventional high probability, p_0, and then define an acceptance region, $A_i = [a_i, b_i]$, such that

$$\int_{a_i}^{b_i} p_i(x)\mathrm{d}x = p_0. \tag{10.2}$$

Then, the test will be passed if, simply, A_i includes x_0:

$$A_i \ni x_0. \tag{10.3}$$

But *we do not know* x_0, so we have to use an appropriate estimate $\hat{x}_0$, obtained from the available data, $\{(\hat{x}_i, u_i), i = 1, \ldots, N\}$.

Intuitively, $\hat{x}_0$ will be some kind of mean of such data, but what kind of mean? The point is that in making this test, for being realistic, we must assume that one, or more, of the available data *may be unreliable*, both in regard to the provided measurement value and to its stated uncertainty. So the estimate should be influenced as low as possible by such possible unreliable values (called *outliers* in statistics): how can we achieve this? For example, consider that we use a weighted mean, using the uncertainties for weighting the data. A common comment that has been made on this choice is that if some biased result has been also provided that declares a low uncertainty, this is expected to strongly adversely affect the result of the comparison[1]! With this in mind, we propose:

- not to use the uncertainties in forming such a mean, so that possible errors in their evaluation do not affect the estimation, and
- to use the *median*, instead of the arithmetic mean, since it is known to be a "robust" estimator.[2]

Remember that a median is described as the numerical value separating the higher half of a sample from the lower half.

So, to sum up, in regard to the assessment of the reliability of each individual result from the participating NMIs, we propose to perform a significance test, according to

$$A_i \ni \hat{x}_0, \tag{10.4}$$

where $\hat{x}_0$ is the median of the provided measurement values

$$\hat{x}_0 = \text{median}\ \{\hat{x}_i,\ i = 1, ..., N\}. \tag{10.5}$$

Then, the non-conformal results should be *excluded* in the continuation of the evaluation. Obviously, a high number of non-conformal results would suggest that the comparison is unreliable in the overall and should therefore be repeated.

[1] In fact, in the weighted-mean procedure, each value is weighted by the *inverse* of its (stated) uncertainty. Thus, a wrong value accompanied by a low stated uncertainty will strongly affect the final mean.

[2] In statistics, an estimator is called *robust* if it is weakly influenced by possible outliers.

After excluding the unreliable results, the point is now how to assign a probability distribution to the value of the travelling standard based on a set of consistent results. In the literature, two main approaches have been proposed for combining the individual distributions to form the final distribution to be assigned to the reference object: one suggests using a "product rule" [7, 9], the other an "addition rule" [8]. For facing this problem, we take advantage of the probabilistic theory that we have developed in Chap. 4.

10.2.3 The Paradigm of the Probabilistic Scale

We *interpret the key comparison as a step in the construction of a reference measurement scale* and proceed in this perspective.

Consider some quantity, x, for example length, and N NMIs participating in a key comparison. Each NMI is in charge of realising a primary reference scale for x: in the case of length, it consists of a highly stable laser for metrology. A ray of such a laser is the basis for the scale, since it is possible to measure accurately the displacement of an object along the laser ray by an optical interferometer, expressing it as a number of wavelengths, or of fractions of wavelength, of the laser radiation. Suppose then that all the NMIs have realised one such experiment on their own, compliant to the relevant specifications and guidelines, so we have *N independent realisations of the primary scale for x, all having the same validity*. The goal of the key comparison is now to establish a *unique* primary "virtual" reference scale, having a worldwide validity.

How is this possible? The basic idea is to compare the various scales to each other by means of a stable *travelling standard*. Suppose that it is an accurately constructed, *very stable*, iron bar having a nominal length x_0, for example $x_0 = 1$ m. By measuring the same object with respect to the different primary scales, it will be possible *to compare those scales amongst them* in a neighbourhood of x_0. It may be instructive to see things in this way: suppose that each NMI has its own one-metre standard, s_{0i}. How do such standards compare to each other? Consider two NMIs, i and j, and let $\hat{x}_i$ and $\hat{x}_j$ be their measurement value for the travelling standard. If, e.g. $\hat{x}_i > \hat{x}_j$, we can conclude that $s_{0i} \prec s_{0j}$.[3] Note that the latter is an *empirical relation*, and this is consistent with the overall framework, since we are now in a scale-construction phase, and a measurement scale, as we now well know, is based on empirical relations. Although in the current practice and language, the results of a key comparison are referred to as measurements, they should be more properly regarded as *evidences concerning empirical relations*.

Let us probe further this subject with the aid of an introductory example. Consider a set of objects, $A = \{a, b, c\}$; suppose that we have just two NMIs, working, respectively, with the subsets of A, A_1 and A_2. Let c be a travelling standard and

[3] Note the inversion of the inequality: since the travelling standard has been compared with standards at the NMIs, when the value obtained is greater, the standard must have been smaller.

$A_1 = \{a, c\}$, $A_2 = \{b, c\}$, so the two subsets have the travelling standard in common and are otherwise independent. The key idea is that, thanks to the element in common, c, we can construct an overall scale on A, without directly comparing the non-common elements, a and b, but rather inferring their mutual relations by the relations that they have with the travelling standard. We can easily recognise that this inference is completely different from the one we have discussed in the previous subsection and that is of the hypothetic-inductive kind, although it does not have a Bayesian structure. Furthermore, we also see how this may be seen as a scale-construction process.

Let us now assume some illustrative numerical values. Let for A_1, the following relational probabilities hold true:

$$\begin{aligned} P(a \succ c) &\triangleq p_1 = 0.1, \\ P(a \prec c) &\triangleq p_2 = 0.6, \\ P(a \sim c) &\triangleq p_3 = 0.3, \end{aligned}$$

and for A_2:

$$\begin{aligned} P(b \succ c) &\triangleq q_1 = 0.8, \\ P(b \prec c) &\triangleq q_2 = 0.1, \\ P(b \sim c) &\triangleq q_3 = 0.1. \end{aligned}$$

Consider the set of numbers $X = \{1, 2, 3\}$. The probabilistic structure associated to A_1 and A_2 are provided in Tables 10.1 and 10.2, respectively. Note that, in contrast with what we have done in Chap. 4, here for each empirical relation, we do not consider just one numerical assignment, but all the assignments that are possible with values in X, and we distribute the probability of the corresponding empirical relation uniformly amongst them. For example, the empirical relation $(a \succ c)$, which has probability p_1, can be represented in X either by $x_a = 2$ and $x'_c = 1$ or by $x_a = 3$ and $x'_c = 2$. Thus, we assign a probability equal to $p_1/2$ to each of them, which is shown in the first two rows of Table 10.1, and so on.[4]

In this way, we obtain the distributions for the probabilistic variables x_a and x'_c in A_1, and to x_b and x''_c in A_2. Note in particular that the probabilistic variable associated to the travelling standard c obtains distinct probability assignments in the two subsets A_1 and A_2: *this precisely models what happens when two NMIs assign different probability distributions to the same standard*. Such distributions are, for x'_c:

[4] We assign an apex to x_c, either x'_c or x''_c, to distinguish between the two ways in which element c, which is common to A_1 and to A_2, is treated in each of them.

Table 10.1 The probabilistic order structure on $A_1 = \{a, c\}$

Weak order $\succsim'_i$	Ordering $\succsim_i$	x_a	x'_c	Probability of the numerical assignment
$\succsim'_1 \triangleq \alpha_1$	$a \succ c$	2	1	$p_1/2$
		3	2	$p_1/2$
$\succsim'_2 \triangleq \alpha_2$	$a \prec c$	1	2	$p_2/2$
		2	3	$p_2/2$
$\succsim'_3 \triangleq \alpha_3$	$a \sim c$	1	1	$p_3/3$
		2	2	$p_3/3$
		3	3	$p_3/3$

Table 10.2 The probabilistic order structure on $A_2 = \{b, c\}$

Weak order $\succsim''_i$	Ordering $\succsim_i$	x_a	x''_c	Probability of the numerical assignment
$\succsim''_1 \triangleq \beta_1$	$b \succ c$	2	1	$q_1/2$
		3	2	$q_1/2$
$\succsim''_2 \triangleq \beta_2$	$b \prec c$	1	2	$q_2/2$
		2	3	$q_2/2$
$\succsim''_3 \triangleq \beta_3$	$b \sim c$	1	1	$q_3/3$
		2	2	$q_3/3$
		3	3	$q_3/3$

$$
\begin{aligned}
P(x'_c = 1) &= 0.15,\\
P(x'_c = 2) &= 0.45,\\
P(x'_c = 3) &= 0.40,
\end{aligned}
$$

and for x''_c:

$$
\begin{aligned}
P(x''_c = 1) &= 0.43,\\
P(x''_c = 2) &= 0.49,\\
P(x''_c = 3) &= 0.08.
\end{aligned}
$$

The point is now how to infer the probabilistic order structure associated to the entire set A, from these two distinct assignments. For doing that we have to compose the two structures: this gives rise to the related *product structure*, whose elements are:

$$
\begin{aligned}
&\alpha_1\beta_1\ \alpha_1\beta_2\ \alpha_1\beta_3,\\
&\alpha_2\beta_1\ \alpha_2\beta_2\ \alpha_2\beta_3,\\
&\alpha_3\beta_1\ \alpha_3\beta_2\ \alpha_3\beta_3.
\end{aligned}
$$

If we assume *independence*, we have

$$
P(\alpha_i\beta_j) = P(\alpha_i)P(\beta_j). \tag{10.6}
$$

Table 10.3 Probabilistic order structure on $A = \{a, b, c\}$

Weak order $\succsim_i$	Ordering $\succsim_i$	x_a	x_b	x_c	$P(\succsim_i)$
$\succsim_1$	$a \succ b \succ c$	3	2	1	$p_1q_1/3$
$\succsim_2$	$a \succ c \succ b$	3	1	2	p_1q_2
$\succsim_3$	$b \succ a \succ c$	2	3	1	$p_1q_1/3$
$\succsim_4$	$b \succ c \succ a$	1	3	2	p_2q_1
$\succsim_5$	$c \succ a \succ b$	2	1	3	$p_2q_2/3$
$\succsim_6$	$c \succ b \succ a$	1	2	3	$p_2q_2/3$
$\succsim_7$	$a \sim b \succ c$	2	2	1	$p_1q_1/3$
$\succsim_8$	$a \sim c \succ b$	2	1	2	p_3q_2
$\succsim_9$	$b \sim c \succ a$	1	2	2	p_2q_3
$\succsim_{10}$	$a \succ b \sim c$	2	1	1	p_1q_3
$\succsim_{11}$	$b \succ a \sim c$	1	2	1	p_3q_1
$\succsim_{12}$	$c \succ a \sim b$	1	1	2	$p_2q_2/3$
$\succsim_{13}$	$a \sim b \sim c$	1	1	1	p_3q_3

What is now the relation between the elements of this product structure and the possible orderings on A? Look at Table 10.3, where all the weak order relations in A are considered and numbered from 1 to 13.

Note that, for example, $\alpha_1\beta_2$ is equivalent to $\succsim_2$, and consequently, it may be given the same probability:

$$P(\succsim_2) = P(\alpha_1\beta_2) = p_1q_2.$$

But what happens, for example, with $\alpha_1\bar{\beta}_2$? It implies that both $a \succ c$ and $b \succ c$, but this is true both with $\succsim_1$, with $\succsim_3$ and with $\succsim_7$. So we will uniformly distribute its probability over these three possibilities. The final result is presented in the table.

In this way, it is possible to calculate the probability distribution of the probabilistic variable x_c that describes the travelling standard in the overall environment A. We obtain

$$P(x_c = 1) = 0.20,$$
$$P(x_c = 2) = 0.69,$$
$$P(x_c = 3) = 0.11.$$

On the other hand, if we apply the "product rule" to the initial distributions $P(x_c')$ and $P(x_c'')$, we obtain

$$P_{\text{prod}}(x_c = 1) = 0.21,$$
$$P_{\text{prod}}(x_c = 2) = 0.69,$$
$$P_{\text{prod}}(x_c = 3) = 0.10,$$

which is very close to what we have found, whilst a very different result is obtained by applying the "addition rule".

So we conclude that this approach essentially confirms the validity of the multiplicative approach and provide an additional argument to it, based on the notion of probabilistic measurement scale.

10.2.4 Summary of the Proposed Approach

To sum up, given a set of results $\{(\hat{x}_i, u_i), i = 1, \ldots, N\}$ from a key comparison over a stable travelling standard, we propose to process the data as follows.

10.2.4.1 Reliability Assessment of the Individual Results

Firstly, it is necessary to assess the reliability of the results presented by the participating NMIs. A reference acceptance probability value p_0 must be fixed prior to executing the exercise. This has to be agreed by the involved scientific and/or technological community. Typical values are 0.95 or 0.99. Then, a significance test can be performed according to formulae (10.4 and 10.5), where the acceptance regions A_i are defined according to (10.2). In the special important case where Gaussian distributions are assumed, and $p_0 = 0.95$ has been agreed, a simplified but reasonable expression for the acceptance regions is simply

$$A_i = [\hat{x}_i - 2u_i, \hat{x}_i + 2u_i]. \tag{10.7}$$

The results that do not pass the test should not be included in the successive evaluation of the distribution for the standard, but a degree of equivalence can be calculated for them also.

10.2.4.2 Assigning a Probability Distribution to the Standard

Let $\{(\hat{x}_i, u_i), i = 1, \ldots, M\}$, with $M \leq N$, be a consistent subset of results from the key comparison. Then, a probability distribution p_{ref} can be assigned to the unknown value of the travelling standard, according to the product rule, as follows:

$$p_{\text{ref}}(x) \propto \prod_{i=1}^{M} p_i(x), \tag{10.8}$$

where the proportionality implies that the resulting distribution must be scaled to integrate to one. In the case of Gaussian assumption, this reduces to the distribution of the weighted mean, that is, we have,

$$\hat{x} = \frac{\sum_{i=1}^{M} \hat{x}_i / u_i^2}{\sum_{i=1}^{M} 1/u_i^2} \tag{10.9}$$

and

$$1/u_{\hat{x}}^2 = \sum_{i=1}^{M} 1/u_i^2. \tag{10.10}$$

10.2.4.3 Determining the Degree of Equivalence

Lastly, the degree of equivalence can be obtained, as required by the MRA, by calculating the differences between each result and the reference value. The ith difference is defined by:

$$d_i = \hat{x}_i - \hat{x}. \tag{10.11}$$

Its standard uncertainty, in the Gaussian case, is [5]

$$u_{d_i}^2 = u_i^2 - u_{\hat{x}}^2, \tag{10.12}$$

and the expanded uncertainty can often be simply assigned as

$$U_{d_i} = 2u_{d_i}. \tag{10.13}$$

10.2.5 A Working Example

We show now an application of the method to key comparison CCL-K1, also discussed in Ref. [9]. This was a length comparison involving ten steel gauge blocks and ten tungsten carbide gauge blocks. As in the mentioned paper, we consider only the results for a tungsten carbide block of nominal length 1 mm. The results, expressed as measured differences with the nominal value of 1 mm, $\hat{x}_i$, with the related standard uncertainty, u_i, are given in Table 10.4, columns 2–3.

We first calculate the median of the measurement value, and we obtain $x_0 = 15.0$. By performing the significance test, with $p_0 = 0.95$, we exclude results 6 and 7. With this reduced set, we calculate the probability distribution for the reference standard, reported in Fig. 10.1.

From this distribution, it is possible to calculate the measurement value and the standard uncertainty for the standard. We obtain

- $\hat{x} = 20.3$ nm and
- $u_x = 2.9$ nm,

and the related distances, d_i, with their uncertainties, U_{d_i}, calculated according to formulae 10.11, 10.12 and10.13 and reported in columns 4–5 in the table. These

Table 10.4 Results from key comparison CCL-K1

NMI index i	$\hat{x}_i$/nm	u_i/nm	d_i/nm	U_{d_i}/nm
1	15.0	9.0	− 5.3	17.0
2	15.0	14.0	− 5.3	24.7
3	30.0	10.0	+ 9.7	19.1
4	18.0	13.0	− 2.3	25.3
5	24.0	9.0	+ 3.7	17.0
6	−9.0	7.0	−29.3	15.2
7	−9.0	8.0	−29.3	17.0
8	33.0	9.0	+12.7	17.0
9	12.5	8.6	− 7.8	16.2
10	8.8	10.0	−11.5	19.1
11	21.0	5.4	+0.7	9.1

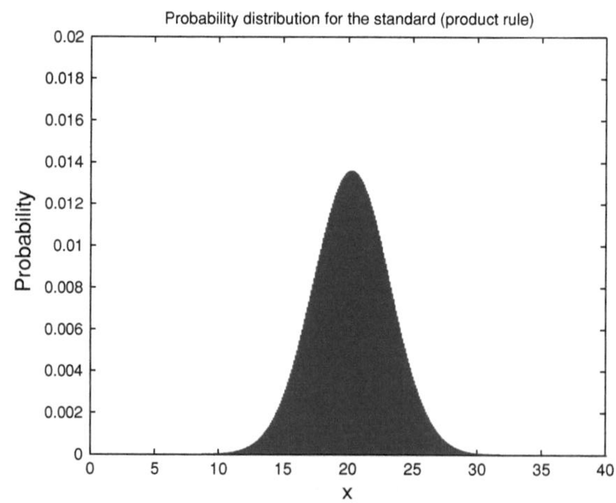

Fig. 10.1 Probability distribution for the standard. Precisely, this is the distribution for the variable x, discretised with quantisation interval $q = 0.1$ nm. Then, the ordinates sum to 1. For obtaining a (more common) probability density function, the ordinates should be divided by q: remember discussion in Sect. 4.1.8

results are in essential agreement with what was established in the quoted paper, apart from the fact that we suggest a different strategy for assessing the reliability of the individual results.

10.3 Calibration

Once the primary reference scales are established by the NMIs, it is necessary that the measuring devices are calibrated with respect to them, either directly or through intermediate steps. Devices that are calibrated directly with respect to the primaryrealisation of the scale are usually called secondary standard, and so forth.

Usually, there is a national network of calibration laboratories that are accredited by the national accreditation system and whose competence is recognised also beyond the national boundaries, thanks to recognition agreements similar to the already mentioned MRA.

Calibration is therefore the second key tool for ensuring the quality of the overall system of metrology and measurement, and thus, it is important to understand at least its basic principles. Noteworthy in this regard is the book by Dietrich [10]; modern textbooks on measurement also feature chapters on calibration [11, 12], and modern monographs on measurement also emphasise this topic [13].

From a theoretic standpoint, it is useful to understand calibration in a wide sense as a complex of modelling and experimentation performed in order to characterise a measuring system. As such, it provides vital information for the measurement processes that are based on that measuring system.[5] Due to the variety of the measuring systems, we can consider static and dynamic calibration, and we can consider one-dimensional or multidimensional systems. For the sake of simplicity, let us focus on one-dimensional static calibration, but most of the ideas we will discuss can be, at least in principle, extended to more complex cases.

In terms of the probabilistic model developed in Chap. 5, we can compactly say that calibration is the complex of operations aimed at obtaining the conditional distribution that characterises observation, that can be written, in its most general form, as

$$p(\mathbf{y}|\mathbf{x}, \boldsymbol{\theta}). \tag{10.14}$$

For static calibration, the basic experiment consists in inputting the system with a series of standard objects[6] that realises a series of states of the quantity under investigation, whose values are known with low (often negligible) uncertainty, and in recording the corresponding steady-state responses of the instrument. Thus, in our usual notation, the data obtainable by calibration consist in a series of pairs, $\{(x_i^*, y_i^*), i = 1, \ldots, n\}$, where the asterisk highlights the fact that such values are *known* at the end of the calibration experiment. Then, some model is assumed for describing the behaviour of the measuring system. Often, but not necessarily, a linear model is adopted. Such a model will depend upon some parameters, and then, the goal of calibration is to obtain an estimate of them. We can formalise all this by considering a special interpretation of expression (10.14) where now both $\mathbf{x} = \mathbf{x}^*$ and $\mathbf{y} = \mathbf{y}^*$ are known and $\boldsymbol{\theta}$ is now the vector of the parameters to be estimated. Then, we can obtain a probabilistic estimate of the required parameters by the Bayes–Laplace rule, as

$$p(\boldsymbol{\theta}|\mathbf{x}^*, \mathbf{y}^*) = \frac{p(\mathbf{y}^*|\mathbf{x}^*, \boldsymbol{\theta})p(\boldsymbol{\theta})}{\int_{\Theta} p(\mathbf{y}^*|\mathbf{x}^*, \boldsymbol{\theta})p(\boldsymbol{\theta})\mathrm{d}\boldsymbol{\theta}}. \tag{10.15}$$

[5] As already noted, we distinguish between a measuring system and a measurement process, since the same measuring system usually can be employed in different measurement conditions, thus giving rise to a plurality of measurement processes.

[6] Remember that the term "object" has to be understood in a wide sense and does not need to be a material object. For example, in the calibration of phonometers, it can be a standard sound.

Once such a probabilistic estimate is obtained, it is possible to fully define the required distribution (10.14), as a function of all pairs $(\mathbf{x}, \mathbf{y})$ in the measuring range. Stated in such general terms, the discourse may sound quite abstract, so let us illustrate it by a simple example.

Consider a linear, measuring device, such as a contact thermometer, whose steady-state behaviour is described by the input–output equation:

$$y = kx + w, \tag{10.16}$$

where x is input temperature, k is the sensitivity of the device, y is the output voltage, and w is the voltage noise in the measuring chain [14]. The system can be calibrated by putting the sensor in a bath where a series of different thermal states are realised; the temperatures of which are accurately measured by a reference platinum thermometer [13]. The data set $\{(x_i, y_i), i = 1, \ldots, n\}$ is acquired, where we have omitted the asterisks for the sake of simplicity. If we assume that the observation noise is normally distributed, introducing as usually the standard Gaussian distribution

$$\varphi(\xi) = (2\pi)^{-1/2} \exp(-\xi^2/2), \tag{10.17}$$

we can express the distribution that characterises observation as

$$p(\mathbf{y}|\mathbf{x}, k, \sigma) = \prod_i \sigma^{-1} \varphi[\sigma^{-1}(y_i - kx_i)], \tag{10.18}$$

where σ denotes the standard deviation of w. Let us now introduce in this expression the following (sufficient) statistics [15]:

$$\hat{k} = \sum_i x_i y_i / \sum_i x_i^2, \tag{10.19}$$

$$\hat{\sigma}^2 = (n-1)^{-1} \sum_i (y_i - \hat{k} x_i)^2. \tag{10.20}$$

After some calculation, we obtain

$$p(\mathbf{y}|\mathbf{x}, k, \sigma) = (2\pi)^{-n/2} \sigma^{-n} \exp\left\{ -\frac{1}{2\sigma^2} \left[(n-1)\hat{\sigma}^2 + (k - \hat{k})^2 \sum_i x_i^2 \right] \right\}. \tag{10.21}$$

After assuming non-informative priors for k and σ, we reach the final joint distribution:

$$p(k, \sigma|\mathbf{y}, \mathbf{x}) \propto \sigma^{-(n+1)} \exp\left\{ -\frac{1}{2\sigma^2} \left[(n-1)\hat{\sigma}^2 + (k - \hat{k})^2 \sum_i x_i^2 \right] \right\}. \tag{10.22}$$

From this distribution, the marginals for k and σ can be obtained. For writing them compactly, we recall the *t-Student* distribution, with ν degrees of freedom:

$$\psi(\xi;\nu) \propto \left(1+\frac{\xi^2}{\nu}\right)^{-\frac{\nu+1}{2}}, \tag{10.23}$$

and the *inverse gamma*:

$$\zeta(\xi;\alpha,\beta) \propto \xi^{-(\alpha+1)} exp(-\beta/\xi), \tag{10.24}$$

and we finally obtain, ignoring from now on the dependence on $(\mathbf{x},\mathbf{y})$ [16],

$$p(k) = \psi\left(\frac{k-\hat{k}}{\hat{\sigma}^2 \sum_i x_i^2}; n-1\right), \tag{10.25}$$

$$p(\sigma) = \zeta\left(\hat{\sigma}^2; \frac{n}{2}-1, \frac{1}{2}(n-1)\hat{\sigma}^2\right). \tag{10.26}$$

Calibration provides all the information required to use the instrument in operating and environmental conditions equivalent to the calibration ones. This happens, e.g. when measurement takes place in the same or in a similar laboratory and when the measurand has a comparable definition uncertainty. Then, the conditional distribution for (a single) observation is [15]:

$$p(y|x) = \int_{K\Sigma} p(y|x,k,\sigma)p(k,\sigma)\mathrm{d}k\mathrm{d}\sigma = \psi\left(\frac{y-kx}{\hat{\sigma}\sqrt{1+x^2/\sum_i x_i^2}}; n-1\right). \tag{10.27}$$

If the operating conditions are instead different, such differences should be properly accounted for according to general guidelines for uncertainty evaluation, as discussed in Chap. 9.[7]

To sum up, the probabilistic framework developed in this book allows treating measurement and calibration in a fully consistent way and information coming from calibration can be immediately transferred in the characterisation of the observation process and consequently used for restitution [14]. A numerical example of this was provided in Sect. 8.2.4, concerning the application of the master scaling method to the measurement of loudness. In that case, the "instrument" to be "calibrated" was a person. Readers are encouraged to try and apply these ideas to cases of their concern. Feedback and comments will be welcome.

[7] See, e.g., Ref. [14] for an example of how to combine information from calibration with information on the measurement environment.

References

1. BIPM.: Mutual Recognition. STEDI, Paris (2008)
2. BIPM.: Guidelines to CIPM Key Comparisons. (2003)
3. EURACHEM.: EURACHEM/CITAC guide CG 4: quantifying uncertainty in analytical measurement. (2000)
4. Rossi, G.B., Berglund, B.: Measurement of quantities involving human perception and interpretation. Measurement **44**, 815–822 (2011)
5. Cox, M.G.: The evaluation of key comparison data: an introduction. Metrologia **39**, 587–588 (2002)
6. Cox, M.G.: The evaluation of key comparison data. Metrologia **39**, 589–595 (2002)
7. Willink, R.: Forming a comparison reference value from different distributions of belief. Metrologia **43**, 12–20 (2006)
8. Duewer, D.: How to combine results having stated uncertainties: to MU or not to MU? In: Fajgelj, A., Belli, M., Sansone, U. (eds.) Combining and Reporting Analytical Results, pp. 127–142. Royal Society of Chemistry, London (2007)
9. Cox, M.G.: The evaluation of key comparison data: determining the largest consistent subset. Metrologia **44**, 187–200 (2007)
10. Dietrich, C.F.: Uncertainty, Calibration and Probability. IOP, Bristol (2000)
11. Bentley, J.P.: Principles of Measurement Systems, 4th edn. Pearson Education Ltd., Harlow, Essex, UK (2005)
12. Morris, A.S., Langari, R.: Measurement and Instrumentation. Academic Press/Elsevier, Waltham (2012)
13. Nicholas, J.V., White, D.R.: Traceable temperatures. Wiley, Chichester (1994)
14. Rossi, G.B.: Measurement modelling: foundations and probabilistic approach. Paper presented at the 14th joint international IMEKO TC1+TC7+TC13 symposium, Jena, 31 August–2 September 2011
15. Press, S.J.: Bayesian Statistics. Wiley, New York (1989)
16. Gill, J.: Bayesian Methods. Chapman and Hall/CRC, Boca Raton (2002)

Chapter 11
Measurement-Based Decisions

11.1 The Inferential Process in Conformance Assessment

Measurement provides objective and reliable support to decision-making [1, 2]. In manufacturing, for example, it is necessary to check workpieces for conformance to their design [3–5]. In mass production, as occurs, for example, in the automotive field, parts are produced independently and then assembled. In order to assemble properly, it is necessary that critical dimensions and forms are kept under control. This often requires length measurement to be taken in the production line, on the basis of which decisions are automatically taken on acceptance or refusal of workpieces [6]. Due to measurement uncertainty, there is the risk of accepting non-conformal items (called *user's or consumer's risk*) or to reject good ones (called *producer's risk*). Both risks can be expressed in probabilistic terms, and their evaluation is a key factor for the design and the management of the production process [7]. Another example is the monitoring and control of physical agents, such as vibration, noise or electromagnetic fields, in working environments, to ensure safe working conditions [8, 9]. For ergonomic reasons, their intensity should be minimised, and in any case it should not exceed safety limits. The related quantities are thus measured, and their values are compared to threshold values established by safety regulations or ergonomic recommendations. Similarly, for health care the presence of pollutants in food, in air or in water must be assessed: this requires chemical analysis for verifying that the amount of pollutant is under safeguard limits. In the general perspective taken in this book, a chemical analysis may be regarded as a measurement. Decision on medical treatment is often based on biochemical analysis: the cholesterol rate in blood is an example known to many! Other critical decisions concern the forensic area, where objective evidence is searched for assessing the responsibility of persons in criminal acts. The list could be continued, but we think that we have given a sufficient feeling of the impact of measurement on critical decision-making.

In the following, we will consider a generic problem of *conformity assessment* that consists in assessing whether the values of some critical quantity, x, in a given situation, is or not in some safe region, A. In particular such a region may be an

G. B. Rossi, *Measurement and Probability*, Springer Series in Measurement Science and Technology, DOI: 10.1007/978-94-017-8825-0_11,

interval centred on a nominal value, x_0, that is, $[x_0 - a, x_0 + a]$, as it happens in dimensional testing in manufacturing, or from zero to a threshold value, $[0, a]$, as it happens in environmental monitoring. Extension to vector measurement will be also addressed.

11.2 A Probabilistic Framework for Risk Analysis

11.2.1 Insight into Conformance Assessment

We discuss now how to evaluate the risks related to decisions to be taken in conformity assessment [10]. Consider a production process $\mathcal{P}$ whose state is characterised by a positive-defined critical parameter, x, that could represent the concentration of a pollutant such as pesticide in bread.[1] Let the safe region be defined by

$$x \leq a, \tag{11.1}$$

that is the concentration of the pollutant must be less than some threshold value a, for example 2 mg kg^{-1}. For maximum simplicity, we assume by now that x can take only a finite number of discrete values: the generalisation to the continuous case is simple and will be presented at a later stage.

In this evaluation, we have to combine information, usually expressed by probability distributions, both on the production and the measurement process: when necessary, we will use the subscript $\mathcal{P}$ for the former and $\mathcal{M}$ for the latter. Often the available information on the process, based on historical records, may be summarised by a probability distribution $P_{\mathcal{P}}(x)$—or $P(x)$ for short, when the interpretation is not dubious—of x taking any possible value: a very simple example, intended for illustrative purposes only, is presented in Fig. 11.1a. Here, x can take just values 1 or 2, in some arbitrary units, and

$$P(x = 1) = 0.9,$$
$$P(x = 2) = 0.1,$$

that is, we have typically 10 % defective items. Let the threshold be $a = 1.5$. If we are able to detect any violation of condition 1, we can take appropriate actions, otherwise we will suffer negative consequences. Let us then discuss the associated risks and how measurement can help to reduce them. We distinguish between *consumer's risk*, the risk of not detecting a violation of condition (11.1), and *producer's risk*, the risk of rejecting an item, when in reality condition (11.1) is not violated. Moreover, in both cases, we consider the *specific risk*, related to a single item, and the *global risk*, related to the entire process: so in total we have four types of risk.

[1] A numerical example will be provided in Sect. 11.4.

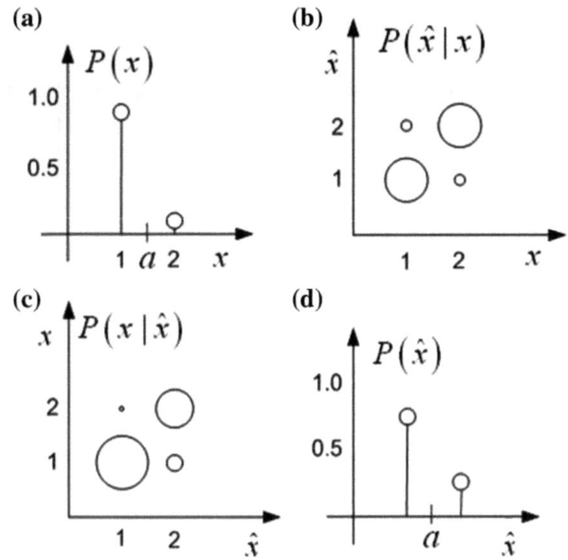

Fig. 11.1 Introductory example. Involved probability distributions: **a** $P(x)$ **b** $P(\hat{x}|x)$ **c** $P(x|\hat{x})$ **d** $P(\hat{x})$

As an extreme case, if we take no measurement and, consequently, we do not reject any item, the global consumer's risk, R_0, that the value of the parameter is greater than the threshold, is, in our example (see Fig. 11.1a),

$$R_0 = P(x > a) = P(x > 1.5) = 0.1,$$

whilst the global producer's risk, R_0^*, is null since no item is rejected: in general the two risks are conflicting. The goal of a selection process based on measurement is to reduce the consumer's risk without raising too much the producer's risk. If we measure the critical parameter x and obtain the measurement value $\hat{x}$, we can adopt the acceptance rule

$$\hat{x} \leq 1.5, \tag{11.2}$$

which constitutes the practical application of (11.1). In this way, the risk will be reduced but not totally eliminated due to uncertainty, that is due to the fact that in general

$$\hat{x} \neq x. \tag{11.3}$$

We can thus consider the following risks. If the measurement result $\hat{x}$ is in the acceptance region, the *(specific) consumer's risk* that the value x of the parameter is outside that region is

$$R(\hat{x}) = P(x > a|\hat{x}). \quad (11.4)$$

Instead, if the measurement result $\hat{x}$ is out of the acceptance region, the (*specific*) *producer's risk* that the value x of the parameter is in that region is

$$R^*(\hat{x}) = P(x \leq a|\hat{x}). \quad (11.5)$$

The *global consumer's risk* is the average consumer's risk associated to the monitoring of the process:

$$R = \sum_{\hat{x} \leq a} R(\hat{x})P(\hat{x}), \quad (11.6)$$

whilst the *global producer's risk* is, similarly:

$$R^* = \sum_{\hat{x} > a} R^*(\hat{x})P(\hat{x}). \quad (11.7)$$

Let us now practice evaluating such risks in our example. For doing so, we need information on the measurement process, which is synthesised by the probability distribution $P(\hat{x}|x)$ that we have defined and discussed in Chap. 5.[2] In our example, we assume the following, very simple, distribution:

$P(\hat{x} = 1|x = 1) = 0.8,$
$P(\hat{x} = 2|x = 1) = 0.2,$
$P(\hat{x} = 1|x = 2) = 0.2,$
$P(\hat{x} = 2|x = 2) = 0.8,$

reported in Fig. 11.1b, and we have to combine it with information on the production process. We may thus calculate the joint probability distribution

$$P(x, \hat{x}) = P(\hat{x}|x)P(x). \quad (11.8)$$

This distribution contains all the needed information for risk assessment. In our example, it is

$P(x = 1, \hat{x} = 1) = 0.72,$
$P(x = 1, \hat{x} = 2) = 0.18,$
$P(x = 2, \hat{x} = 1) = 0.02,$
$P(x = 2, \hat{x} = 2) = 0.08.$

It is also useful to calculate the marginal with respect to $\hat{x}$:

$P(\hat{x} = 1) = 0.74,$
$P(\hat{x} = 2) = 0.26,$

[2] Recall in particular Sects. 5.3, 5.6 and 5.7.

shown in Fig. 11.1d. We are now ready to evaluate the risks. Suppose for example that we obtain $\hat{x} = 1$. In this case, we will accept the item and the related consumer's risk is given by (11.4). Since

$$P(x > a|\hat{x}) = \sum_{x>a} P(x|\hat{x}), \tag{11.9}$$

for this calculation, we need the conditional distribution $P(x|\hat{x})$ that can be obtained as:

$$P(x|\hat{x}) = \frac{P(x, \hat{x})}{P(\hat{x})} \tag{11.10}$$

and results (approximately) as follows:

$$\begin{aligned}
&P(x = 1|\hat{x} = 1) = 0.97,\\
&P(x = 2|\hat{x} = 1) = 0.03,\\
&P(x = 1|\hat{x} = 2) = 0.31,\\
&P(x = 2|\hat{x} = 2) = 0.69,
\end{aligned}$$

and is reported in Fig. 11.1c. So, for $\hat{x} = 1$, we obtain

$$R(\hat{x} = 1) = P(x > 1.5|\hat{x} = 1) = P(x = 2|\hat{x} = 1) = 0.03.$$

The global consumer's risk is

$$R = \sum_{\hat{x} \le a} R(\hat{x})P(\hat{x}) = R(\hat{x} = 1)P(\hat{x} = 1) = 0.03 \times 0.74 = 0.02.$$

The calculation of the producer's risks proceeds in a similar way. Suppose that we obtain $\hat{x} = 2$: then the specific risk is

$$R^*(\hat{x} = 2) = P(x = 1|\hat{x} = 2) = 0.69$$

and the global risk is

$$R^* = \sum_{\hat{x} > a} R^*(\hat{x})P(\hat{x}) = P(x = 1|\hat{x} = 2)P(\hat{x} = 2) = 0.69 \times 0.26 = 0.18.$$

It is interesting to consider what happens if the measurement process is more accurate, for example if it is characterised by the following conditional distribution:

$$\begin{aligned}
&P(\hat{x} = 1|x = 1) = 0.9,\\
&P(\hat{x} = 2|x = 1) = 0.1,\\
&P(\hat{x} = 1|x = 2) = 0.1,\\
&P(\hat{x} = 2|x = 2) = 0.9.
\end{aligned}$$

Proceeding in the same way, we obtain:

$$\begin{aligned}
&R(\hat{x} = 1) = 0.01,\\
&R = 0.01,\\
&R^*(\hat{x} = 2) = 0.5,\\
&R^* = 0.09.
\end{aligned}$$

We note that there is an improvement of all the risk figures. On the other hand, we most probably have an increase in the cost of the measurement process, so some trade-off is usually required. A brief introduction to cost analysis will be provided at a later stage. Prior to that, we have to generalise what we have so far presented.

11.2.2 Probabilistic Framework

For maximum generality, we consider continuous vector measurement.

Let then $\mathcal{P}$ be a production process characterised by the parameter vector $\mathbf{x} \in \mathbf{X}$, governed by the probability distribution $p(\mathbf{x})$. Let $\mathbf{A} \subset \mathbf{X}$ be a safe region in the parameter space and $\mathbf{B} \subseteq \mathbf{A}$ an acceptance region: in fact it is common practice to restrict the acceptance region with respect to the safe region, in order to reduce the consumer's risk. For example, in the case of a scalar positive-defined parameter, subject to a threshold condition, $A = \{x|x \leq a\}$ and $B = \{x|x \leq b\}$, with $b = \kappa a$, where $\kappa \leq 1$ is a safeguard factor.

Let then $\mathcal{M}$ be a measurement process for $\mathbf{x}$, characterised by the distribution $p(\hat{\mathbf{x}}|\mathbf{x})$. The reference distribution for risk evaluation is the joint distribution for $\mathbf{x}$ and $\hat{\mathbf{x}}$ that *combines information on the production process and information on the measurement process*:

$$p(\mathbf{x}, \hat{\mathbf{x}}) = p(\hat{\mathbf{x}}|\mathbf{x})p(\mathbf{x}). \tag{11.11}$$

It is then possible to calculate

$$p(\hat{\mathbf{x}}) = \int_{\mathbf{X}} p(\mathbf{x}, \hat{\mathbf{x}})\mathrm{d}\mathbf{x}, \tag{11.12}$$

and

$$p(\mathbf{x}|\hat{\mathbf{x}}) = p(\mathbf{x}, \hat{\mathbf{x}})[p(\hat{\mathbf{x}})]^{-1}. \tag{11.13}$$

Then all the required risks can be calculated on the basis of this probabilistic framework. For $\hat{\mathbf{x}} \in \mathbf{B}$, the *specific consumer's risk* is

$$R(\hat{\mathbf{x}}) = P(\mathbf{x} \in \mathbf{X} - \mathbf{A}|\hat{\mathbf{x}}) = \int_{\mathbf{X}-\mathbf{A}} p(\mathbf{x}|\hat{\mathbf{x}})\mathrm{d}\mathbf{x}, \tag{11.14}$$

whilst for $\hat{\mathbf{x}} \in \mathbf{X} - \mathbf{B}$ such risk is null.

The *global consumer's risk* is

$$R = P(\mathbf{x} \in \mathbf{X} - \mathbf{A}, \hat{\mathbf{x}} \in \mathbf{B}) = \int_{\mathbf{B}} \int_{\mathbf{X}-\mathbf{A}} p(\mathbf{x}, \hat{\mathbf{x}})\mathrm{d}\mathbf{x}\mathrm{d}\hat{\mathbf{x}}. \tag{11.15}$$

On the other hand, for $\hat{\mathbf{x}} \in \mathbf{X} - \mathbf{B}$, the *specific producer's risk* is

$$R^*(\hat{\mathbf{x}}) = P(\mathbf{x} \in \mathbf{A}|\hat{\mathbf{x}}) = \int_{\mathbf{A}} p(\mathbf{x}|\hat{\mathbf{x}})\mathrm{d}\mathbf{x}, \quad (11.16)$$

whilst for $\hat{\mathbf{x}} \in \mathbf{B}$ such risk is null.

Lastly, the *global producer's risk* is

$$R^* = P(\mathbf{x} \in \mathbf{A}, \hat{\mathbf{x}} \in \mathbf{X} - \mathbf{B}) = \int_{\mathbf{X}-\mathbf{B}} \int_{\mathbf{A}} p(\mathbf{x}, \hat{\mathbf{x}})\mathrm{d}\mathbf{x}\mathrm{d}\hat{\mathbf{x}}. \quad (11.17)$$

Let us practice applying these formulae to a simple but important example.

11.2.3 Illustrative Example

Consider the inspection of a workpiece having some critical length x to be checked for conformance with a tolerance region $A = [x_0 - a, x_0 + a]$, $a > 0$, and let the acceptance region be $B = [x_0 - b, x_0 + b]$, $0 < b \leq a$.[3] Suppose also that the production process may be described by a Gaussian probability distribution centred on the nominal value x_0, with variance σ^2. Let the measurement process be Gaussian also, with standard uncertainty u. Let φ denote, as usual, the normalised Gaussian distribution.

Then

$$p(x) = \sigma^{-1}\varphi[\sigma^{-1}(x - x_0)] \quad (11.18)$$

and

$$p(\hat{x}|x) = u^{-1}\varphi[u^{-1}(\hat{x} - x)]. \quad (11.19)$$

It is immediate to obtain the joint distribution $p(x, \hat{x})$ by the factorisation [12]

$$p(x, \hat{x}) = p(\hat{x}|x)p(x) = \sigma^{-1}u^{-1}\varphi[\sigma^{-1}(x - x_0)]\varphi[u^{-1}(\hat{x} - x)]. \quad (11.20)$$

It is instructive to consider the alternative factorisation $p(x, \hat{x}) = p(x)p(x|\hat{x})$ also. Firstly, we obtain

$$p(\hat{x}) = \sigma'^{-1}\varphi[\sigma'^{-1}(\hat{x} - x_0)], \quad (11.21)$$

[3] In this and in the following numerical examples, we mention some of the results that we extensively presented in Ref. [10]. Readers are referred to that paper for probing this subject further. The basic assumptions for this first example are taken from Ref. [11], a well and informative paper that we also recommend to read in full.

where $\sigma'^2 = u^2 + \sigma^2$ [11]. This formula shows that the distribution of the measurement value, $\hat{x}$, is the convolution of the distribution of the parameter and of the measurement "error",[4] $e = \hat{x} - x$, so that its variance, σ'^2, is the sum of the variance of the process, σ^2, and of the variance associated to measurement uncertainty, u^2, centred on the average value of the parameter, x_0. Furthermore, let us introduce the weighted mean of x:

$$\tilde{x} = \frac{\sigma^2}{u^2 + \sigma^2}\hat{x} + \frac{u^2}{u^2 + \sigma^2}x_0 \tag{11.22}$$

which is a function of $\hat{x}$, and its variance,

$$\tilde{\sigma}^2 = \frac{u^2\sigma^2}{u^2 + \sigma^2}. \tag{11.23}$$

After some calculation, we obtain

$$p(x|\hat{x}) = \tilde{\sigma}^{-1}\varphi[\tilde{\sigma}^{-1}(x - \tilde{x})]. \tag{11.24}$$

Thus the final distribution for x, given $\hat{x}$, is a Gaussian one, centred in the weighted mean of x, $\tilde{x}$, and having the variance of the weighted mean, $\tilde{\sigma}^2$. This is not surprising, since this estimate accounts both for the initial distribution, $p(x)$ and for the distribution related to measurement uncertainty, which are both Gaussian, with a different variance. So the resulting final distribution is properly associated to the weighted mean. The joint distribution can be then also factorised as follows:

$$p(x, \hat{x}) = p(x|\hat{x})p(\hat{x}) = \tilde{\sigma}^{-1}\sigma'^{-1}\varphi[\tilde{\sigma}^{-1}(x - \tilde{x})]\varphi[\sigma'^{-1}(\hat{x} - x_0)]. \tag{11.25}$$

It is now possible to calculate the various risks. For the specific consumer's risk, we obtain:

$$\begin{aligned} R(\hat{x}) &= \int_{-\infty}^{x_0-a} p(x|\hat{x})\mathrm{d}x + \int_{x_0+a}^{\infty} p(x|\hat{x})\mathrm{d}x \\ &= \int_{-\infty}^{x_0-a} \tilde{\sigma}^{-1}\varphi[\tilde{\sigma}^{-1}(x - \tilde{x})]\mathrm{d}x + \int_{x_0+a}^{\infty} \tilde{\sigma}^{-1}\varphi[\tilde{\sigma}^{-1}(x - \tilde{x})]\mathrm{d}x. \end{aligned} \tag{11.26}$$

If we consider the cumulative standard Gaussian distribution, Φ, we can develop this further as

$$R(\hat{x}) = \Phi[\tilde{\sigma}^{-1}(x_0 - a - \tilde{x})] + 1 - \Phi[\tilde{\sigma}^{-1}(x_0 + a - \tilde{x})]. \tag{11.27}$$

---

[4] See Ref. [12] for a discussion of the interpretation of $e = \hat{x} - x$ as the measurement error, in particular Sect. 3 and footnote 10 in that paper.

Table 11.1 Quality control of workpieces: risk analysis

Case	$\frac{\sigma}{a}$	$\frac{u}{a}$	$R\ (b = a)$	$R\ (b = a - 2u)$	$R^*\ (b = a)$	$R^*\ (b = a - 2u)$
A	$\frac{\sqrt{15}}{8}$	$\frac{1}{8}$	7.1×10^{-3}	1.7×10^{-4}	1.4×10^{-2}	9.5×10^{-2}
B	$\frac{\sqrt{3}}{6}$	$\frac{1}{6}$	1.8×10^{-4}	5.7×10^{-6}	2.4×10^{-3}	4.5×10^{-2}

For the global risk, we obtain

$$R(\hat{x}) = \int_{x_0-b}^{x_0+b} \left[\int_{-\infty}^{x_0-a} p(x, \hat{x}) \mathrm{d}x + \int_{x_0+a}^{\infty} p(x, \hat{x}) \mathrm{d}x \right] \mathrm{d}\hat{x}. \tag{11.28}$$

Similar equations can be obtained for producer's risks.

Let us now demonstrate the application of these results to a simple numerical example [11]. We consider two cases:

(a) 95 % of workpieces fall in the tolerance (safe) region: this means that the semi-width of the tolerance region, a, equals twice the standard deviation of the production process: $a = 2\sigma'$; we also assume a "gauging ratio"[5] 4:1, that is $U = 2u = \frac{1}{4}a$;
(b) 99 % of workpieces in the tolerance region, that is, $a = 3\sigma'$ and a "gauging ratio" 3:1, that is $U = 2u = \frac{1}{3}a$.

With these assumptions, we can calculate both the ratio σ/a that describes the behaviour of the production process and the ratio u/a that characterises the measurement process. In case A, for example, we have

- $\sigma' = a/2$,
- $u = a/8$,
- $\sigma = (\sigma'^2 + u^2)^{1/2} = (15)^{1/2}/8$,

and similar calculations hold true for case B. Qualitatively, in case A the degree of control of the production process is lower than in case B. To partially compensate for that a higher gauging factor is established, in order to make the inspection process more effective.

For both cases, we considered two inspection strategies: one in which no safeguard factor is assumed, that is $b = a$, the other, more cautionary, in which the acceptance region is reduced according to the expanded uncertainty, $U \cong 2u$, that is $b = a - 2u$. The corresponding numerical results are summarised in Table 11.1, where it is possible to appreciate quantitatively the combined effect of all these parameters, in terms of global consumer's and producer's risks.

[5] The gauging ratio is a parameter that relates measurement uncertainty to the characteristics of the production process, here summarised by parameter a. A high gauging factor is typical of an accurate inspection process.

Table 11.2 Quality control of workpieces: cost analysis

Case	Cost, for $b = a$ (%)	Cost, for $b = a - 2u$ (%)
A	12.0	9.8
B	0.5	4.5

The comparison of different strategies is even more effective if a cost analysis is performed [5]. Let C_1 be the cost of accepting a wrong item, whilst C_2 is the cost of rejecting a good one. Then the expected cost per item is

$$C = C_1 R + C_2 R^*. \tag{11.29}$$

Basically, C_2 is the cost of the item, whilst C_1 is more difficult to evaluate, since it includes a provision of all the consequences that may derive from this wrong decision. Anyway C_2 may be in general expected to be much greater than C_1. To carry out our example, let us assume the ratio $C_1/C_2 = 15$ as in [11]. The costs for different strategies are presented in Table 11.2. Now the two kinds of risk are merged in terms of overall cost, so that each situation is characterised by a single feature and comparison is even more apparent.

11.3 Software for Risk Analysis

The application of this probabilistic approach to complex real-life cases requires some software. In our Measurement Laboratory, we developed a package called MEAS RISK [10]. It basically implements formulae 11.14–11.17, for *vector quantities*. Similarly to the code UNCERT, discussed in Chap. 9, it considers such quantities as properly discretised and thus treatable as discrete probabilistic variables. Thus, for a vector quantity $x = [x_1, x_2, ...x_M]$, the code expects, as input, for each component x_i, the production process probability distribution, $P_{\mathcal{P}}(x_i)$, the related measurement probability conditional distribution $P_{\mathcal{M}}(\hat{x}_i|x_i)$ and the parameters, a_i and b_i, which define the safe regions. Then it calculates the required vector probability distributions, $P(\mathbf{x}, \hat{\mathbf{x}})$ and $P(\mathbf{x}|\hat{\mathbf{x}})$, assuming the independence of the individual components, and then the local and global risks, according to a discrete version of formulae (11.14)–(11.17) as illustrated in Fig. 11.2.

11.4 Chemical Analysis

A very important case of vector measurement is chemical analysis. In such analysis, the concentration of one or more components in some substance is measured. If such components are pollutants, it is necessary to test whether their content is or is not

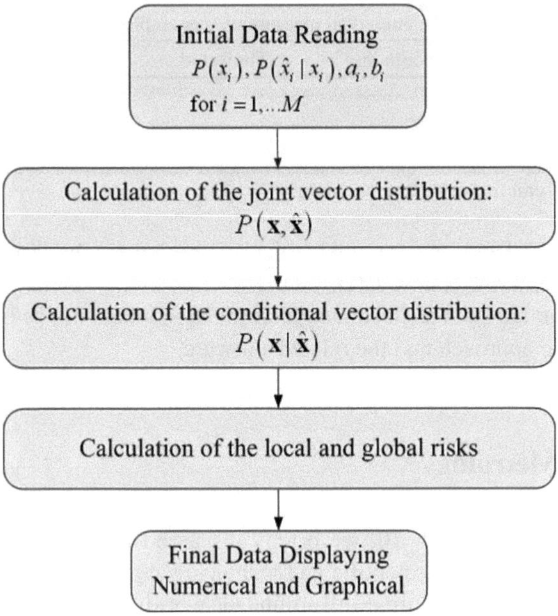

Fig. 11.2 Flowchart of the Code MEAS RISK

below some safe threshold. Since such concentrations are usually small, the impact of measurement uncertainty on decisions is often non-negligible. In such cases, the probabilistic approach is particularly effective. To give a feeling of this, we now briefly discuss the determination of organo-phosphorous pesticides in bread, taking numerical data from the Eurachem Guide on uncertainty in analytical measurement [13]. A complete account of this study of ours can be found in Ref. [10], to which the reader is referred for probing this subject further.

Consider the simultaneous measurement of the concentration of three pesticides:

(a) Clorpirifos-Metile,
(b) Pirimifos-Metile,
(c) Malation.

The vector measurand is thus denoted by $\mathbf{x} = [x_a, x_b, x_c]$. We assume for the production process a uniform distribution in the neighbourhood of the threshold. Let the measurement uncertainties, thresholds and probability distributions for the three pollutants be as in Table 11.3.

By performing calculations with the Code MEAS RISK, we obtain, for the global user's risk, $R = 0.01$ and, for the global consumer's risk, $R^* = 0.01$.

Suppose now that we perform an analysis and we obtain $\hat{\mathbf{x}} = [1.7\ 1.7\ 0.7]$. This result falls inside the safe region. We thus calculate the specific consumer's risk for this result, and we obtain $R(\hat{\mathbf{x}}) = 0.16$. On the other hand, suppose the result is

Table 11.3 Data concerning the analytical measurement example

Component	Standard uncertainty	Threshold	Probability distribution
a	0.25	2	Uniform over the range [0, 4]
b	0.21	2	Uniform over the range [0, 4]
c	0.21	1	Uniform over the range [0, 2]

The measurement unit for concentration is mg kg^{-1}

$\hat{\mathbf{x}} = [1.7\ 1.7\ 1.3]$. Here, we are outside the safe region for the last component: the specific producer's risk is now $R^*(\hat{\mathbf{x}}) = 0.10$.

It is apparent that such an accurate risk evaluation would not be possible without the probabilistic approach and the related software.

11.5 Legal Metrology

In Sect. 3.7.4 and in Chap. 10, we briefly discussed the International system of metrology which provides scientific and technical support virtually to all the measurements that are daily performed all around the world. An important subset of these measurements is also of legal concern. They include "those responding for reasons of public interest, public health, safety and order, protection of the environment and the consumer, of levying taxes and duties and of fair trading, which directly and indirectly affect the daily life of citizens in many ways" [14]. Examples of instruments involved in such measurements are water and gas meters, electrical-energy meters, heat meters, automatic weighing instruments, taximeters and exhaust gas analysers. This is the area of *legal metrology* [15]. For supporting such measurements and their legal control, there is a somewhat parallel international structure, the International Organization of Legal Metrology (OIML) which is an intergovernmental treaty organization, established in 1955 in order to promote the global harmonisation of legal metrology procedures. The organisational structure includes the Bureau International de Métrologie Légale (BIML), based in Paris, which is the Secretariat and Headquarters of the OIML, the International Committee (CIML), which is composed of one appointed representative from each Member State and acts as the steering committee, meeting annually to review the Organization's technical progress and administrative operations, and Technical Committees and Subcommittees.

The general philosophy of legal metrological control has evolved over the years. In Europe, for example, the quite recent Directive 2004/22/EC on measuring instruments (MID) [14] introduced new ideas and criteria.

The previous legislation included a Framework Directive (71/316/CEE) and a set of specific directives, for the various classes of instruments. The need for a change emerged for two main reasons. Firstly, the requirements were often formulated in terms of existing technologies and were thus unable to follow the, sometimes rapid, evolution of instrumentation. Secondly, conformity assessment was essentially based on a third party verification on the final product, whilst the current wide diffusion of quality assurance in production processes allows other strategies to be considered. In the new approach instead the legislation only considers essential requirements, which are formulated as functional requirements independent from the measurement technology adopted. Furthermore, manufacturers can choose amongst different conformance assessment procedures, those best suited to their production organisation.

In general, this approach assigns a greater responsibility to instrument producers, in view of a more competitive market. Thus the use of sophisticated evaluation methods such as that presented in this book, can be particularly rewarding [15, 16]. Let us the illustrate this point in a practical test case.

11.6 A Working Example

The essential requirements established by the MID consider that "a measuring instrument shall provide a high level of metrological protection in order that any party affected can have confidence in the result of measurement, and shall be designed and manufactured to a high level of quality in respect of the measurement technology and security of the measurement data". As a consequence of this, a major concern is devoted to "allowable errors".[6] The MID requires that:

(a) "Under rated operating conditions and in the absence of a disturbance, the error of measurement shall not exceed the maximum permissible error (MPE) value as laid down in the appropriate instrument-specific requirements.
(b) Under rated operating conditions and in the presence of a disturbance, the performance requirement shall be as laid down in the appropriate instrument-specific requirements.
(c) The manufacturer shall specify the climatic, mechanical and electromagnetic environments in which the instrument is intended to be used, power supply and other influence quantities likely to affect its accuracy, taking account of the requirements laid down in the appropriate instrument-specific annexes".

Consider now the application of this approach to *water meters*, that is "instruments designed to measure, memorise and display the volume at metering conditions of water passing through the measurement transducer". Such devices are very common in daily life since they are used to measure water consumption and to bill the consumer accordingly. The main functional features for such devices include the flowrate, Q, which is further specified as minimum, Q_1, transitional, Q_2, permanent, Q_3, and overload, Q_4, and the water temperature range, typically below or beyond 30 °C. Requirements are specified according to operational conditions. Namely, the maximum permissible error, positive or negative, on volumes delivered at flowrates between Q_2 (included) and Q_4 is:

[6] In a more appropriate language, this concept should be expressed as "allowable (measurement) uncertainty". Otherwise, apart from language subtleties, the metrological requirements are stated in a sound way.

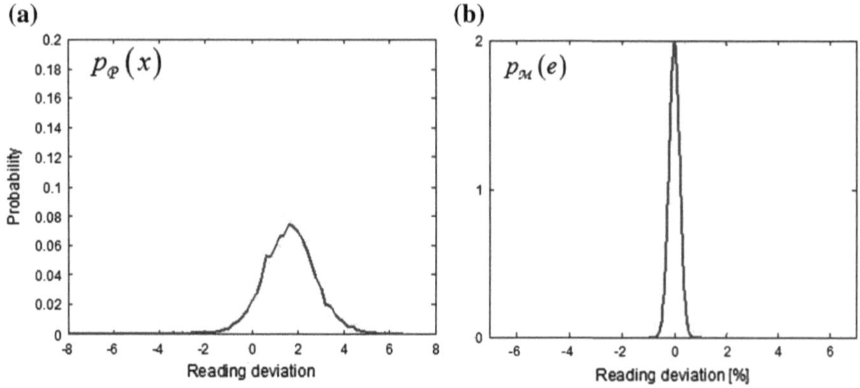

Fig. 11.3 Probability distribution for water meter conformity assessment, for the reading deviation (**a**) and for the testing error (**b**) [16]

2 % for water having a temperature $\leq$30 °C,
3 % for water having a temperature $>$30 °C,

whilst on volumes delivered at flowrates between Q_1 and Q_2 (excluded) is 5 % for water having any temperature.

In 2008, our Laboratory participated in a National research programme aimed at supporting the implementation of the MID. We studied water meters in particular and applied the probabilistic approach here considered. Results were published in Ref. [16]. We present here one such result to give a feeling for what can be achieved. The interested reader can consult the full paper for additional information.

The basis for risk evaluation includes the distribution that characterises the process,[7] $p_{\mathcal{P}}(x)$, where here x represents the *reading deviation* of the water meters, and the distribution $p_{\mathcal{M}}(\hat{x}|x)$ that characterises the testing process, where $\hat{x}$ denotes the measured deviation. Often the *testing error*,[8] $e \triangleq \hat{x} - x$ may be assumed to be independent from x, so that the (two-arguments) distribution $p_{\mathcal{M}}(\hat{x}|x)$ can be replaced by the simpler (one-argument) distribution $p_{\mathcal{M}}(e)$. In the study we performed, the distributions for the reading deviation were assigned based on historical data, obtained from testing boards, whilst those for the testing error were obtained from (testing) instrument data sheets complemented with information from the technical literature. An example of such distributions, for one class of water meters at minimum flow rate Q_1, is reported in Fig. 11.3.

The corresponding estimated global consumer's and producer's risks were 0.07 and 0.08 %, respectively. Such results are useful for optimising the testing process by setting appropriate guard factors and thus validly support decision-making.

---

[7] Note that here the "product" is a measuring device, so the production process is characterised by the measurement "error" of such devices, as detected in the testing process, and the measurement process is that performed by the testing device(s).

[8] Remember footnote 4.

References

1. BIPM: Guide to the expression of uncertainty in measurement—supplement 2: measurement uncertainty and conformance testing: risk analysis (2005)
2. Pendrill, L.R.: Risk assessment and decision-making. In: Berglund, B., Rossi, G.B., Townsend, J., Pendrill, L. (eds.) Measurement with persons, pp. 353–368. Taylor and Francis, London (2012)
3. Yano, H.: Metrological control: Industrial measurement menagement. Asian Production Organisation (1991)
4. ISO: ISO 14253–1: Geometrical Product Specification (GPS)—Inspection by measurement of workpieces and measuring instruments—Part I: Decision rules for proving conformance or non-conformance with specifications (1998)
5. Pendrill, L.R.: Optimised measurement uncertainty and decision-making when sampling by variables or by attributes. Measurement **39**, 829–840 (2006)
6. Estler, W.T.: Measurement as inference: fundamental ideas. Ann. CIRP **48**, 1–22 (1999)
7. Lira, I.: A Bayesian approach to consumer's and user's risk in measurement. Metrologia **36**, 397–402 (1999)
8. IEC: IEC CISPR/A/204/CD: Accounting for measurement uncertainty when determining compliance with a limit (1997)
9. CENELEC: CENELEC—Draft prEN 50222: Standard for the evaluation of measurement results taking measurement uncertainty into account (1997)
10. Rossi, G.B., Crenna, F.: A probabilistic approach to measurement-based decisions. Measurement **39**, 101–119 (2006)
11. Phillips, S.D., Estler, W.T., Levenson, M.S., Eberhart, K.R.: Calculation of measurement uncertainty using prior information. J Res Natl Inst Stand Technol **103**, 625–632 (1998)
12. Cox, M.G., Rossi, G.B., Harris, P.M., Forbes, A.: A probabilistic approach to the analysis of measurement processes. Metrologia **45**, 493–502 (2008)
13. EURACHEM: EURACHEM/CITAC Guide CG 4: Quantifying uncertainty in analytical measurement (2000)
14. EU; Directive 2004/22/EC of the European Parliament and of the Council of the 31 March 2004 on measuring instruments, Official Journal of the European Union, L 135 (2004)
15. Sommer, K.D., Kochsiek, M., Schultz, W.: Error limits and measurement uncertainty in legal metrology. In: Proceedings of the XVI IMEKO World Congress, Vienna, 2000 (2000)
16. Crenna, F., Rossi, G.B.: Probabilistic measurement evaluation for the implementation of the measuring instrument directive. Measurement **42**, 1522–1531 (2009)

Chapter 12
Dynamic Measurement

12.1 Dynamic Measurement: An Introduction

Dynamic measurement sets out to measure the variations in the values of a quantity over time [1–4]. It covers a wide application area, since it is very common that a quantity varies over time: examples are vibration, sound and electromagnetic radiations. Examples of application areas include the monitoring of continuous industrial processes, the sensing function in automatic control, vehicles guidance, experimental biomechanics, psychophysics and perception. Such measurements are typically affected by the dynamic characteristics of the measuring device and by noise in the measuring chain. A proper analysis often requires quite sophisticated measurement tools, including signal and system theory. Much of such theory has been developed over the years, especially in the areas of telecommunications and automatic control, and important contributions have also come for the statistical community, in particular for discrete-time series, as those occurring in econometrics [5–10]. Despite this noteworthy literature, approaching the problem from a measurement standpoint requires additional works, in consideration of problems such as calibration or uncertainty evaluation, often not explicitly considered in the above disciplines, as well as, on a more theoretical side, for regarding dynamic measurement as a special kind of multidimensional, or vector, measurement [3, 4, 11]. This last perspective will be particularly pursued here [12].

As an overview, we can distinguish two main classes of problems, related to the classical distinction between direct and indirect measurements. *Direct* here means that *the goal is to measure the variation of some quantity with time*. So the final result is basically a function of time. *Indirect* refers to the case where *the final goal is to estimate some features of the dynamic process*, other than its time profile: *spectrum measurement* is a good example. Here, the acquisition of a time signal is just a first step for achieving the final result, which is expressed as a function of frequency. Interestingly enough, the general theory for vector measurement presented in this book can provide an elegant approach to both problems, through a different interpretation of the measurand vector [12].

G. B. Rossi, *Measurement and Probability*, Springer Series in Measurement Science and Technology, DOI: 10.1007/978-94-017-8825-0_12,

12.2 Direct Dynamic Measurement

12.2.1 A Probabilistic Framework for Direct Dynamic Measurement

Although continuous variations in time are often assumed for the quantities involved, a discrete-time representation is also appropriate, provided that the sampling rate is sufficiently high (Nyquist condition [6, 13]). In this perspective, the measurand is a vector, $\mathbf{x}$, that collects the values assumed by the quantity x_t, in a time interval $T = N\Delta t$, where t is a discrete-time index, Δt is the sampling interval , and $\mathbf{y}$ is the corresponding vector of instrument indications.

Then, the reference equations for our study are those presented in Sect. 5.7, Table 5.6. In particular, we will use the formula concerning restitution, which we re-write here in a slightly simplified notation. In fact, in Chap. 5, we denoted the *probabilistic measurement value* by x' and the expected measurement value by $\hat{x}$, and the related vector versions by $\mathbf{x}'$ and $\hat{\mathbf{x}}$, respectively. Here, we drop the "prime" ($'$) from the former, whilst we leave unchanged the latter. In fact, applying the prime is necessary when we want to model the entire measurement process by chaining observation and restitution. Since here we will not do that, the simplified notation will be, hopefully, clear enough. Thus, we model observation by a vector parametric conditional distribution that we will call *characteristic distribution* for ease of reference:

$$p(\mathbf{y}|\mathbf{x}, \boldsymbol{\theta}), \tag{12.1}$$

and restitution by the equation

$$p(\mathbf{x}|\mathbf{y}) = \int_{\boldsymbol{\Theta}} \frac{p(\mathbf{y}|\mathbf{x}, \boldsymbol{\theta})}{\int_{\mathbf{X}} p(\mathbf{y}|\mathbf{x}, \boldsymbol{\theta}) \mathrm{d}\boldsymbol{\theta}} p(\boldsymbol{\theta}) \mathrm{d}\boldsymbol{\theta}. \tag{12.2}$$

As an additional simplification, we will also ignore the dependence of the characteristic distribution upon $\boldsymbol{\theta}$, when not explicitly needed. As usually, the key modelling point is to develop the observation equation, restitution being essentially a calculation concern. Let us then discuss the structure of the characteristic distribution, $p(\mathbf{y}|\mathbf{x}, \boldsymbol{\theta})$. Consider the following convenient notation:

$$x^t = (x_1, x_2, ..., x_t),$$
$$y^t = (y_1, y_2, ..., y_t).$$

Then, ignoring, for the sake of simplicity, the dependence on $\boldsymbol{\theta}$, the characteristic distribution (12.1), can be factorised as

$$p(\mathbf{y}|\mathbf{x}) = p(y_1|\mathbf{x})...p(y_t|y^{t-1}, \mathbf{x})...p(y_N|y^{N-1}, \mathbf{x}). \tag{12.3}$$

This results from a general property of joint probability distributions [8]. Furthermore, if we add a causality assumption, that is we assume that the indication y at instant t only depends upon the value of the measurand up to instant $t-1$, we have a further simplification, and we can write:

$$p(\mathbf{y}|\mathbf{x}) = p(y_1|x_1)...p(y_t|y^{t-1}, x^{t-1})...p(y_N|y^{N-1}, x^{N-1}). \tag{12.4}$$

The point is now how to calculate the required sequence of conditional probability distributions, which may anyway look like a rather formidable task!

This requires adopting an *internal* model[1] describing the observation process. We will consider and treat, in general terms, a wide class of models and will show how the model allows a step-by-step calculation of the characteristic distribution. But prior to that, let us discuss a simple, but in reality quite general, introductory example.

Consider a contact thermometer having a linear behaviour. The dynamics of such a device depends on the characteristics of the thermal interface. In the simple case of the measurement of the temperature of a fluid, where the sensing element is immersed in the fluid and in direct contact with it, we may derive the dynamic equation by the heat transfer balance condition:

$$k_s S(T - T_s) = mc\frac{\mathrm{d}T_s}{\mathrm{d}t}, \tag{12.5}$$

where

k_s heat transfer coefficient ($\mathrm{W\,m^{-2}\,K^{-1}}$),
c specific heat of the sensor ($\mathrm{J\,kg^{-1}\,K^{-1}}$),
m sensor's mass (kg),
S sensor's surface ($\mathrm{m^2}$),
T_s sensor's temperature (K),
T temperature to be measured (K).

This is a first-order differential equation. Introducing the *time constant* $\theta = mc/k_s S$ and the *state variable* $z = T_s$, and denoting the measurand by x as usually, we may compactly re-write it as

$$\theta\dot{z} + z = x, \tag{12.6}$$

that we will call the (*dynamic*) *state equation* of the system. Let then k (mV/K) be the overall sensitivity of the measuring chain, w (mV) a random variable describing some output noise, and y (mV), as usually, the instrument indication; we can also write the *observation equation*:

$$y = kx + w. \tag{12.7}$$

[1] An internal model is one in which (internal) state variables appear in addition to input/output variables that solely appear in input–output models.

These two equations together constitute an *internal model* of the observation process, that is a model where the relation between the input and the output is explicated in reference to the evolution of the state of the process.

Since we are going to work with discrete, time-sampled, observations, we have first to derive a suitable correspondent discrete-time model.

There is no unique way to do such a transformation; rather, some reference criteria have to be adopted. An advantageous one is based on the invariance of impulse response, for trait-constant output. Let us practice applying it to our example. The response due to an input $x(\tau)$ is

$$z(\tau) = z_0 \exp\left(-\frac{\tau - \tau_0}{\theta}\right) + \frac{1}{\theta}\int_{\tau_0}^{\tau} \exp\left(-\frac{\tau - \xi}{\theta}\right) x(\xi)\mathrm{d}\xi, \tag{12.8}$$

where $z_0 = z(\tau_0)$. Assume now that the input x is constant over each sampling interval, that is $x(\tau) = x(t\Delta t)$ for $t\Delta t \leq \tau < (t+1)\Delta t$. Then, by applying formula (12.8) between $(t-1)\Delta t$ and $t\Delta t$, we obtain

$$\begin{aligned} z_t &= z_{t-1} \exp\left(-\frac{\Delta t}{\theta}\right) + \frac{1}{\theta}\int_{(t-1)\Delta t}^{t\Delta t} \exp\left(-\frac{t\Delta t - \xi}{\theta}\right) x(\xi)\mathrm{d}\xi \\ &= z_{t-1} \exp\left(-\frac{\Delta t}{\theta}\right) + \left[1 - \exp\left(-\frac{\Delta t}{\theta}\right)\right] x_{t-1}. \end{aligned} \tag{12.9}$$

So the difference equation we seek is

$$z_t = az_{t-1} + bx_{t-1}, \tag{12.10}$$

where $a = \exp\left(-\frac{\Delta t}{\theta}\right)$ and $b = (1-a)$, that, together with the discretised observation equation,

$$y_t = kz_t + w_t, \tag{12.11}$$

form the discrete internal model of the process.

Let us now derive an "external", that is input–output, model.

By introducing the random variable $v_t = w_t - aw_{t-1}$, we obtain

$$y_t = ay_{t-1} + kbx_{t-1} + v_t, \tag{12.12}$$

from which we can calculate the tth term in formula (12.4), as

$$p(y_t|y^{t-1}, x^{t-1}) = p(y_t|y_{t-1}, x_{t-1}) = p_v(y_t - ay_{t-1} - kbx_{t-1}), \tag{12.13}$$

that may be initialised by setting

$$p(y_1|x_1) = p(y_1). \tag{12.14}$$

In this way, it is possible to factorise the characteristic distribution as:

$$p(\mathbf{y}|\mathbf{x}) = p(y_1) \prod_{t=1}^{N-1} p_v(y_t - ay_{t-1} - kbx_{t-1}), \tag{12.15}$$

and restitution is given by

$$p(\mathbf{x}|\mathbf{y}) \propto \prod_{t=2}^{N} p_v(y_t - ay_{t-1} - kbx_{t-1}). \tag{12.16}$$

From this expression, it is possible to obtain the marginal distribution $p(x_t|\mathbf{y})$ that provides restitution at time t. If we now assume that the w_t are independent realisations of a zero-mean Gaussian random variable, with variance σ_w^2, we find that the measurement value at time t is

$$\hat{x}_t = \frac{y_{t+1} - ay_t}{kb} \tag{12.17}$$

and, introducing the variable

$$c_t \triangleq x_t - \hat{x}_t = -\frac{w_{t+1} - aw_t}{kb}, \tag{12.18}$$

whose variance is

$$\sigma_c^2 = \frac{1 + a^2}{k^2 b^2}, \tag{12.19}$$

we may rewrite the above formula as

$$p(\mathbf{x}|\mathbf{y}) = \prod_{t=2}^{N} \sigma_c^{-1} \varphi(\sigma_c^{-1}(x_t - \hat{x}_t)). \tag{12.20}$$

From this, we can readily obtain the marginal distribution with respect to x_t

$$\begin{aligned} p(x_t|\mathbf{y}) &= \int \prod_{t=2}^{N} \sigma_c^{-1} \varphi(\sigma_c^{-1}(x_t - \hat{x}_t)) \mathrm{d}x_1 \ldots \mathrm{d}x_{t-1} \mathrm{d}x_{t+1} \ldots \mathrm{d}x_{N-1} \\ &= \sigma_c^{-1} \varphi(\sigma_c^{-1}(x_t - \hat{x}_t)), \end{aligned} \tag{12.21}$$

which constitute the final result, at each time instant.

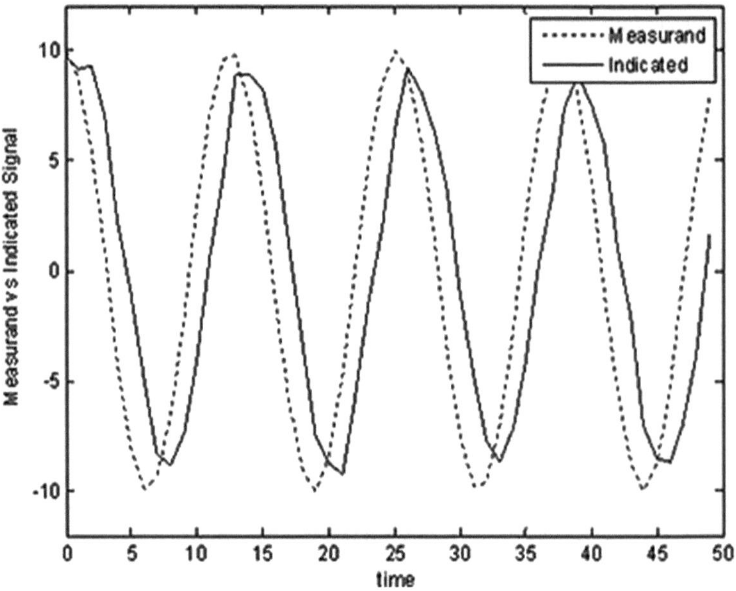

Fig. 12.1 Measurement of a simple cosine process

Let us now illustrate this by a numerical example. Consider the measurement of a simple cosine signal, where the indication is affected by dynamic effect and noise, as illustrated in Fig. 12.1.

If we perform restitution according to the above, we obtain a set of probability distributions, ordered according to time, that appear as in Fig. 12.2.

We may also visualise the result in another way, as in Fig. 12.3.

Here, the measured signal is compared with the measurand and uncertainty intervals of $2u$ semi-width are also plotted. We can see that, as the result of restitution, the measured signal is now in phase with the original: compare with Fig. 12.1. Furthermore, the two signals are compatible to each other within the uncertainty band. The operation of compensating dynamic effects is, technically, a de-convolution [14]. So we can say that the restitution algorithm actually performs a *probabilistic de-convolution* since it, at the same time, compensates dynamic effects and applies probability distributions to the compensated (corrected) signal [11]. All this happens without any heuristics; rather, it results from the application of the general framework, without requiring ad hoc considerations!

Assuming that this example has been a good illustration of what can be done, we are now ready to discuss how the general framework can be applied to a wide class of models [5]. We thus consider the following, generally nonlinear, Markovian stochastic process:

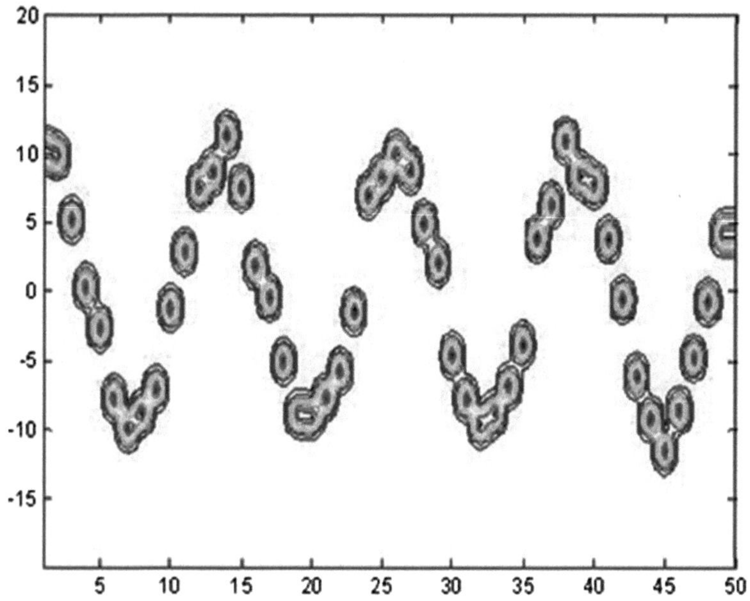

Fig. 12.2 Restitution expressed as probability distributions ordered with time

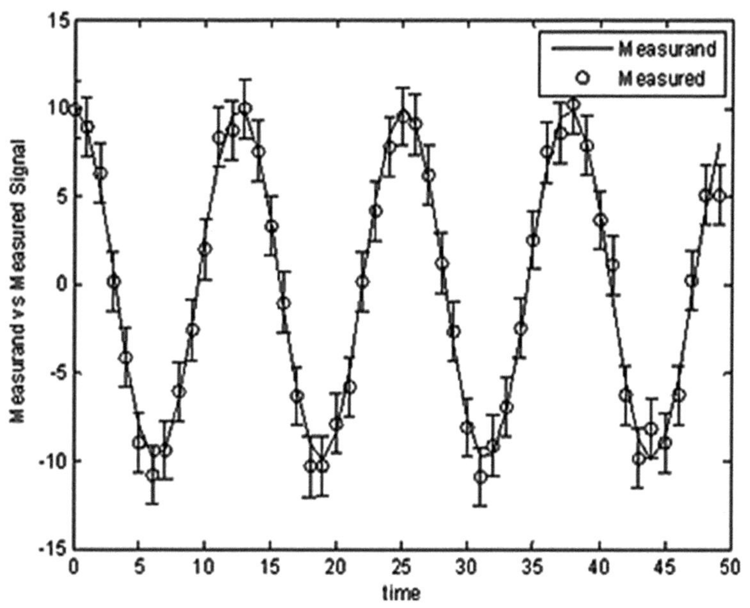

Fig. 12.3 Plot of the measured signal compared with the measurand and also showing uncertainty intervals

$$\mathbf{z}_t = g(\mathbf{z}_{t-1}, \xi_{t-1}, x_t),$$
$$y_t = f(\mathbf{z}_t, w_t), \tag{12.22}$$

where

$\mathbf{z}_t$ is the time sequence of the (vector) state of the process,
ξ_t is an uncorrelated sequence of driving vector random variables,
w_t is a random observation noise and
f and g are two functions that may be generally nonlinear.

This model is indeed very general since the vector state $\mathbf{z}_t$ allows representing a whatever complex internal dynamics. Furthermore, random variability may be flexibly expressed thanks to a combination of the driving (vector) random variables ξ_t and the (scalar) observation noise w_t. Lastly, even nonlinear behaviours can be properly represented by nonlinear f and g functions.

We will now show that, in spite of its complexity, this model still allows, as the example, a step-by-step calculation of the factors of the characteristic distribution that appear in formula (12.4), which makes the overall framework manageable.

More explicitly, we *show now how to calculate* $p(y_t|y^{t-1}, x^{t-1})$ *as a function of the previous term,* $p(y_{t-1}|y^{t-2}, x^{t-2})$, *and of the model* (12.22).

For doing so, we need to calculate $p(\mathbf{z}_t|y^t, x^{t-1})$ as a function of $p(\mathbf{z}_{t-1}|y^{t-1}, x^{t-2})$ first.

At Step 1, by setting $x_0 = 0$, we have

$$p(\mathbf{z}_1|y^1, x^0) = p(\mathbf{z}_1|y_1, x_0) \propto p(y_1|\mathbf{z}_1)p(\mathbf{z}_1), \tag{12.23}$$

where

$$p(y_1|\mathbf{z}_1) = \int \delta(y_1 - f(\mathbf{z}_1, w_1))p(w_1)\mathrm{d}w_1, \tag{12.24}$$

where $\delta(\cdot)$ is the (continuous) Dirac delta operator.

At step t, instead we obtain,

$$\begin{aligned} p(\mathbf{z}_t|y^t, x^{t-1}) &= p(\mathbf{z}_t|y_{t-1}, x_{t-1}) \\ &\propto p(y_t|\mathbf{z}_t, y_{t-1}, x_{t-1})p(\mathbf{z}_t|y_{t-1}, x_{t-1}) \\ &= p(y_t|\mathbf{z}_t)\int p(\mathbf{z}_t|\mathbf{z}_{t-1}, x_{t-1})p(\mathbf{z}_{t-1}|y^{t-1}, x^{t-2})\mathrm{d}\mathbf{z}_{t-1}. \end{aligned} \tag{12.25}$$

Now, since

$$p(y_t|\mathbf{z}_t) = \int \delta(y_t - f(\mathbf{z}_t, w_t))p(w_t)\mathrm{d}w_t \tag{12.26}$$

and

$$p(\mathbf{z}_t|\mathbf{z}_{t-1}, x_{t-2}) = \int \delta[\mathbf{z}_t - g(\mathbf{z}_{t-1}, \xi_{t-1})]p(\xi_{t-1})\mathrm{d}\xi_{t-1}, \tag{12.27}$$

formula (12.25) allows $p(\mathbf{z}_t|y^t, x^{t-1})$ to be calculated, once $p(\mathbf{z}_{t-1}|y^{t-1}, x^{t-2})$ is known.

Let us now consider the calculation of $p(y_t|y^{t-1}, x^{t-1})$ as a function of $p(y_{t-1}| y^{t-2}, x^{t-2})$, which is our final goal.

Note that

$$p(y_t|y^{t-1}, x^{t-1}) = \int p(y_t|\mathbf{z}_t, y^{t-1}, x^{t-1})p(\mathbf{z}_t|y^{t-1}, x^{t-1})\mathrm{d}\mathbf{z}_t. \tag{12.28}$$

But the right-hand side of this equation is the integral of the right-hand side of (12.25); so, with regard to our observation, we attain

$$p(y_t|y^{t-1}, x^{t-1}) = \int \left[p(y_t|\mathbf{z}_t) \int p(\mathbf{z}_t|\mathbf{z}_{t-1}, x_{t-1})p(\mathbf{z}_{t-1}|y^{t-1}, x^{t-2})\mathrm{d}\mathbf{z}_{t-1} \right] \mathrm{d}\mathbf{z}_t. \tag{12.29}$$

To sum up, at step t, we know, from the previous step, $p(\mathbf{z}_{t-1}|y^{t-1}, x^{t-2})$ and $p(y_{t-1}|y^{t-2}, x^{t-2})$. Then, by formulae (12.25, 12.26 and 12.27), we calculate $p(\mathbf{z}_t|y^t, x^{t-1})$ and, by formula (12.29), we calculate $p(y_t|y^{t-1}, x^{t-1})$. In this way, by assuming a proper distribution $p(w)$ for w, which may be the same for all the time instants, and an initial, even vague, distribution for $\mathbf{z}$, $p(\mathbf{z}_1)$, initialising the procedure by formulae (12.23, 12.24), *we can calculate step by step the characteristic distribution* (12.4) *which fully describes the observation process and is the core of the probabilistic model* [12].

We can note now that the introductory example is a special case of what we just discussed, in which the state at each instant reduces to a scalar, z_t, the driving input ξ_t is absent, and f and g are simple linear functions. This example may be generalised in a quite straightforward way into a general linear model, of order n, which allows a proper description of a wide class of real measurement processes. Furthermore, if the random inputs are assumed to be Gaussian, an explicit solution can be found [5].

However, these developments are outside the scope of this introductory exposition, in which we are more interested in understanding the basic philosophy rather than in going into detail.

12.2.2 Evaluation of the Uncertainty Generated by Dynamic Effects in Instrumentation

In the previous section, we have seen how the probabilistic model allows restitution of a direct dynamic measurement in terms of a probability distribution. Apart from its theoretical value, this may be also of practical interest for sophisticated high-precision measurement. Yet the probabilistic model also allows us to obtain simple formulae for the evaluation of uncertainty generated by dynamic effects, for practical use in everyday measurement.

To illustrate this, let us start from a simple example. Suppose that we want to measure a dynamic phenomenon, like a temperature variation, having a simple cosine behaviour:

$$x(\tau) = x_0 \cos(2\pi f \tau), \tag{12.30}$$

where f denotes frequency. The corresponding instrument indication will include dynamic effects, both of transient and of steady-state nature. Since we are measuring a *stationary* phenomenon, that is, one that maintains its characteristics constant with time, usually we can neglect transient effects and consider only steady-state ones [15]. These latter are characterised by the *frequency response* of the instrument, which is a complex function of frequency:

$$\tilde{q}(f) = k\alpha(f) \exp(j\varphi(f)), \tag{12.31}$$

where k is, as usually, the sensitivity of the instrument, $\tilde{q}(f)$ is the frequency response, $\alpha(f)$ its *modulus*, $\varphi(f)$ its *phase*, both real functions of the frequency, and j is the *imaginary unit*, that is $j^2 = -1$.

For example, in the case of the thermometer of the previous subsection, the modulus would be

$$\alpha(f) = (1 + (2\pi f T)^2)^{-1/2} \tag{12.32}$$

and the phase

$$\varphi(f) = -\arctan(2\pi f T). \tag{12.33}$$

Note that, having put in evidence the sensitivity, k, in formula (12.31), the modulus has a unitary dimension: this allows a more elegant presentation of what follows.

Such dynamic effects may be noted in Fig. 12.1, where

- the effect related to the modulus consists in a reduction of the amplitude of the indicated signal, in comparison with the measurand;
- the effect related to the phase consists in some delay.

Note that the combined effect may result in severe errors at some time instants, especially near to zero-crossing points.

Such steady-state response may be expressed, in the general case as

$$y(\tau) = k\alpha(f)x_0 \cos(2\pi f \tau + \varphi(f)). \tag{12.34}$$

The measured signal, without dynamic compensation, will be

$$\hat{x}(\tau) = k^{-1} y(\tau) = \alpha(f)x_0 \cos(2\pi f \tau + \varphi(f)). \tag{12.35}$$

The comparison of $x(\tau)$ and $\hat{x}(\tau)$ suggests the so-called *non-distortion conditions*: in the ideal case, if

$\alpha(f) = 1$ and
$\varphi(f) = 0$,

we would have no dynamic effect and $\hat{x}(\tau) = x(\tau)$. So measuring systems are designed to approach as much as possible these non-distortion conditions. If we have reliable knowledge of the frequency response, we may compensate for such effects, as shown in the previous subsection. But if, as happens in many practical cases, we simply know that the frequency response is close to the ideal one up to some tolerance, that is

$$\alpha(f) = 1 \pm \Delta\alpha \tag{12.36}$$

and the phase

$$\varphi(f) = 0 \pm \Delta\varphi, \tag{12.37}$$

we cannot make any compensation and we rather have to evaluate the uncertainty due to such uncompensated effects. For doing so, let us put

$\alpha(f) = 1 + \delta\alpha$ and
$\varphi(f) = 0 + \delta\varphi$.

Here, $\delta\alpha$ and $\delta\varphi$ are unknown constant values. Since all we know about them is expressed in (12.36, 12.37), we can treat then as zero-mean uniform independent probabilistic variables. Proceeding further, note that the "error" signal is

$$\begin{aligned} e(\tau) &= \hat{x}(\tau) - x(\tau) \\ &= (1 + \delta\alpha)x_0 \cos(2\pi f\tau + \delta\varphi) - x_0 \cos(2\pi f\tau). \end{aligned} \tag{12.38}$$

Remember now that, from trigonometry,

$$\cos(a + b) = \cos(a)\cos(b) - \sin(a)\sin(b) \tag{12.39}$$

and set $a = 2\pi f\tau$ and $b = \delta\varphi$. Furthermore, since $\delta\alpha$ and $\delta\varphi$ can be assumed to be "small", we may also use the approximations $\cos(\delta\varphi) \cong 1$ and $\sin(\delta\varphi) \cong \delta\varphi$. Substituting in formula (12.38), after some manipulations, we obtain:

$$e(\tau) = \delta\alpha x_0 \cos(2\pi f\tau) + \delta\varphi x_0 \sin(2\pi f\tau). \tag{12.40}$$

Now, treating $\delta\alpha$ and $\delta\varphi$ as (zero-mean) independent probabilistic variables, we can calculate the variance of the error, at each time instant τ. Observing that the expected value of the error is zero, we obtain

$$\begin{aligned}\sigma_e^2(\tau) &= \mathrm{var}[e(\tau)] = \mathrm{E}[e^2(\tau)] \\ &= \mathrm{E}[\delta\alpha^2 x_0^2 \cos^2(2\pi f \tau) \\ &\quad + \delta\varphi^2 x_0^2 \sin^2(2\pi f \tau) + 2\delta\alpha\delta\varphi x_0^2 \cos^2(2\pi f \tau) \sin^2(2\pi f \tau)] \\ &= x_0^2[\sigma_\alpha^2 \cos^2(2\pi f \tau) + \sigma_\varphi^2 \sin^2(2\pi f \tau)]. \end{aligned} \tag{12.41}$$

In practical uncertainty evaluation, we would be hardly interested in a time-dependent uncertainty. So it makes sense to average over one period T_p:

$$\begin{aligned}\sigma_e^2 &= \int_0^{T_p} \sigma_e^2 \mathrm{d}(\tau) \\ &= x_0^2 \int_0^{T_p} [\sigma_\alpha^2 \cos^2(2\pi f \tau) + \sigma_\varphi^2 \sin^2(2\pi f \tau)] \\ &= \frac{x_0^2}{2}(\sigma_\alpha^2 + \sigma_\varphi^2). \end{aligned} \tag{12.42}$$

Remembering that $\delta\alpha$ and $\delta\varphi$ are uniform random variables, we obtain

$$\sigma_e^2 = \frac{x_0^2}{2}\left(\frac{\Delta\alpha^2}{3} + \frac{\Delta\varphi^2}{3}\right). \tag{12.43}$$

Lastly, considering that

$$\frac{x_0^2}{2} = x_{\mathrm{rms}}^2, \tag{12.44}$$

we obtain a simple and elegant formula for the evaluation of relative standard uncertainty, of high practical value:

$$\frac{u}{x_{\mathrm{rms}}} = \left(\frac{\Delta\alpha^2}{3} + \frac{\Delta\varphi^2}{3}\right)^{1/2}. \tag{12.45}$$

Note again that the probabilistic approach developed in this book allows us both to obtain a sophisticated restitution, in term of probability distributions, and to obtain simple evaluation formulae for simple everyday measurements. It thus enables a proper treatment of measurement with the degree of sophistication required by any application.

12.3 Indirect Dynamic Measurement: Spectrum Measurement

In dynamic measurement, often the acquisition of a signal that reproduces the *time behaviour* of (some quantity related to) the phenomenon under investigation is not the final goal. Rather some other characteristic of the phenomenon is sought, such as its *spectrum* [7, 9, 10, 16]. A spectrum is a representation of a phenomenon in the frequency domain. In the case of a periodic phenomenon, for instance, that is a phenomenon described by a time-variable quantity $x(\tau)$ whose value repeats over a period T_p:

$$x(\tau) = x(\tau - T_p), \tag{12.46}$$

what a spectrum is can be understood by remembering that quite generally a periodic zero-mean phenomenon $x(\tau)$ can be expressed by a finite sum of cosine functions, of frequency being an integer multiple of a fundamental frequency $f_0 = 1/T_p$, each of them having a proper modulus (amplitude) and phase:

$$x(\tau) = \sum_{i=1}^{n} c_i \cos(2\pi i f_0 \tau + \varphi_i). \tag{12.47}$$

The spectrum in this case consists in two discrete functions of the frequency: the *amplitude (or modulus) spectrum*, that is, the function $f_i \mapsto c_i$, where $f_i = i f_0$, and the phase spectrum, $f_i \mapsto \varphi_i$, with $i = 1, ..., n$.

In many cases, the spectrum provides the information of direct practical interest. For example, in ergonomics, we may want to evaluate the exposure of workers to vibration, to prevent related diseases. For doing so, we have to consider that human response to vibration depends upon frequency. Thus, a proper procedure for such an evaluation includes the measurement of vibration spectra in typical working conditions and their processing, by means of standard weighting functions that account for the frequency dependence of human response. Furthermore, as we have discussed in Chap. 8, spectrum measurement is also a first step in loudness measurement for the assessment of noise exposure.

As another example, consider the signal in Fig. 12.4.

It is a record of shaft vibration in a steam turbine, which is a machine that extracts thermal energy from pressurised steam and converts it into rotary motion.[2] It usually drives an electrical generator, for energy production. Continuous vibration monitoring is a powerful tool for early fault detection, allowing immediate intervention. The record in the figure refers to a turbine in the start-up phase, at a rotation speed of 756 rev/min. Since here we are interested in the spectral distribution and not in the overall value, the signal has been normalised to a unitary root mean square value. For diagnostic purposes, it is important to monitor the spectral components of

[2] It was acquired in the 1980s as a part of a collaboration of our Measurement Laboratory with Politecnico di Milano and ENEL (the Italian National Energy Board) [17].

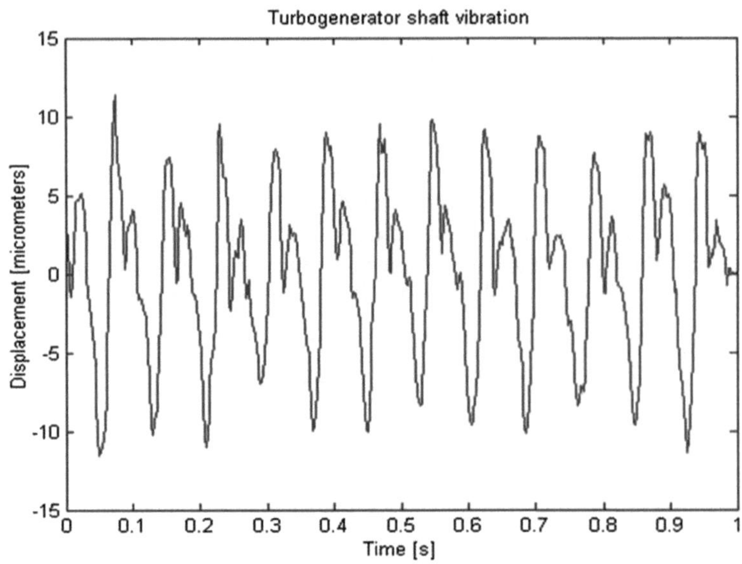

Fig. 12.4 Vibration of a turbine during the start-up phase [17]

vibration at the fundamental frequency, f_0, at second and third harmonics ($2f_0$ and $3f_0$). Sometimes a subharmonic component, around $f_0/2$, may also appear. Note that the signal, being based on real data, is not exactly periodic, mainly due to the presence of noise; so, an appropriate model for the *indicated signal* is

$$y_t = k \sum_{i=1}^{n} c_i \cos(2\pi i f_0 t \Delta t + \varphi_i) + w_t, \tag{12.48}$$

where w_t is, as usually, a random noise sequence. In the following, we will assume, without loss of generality, $k = 1$, since if k is different from one, we simply divide the signal by it. In our example, assuming that no subharmonic is present, we can assume $n = 3$. Let us also assume that we can independently measure f_0, since it corresponds to the rotation speed: if this is 756 rev/min, $f_0 = 12.6$ Hz. Interestingly enough, we can approach this problem via the proposed general framework, that is by Eqs. 12.1 and 12.2, by simply reinterpreting the measurand vector, $\mathbf{x}$, in accordance with the present indirect-measurement situation [16].

In our example, we have assumed to know the fundamental frequency f_0, since we can measure independently the rotation speed, with negligible uncertainty. Let us further assume that the variance of the noise, σ^2 is known, for example by previous experience. Then, we do not have influence parameters and we do not need the vector $\boldsymbol{\theta}$. The reference equation thus simplifies: we have to model observation by the characteristic distribution $p(\mathbf{y}|\mathbf{x})$ and we can perform restitution by

$$p(\mathbf{x}|\mathbf{y}) = \frac{p(\mathbf{y}|\mathbf{x})}{\int_{\mathbf{X}} p(\mathbf{y}|\mathbf{x})\mathrm{d}\mathbf{x}}, \tag{12.49}$$

where we have denoted the measured vector by $\mathbf{x}$, instead of by $\mathbf{x}'$, as we have done in the previous section, to simplify notation. In the characteristic distribution, $\mathbf{y}$ still represents the time-sampled indication, $\mathbf{y} = (y_1, \ldots, y_N)$, whilst $\mathbf{x}$ now collects the parameters of the Fourier series (12.47), c_i and φ_i. In reality, for simplifying things, it is convenient to express the Fourier series in another equivalent way:

$$x_t = \sum_{i=1}^{n} a_i \cos(2\pi i f_0 t \Delta t) + b_i \sin(2\pi i f_0 t \Delta t). \tag{12.50}$$

In fact, in this way, we obtain an expression that is linear in the unknown parameters, a_i and b_i, whilst the previous one was nonlinear in the parameters c_i and φ_i. The relation between the two is established by the trigonometric identity

$$c_i \cos(2\pi i f_0 t \Delta t + \varphi_i) = +c_i \cos(\varphi_i) \cos(2\pi i f_0 t \Delta t) - c_i \sin(\varphi_i) \sin(2\pi i f_0 t \Delta t). \tag{12.51}$$

Then, the measurand vector is $\mathbf{x} = (a_1, b_1 \ldots, a_i, b_i, \ldots, a_n, b_n)$. For the procedure to work, it is necessary that $2n < N$. Once these parameters have been obtained, it is easy to calculate the c_i as

$$c_i = \sqrt{a_i^2 + b_i^2} \tag{12.52}$$

and the φ_i that satisfy

$$\begin{aligned} \cos(\varphi_i) &= +\frac{a_i}{c_i}, \\ \sin(\varphi_i) &= -\frac{b_i}{c_i}, \end{aligned} \tag{12.53}$$

which implies

- for $a_i > 0$, $\varphi_i = \arctan\left(-\frac{b_i}{a_i}\right)$,
- for $a_i < 0$, $\varphi_i = \arctan\left(-\frac{b_i}{a_i}\right) + \pi$ and
- for $a_i = 0$:
  - for $b_i > 0$, $\varphi_i = +\frac{\pi}{2}$,
  - for $b_i < 0$, $\varphi_i = -\frac{\pi}{2}$ and
  - for $b_i = 0$, $\varphi_i = 0$ (this is a conventional value, since, when the modulus is zero, the phase is undefined).

Let us then develop the example. To further simplify the development and especially for making the formulae more understandable, let us first assume $n = 1$, since the generalisation to $n > 1$ is an immediate extension.

Then, the discretised internal model is (remember that we have also assumed $k = 1$):

$$y_t = a\cos(2\pi f_0 t\Delta t) + b\sin(2\pi f_0 t\Delta t) + w_t, \tag{12.54}$$

and the observation distribution is

$$\begin{aligned} p(\mathbf{y}|a,b) &= \prod_{t=1}^{N} \sigma^{-1}\exp\left\{-\frac{1}{2\sigma^2}[y_t - (a\cos(2\pi f_0 t\Delta t) + b\sin(2\pi f_0 t\Delta t)]^2\right\} \\ &= (2\pi)^{-N/2}\sigma^{-N}\exp\left\{-\frac{1}{2\sigma^2}\sum_{t=1}^{N}[y_t - (a\cos(2\pi f_0 t\Delta t) \right. \\ &\quad \left. + b\sin(2\pi f_0 t\Delta t))]^2\right\} \\ &= (2\pi)^{-N/2}\sigma^{-N}\exp\left\{-\frac{1}{2\sigma^2}\sum_{t=1}^{N}[y_t^2 - a^2\cos^2(2\pi f_0 t\Delta t) \right. \\ &\quad - b^2\sin^2(2\pi f_0 t\Delta t) \\ &\quad + 2ab\sin(2\pi f_0 t\Delta t)\cos(2\pi f_0 t\Delta t) - 2ay_t\cos(2\pi f_0 t\Delta t) \\ &\quad \left. - 2by_t\sin(2\pi f_0 t\Delta t)]\right\}. \end{aligned} \tag{12.55}$$

This equation simplifies if the observation time is an integer multiple of the period, $T = mT_p$, and the period is an integer multiple of the sampling interval, $T_p = p\Delta t$. With these assumptions, we obtain

$$\sum_{t=1}^{N}\cos^2(2\pi f_0 t\Delta t) = \sum_{t=1}^{N}\cos^2\left(2\pi\tfrac{t}{N}\right) = \tfrac{N}{2};$$
$$\sum_{t=1}^{N}\sin^2(2\pi f_0 t\Delta t) = \sum_{t=1}^{N}\sin^2\left(2\pi\tfrac{t}{N}\right) = \tfrac{N}{2};$$
$$\sum_{t=1}^{N}\cos(2\pi f_0 t\Delta t)\sin(2\pi f_0 t\Delta t) = \sum_{t=1}^{N}\cos(2\pi\tfrac{t}{N})\sin\left(2\pi\tfrac{t}{N}\right) = 0.$$

These conditions are fulfilled, for example, in the *synchronous monitoring* of rotating machines, where a fixed number of samples is acquired for each rotation cycle [17]. Furthermore, they are also approximately true whenever $T \gg T_p$ and $T_p \gg \Delta t$, for example, for $T \geq 10T_p$ and $T_p \geq 10\Delta t$, which is often the case in typical measuring conditions. Substituting in the previous formula, we obtain:

$$p(\mathbf{y}|a,b) = (2\pi)^{-N/2}\sigma^{-N}\exp\left\{-\frac{1}{2\sigma^2}\left[\sum_{t=1}^{N} y_t^2 + \frac{N}{2}a^2 + \frac{N}{2}b^2 \right.\right.$$
$$\left.\left. -2a\sum_{t=1}^{N} y_t\cos(2\pi f_0 t\Delta t) - 2b\sum_{t=1}^{N} y_t\sin(2\pi f_0 t\Delta t)\right]\right\}. \quad (12.56)$$

So far for the characteristic distribution. Assuming a vague (constant) prior for a and b, restitution is simply given by

$$p(a,b|\mathbf{y}) \propto \exp\left\{-\frac{1}{2\sigma^2}\left[\sum_{t=1}^{N} y_t^2 + \frac{N}{2}a^2 + \frac{N}{2}b^2 \right.\right.$$
$$\left.\left. -2a\sum_{t=1}^{N} y_t\cos(2\pi f_0 t\Delta t) - 2b\sum_{t=1}^{N} y_t\sin(2\pi f_0 t\Delta t)\right]\right\}. \quad (12.57)$$

Let us now introduce the following parameters:

$$\hat{a} = \frac{2}{N}\sum_{t=1}^{N} y_t\cos(2\pi f_0 t\Delta t) \quad (12.58)$$

and

$$\hat{b} = \frac{2}{N}\sum_{t=1}^{N} y_t\sin(2\pi f_0 t\Delta t). \quad (12.59)$$

Then, restitution can be restated as

$$p(a,b|\mathbf{y}) \propto \exp\left\{-\frac{1}{2\sigma^2}\frac{N}{2}\left[\frac{2}{N}\sum_{t=1}^{N} y_t^2 + a^2 + b^2 - 2a\hat{a} - 2b\hat{b}\right]\right\}$$
$$= \exp\left\{-\frac{1}{2\sigma^2}\frac{N}{2}\left[\frac{2}{N}\sum_{t=1}^{N} y_t^2 + a^2 + b^2 - 2a\hat{a} - 2b\hat{b} + \hat{a}^2 + \hat{b}^2 - \hat{a}^2 - \hat{b}^2\right]\right\}$$
$$= \exp\left\{-\frac{1}{2\sigma^2}\frac{N}{2}\left[\left(\frac{1}{N}\sum_{t=1}^{N} y_t^2 - \hat{a}^2\right) + \left(\frac{1}{N}\sum_{t=1}^{N} y_t^2 - \hat{b}^2\right) + (a-\hat{a})^2 + (b-\hat{b})^2\right]\right\}. \quad (12.60)$$

Including the terms independent from the variables a and b in the proportionality factor, we finally attain

$$p(a,b|\mathbf{y}) \propto \exp\left\{-\frac{N/2}{2\sigma^2}\left[(a-\hat{a})^2 + (b-\hat{b})^2\right]\right\}, \quad (12.61)$$

Table 12.1 Results of spectrum measurement for turbogenerator (normalised) vibration

Frequency (Hz)	Modulus	$u(c)$	Phase (rad)	$u(\varphi)$ (rad)
12.6	1.07	0.02	+1.54	0.017
25.2	0.75	0.02	−1.01	0.025
37.8	0.35	0.02	+1.74	0.052

which is a bivariate Gaussian distribution in the independent variables a and b, with expected value $(\hat{a}, \hat{b})$. Their marginal distributions have variance $\frac{2}{N}\sigma^2$. This result can be easily generalised to the case of n spectral components, obtaining a $2n$-variate Gaussian distribution. The marginal distributions of the ith spectral components are thus Gaussian with expected values:

$$\hat{a}_i = \frac{2}{N}\sum_{t=1}^{N} y_t \cos(2\pi i f_0 t \Delta t) \tag{12.62}$$

and

$$\hat{b}_i = \frac{2}{N}\sum_{t=1}^{N} y_t \sin(2\pi i f_0 t \Delta t), \tag{12.63}$$

and variance still equal to $\frac{N}{2}\sigma^2$ [7].

The case of unknown variance can also be treated in a similar way as in Chap. 9, obtaining a t-Student multivariate distribution, with variance

$$\hat{\sigma}^2 = \frac{1}{N-n}\sum_{t=1}^{N}\left[y_t - \sum_{i=1}^{n} \hat{a}_i \cos(2\pi i f_0 t \Delta t) + \hat{b}_i \sin(2\pi i f_0 t \Delta t)\right]^2. \tag{12.64}$$

Once probability distributions have been assigned to the a_i and b_i parameters, the distributions for the c_i and φ_i parameters, which are more widely used, can also be obtained through the corresponding formulae (12.52, 12.53). This involves some numerical calculations. Alternatively, it is possible to obtain an approximate result, by propagating the variance through linearised transformations. Omitting, for greater clarity, the index i, we obtain:

$$\begin{aligned} \sigma_c^2 &= \left(\frac{\partial c}{\partial a}\right)^2 \sigma_a^2 + \left(\frac{\partial c}{\partial b}\right)^2 \sigma_b^2 = \frac{N}{2}\sigma^2, \\ \sigma_\varphi^2 &= \left(\frac{\partial \varphi}{\partial a}\right)^2 \sigma_a^2 + \left(\frac{\partial \varphi}{\partial b}\right)^2 \sigma_b^2 = \frac{N}{2}\frac{\sigma^2}{c^2}. \end{aligned} \tag{12.65}$$

The case where the fundamental frequency is unknown can also be treated, but this is beyond the scope of this introductory presentation [16].

Consider now the application of the above procedure to the turbine vibration example. Here, three spectral components need considering, the fundamental being $f_0 = 12.6$ Hz, corresponding to a rotation speed of 756 rev/min. We obtain the results in Table 12.1. Note that standard uncertainty, u, calculated in this way, only accounts for the effect of measurement noise; other uncertainty sources, if present, should be properly included in the model.

To sum up, both direct and indirect dynamic measurements can be elegantly treated, in the probabilistic approach develop in this book, as a special case of vector measurement, without additional assumptions [18].

References

1. Crenna, F., Michelini, R.C., Rossi, G.B.: Hierarchical intelligent measurement set-up for characterizing dynamical systems. Measurement **21**, 91–106 (1997)
2. Morawski, R.Z.: Unified approach to measurand reconstruction. IEEE Trans. IM **43**, 226–231 (1994)
3. Hessling, J.P.: Dynamic metrology. Measur. Sci. Technol. **19**, 084008 (2008) (7p)
4. Sommer, K.D.: Modelling of measurements, system theory, and uncertanty evaluation. In: Pavese, F., Forbes, A. (eds.) Data Modeling for Metrology and Testing in Measurement Science, pp. 275–298. Birkhauser-Springer, Boston (2009)
5. Kwakernaak, H., Sivan, R.: Linear Optimal Control Systems. Wiley, New York (1972)
6. Oppenheim, A.V., Shafer, R.W.: Digital Signal Processing. Prentice Hall, Englewood Cliffs (1975)
7. Priestley, M.B.: Spectral Analysis and Time Series. Academic, London (1982)
8. Papoulis, A.: Probability, Random Variables and Stochastic Processes, 2nd edn. McGraw-Hill, Singapore (1984)
9. Marple, S.L.: Digital Spectral Analysis. Prentice-Hall, Englewood Cliffs (1987)
10. Kay, S.M.: Modern Spectral Estimation. Prentice Hall, Englewood Cliffs (1988)
11. Rossi, G.B.: A probabilistic model for measurement processes. Measurement **34**, 85–99 (2003)
12. Rossi, G.B.: Measurement modelling: Foundations and probabilistic approach. Paper presented at the 14th joint internatinal IMEKO TC1+TC7+TC13 symposium, Jena, 31 August–2 September, 2011 (2011)
13. Aumala, O.: Fundamentals and trends of digital measurement. Measurement **26**, 45–54 (1999)
14. Eichstädt, S., Elster, C., Esward, T.J., Hessling, J.P.: Deconvolution filters for the analysis of dynamic measurement processes: A tutorial. Metrologia **47**, 522–533 (2010)
15. Bentley, J.P.: Principles of Measurement Systems, 4th edn. Pearson Education Ltd., Harlow (2005)
16. Larry Bretthorst, G.: Bayesian spectrum analysis and parameter estimation. Springer, New York (1988)
17. Diana, G. (ed.): Diagnostics of Rotating Machines in Power Plants. Springer, Berlin (1994)
18. Rossi, G.B.: Toward an interdisciplinary probabilistic theory of measurement. IEEE Trans. Instrum. Meas. **61**, 2097–2106 (2012)

Appendix A
Glossary and Notation

A.1 Glossary

Some key terms used throughout the book are defined in Table A.1.

A.2 Symbols and Notation

As a general criterion, I have tried to keep the notation as lean as possible. Due to the broadness of the subject, some symbols are poly-semantic: for example "m" can denote either "mass" or the "measure" function. In fact, I have preferred to use mnemonic symbols rather than resorting to a large number of difficult-to-relate symbols.

Sets are usually denoted by capital letters, $A, B, C, \ldots$, and their elements by small letters, $a, b, c, d, \ldots$. Parameters of functions are usually expressed by Greek letters, $\alpha, \beta, \gamma, \delta, \ldots$. Scalar variables are denoted by normal characters, $x, y, u, v, w, z, \ldots$, and vectors are in bold, $\mathbf{x}, \mathbf{y}, \mathbf{z}, \ldots$. Typical symbols for functions are $f, g, h, l, \phi, \ldots$; for integers $i, j, k, l, m, n, p, q, \ldots$; when they denote the number of elements in a set or in a series, often they are in capital letters, $M, N, \ldots$, but there are exceptions, such as $m, n, \ldots$.

No special convention (such as capital or bold characters) has been adopted for probabilistic (or random) variables. So, the same symbol may be used for denoting a random variable or its specific value. For example, the probability distribution of v can be denoted either as $p_v(\cdot)$ or, in a shorthand notation, as $p(v)$.

A list of the main symbols used throughout the book, organised by categories of subjects, is provided in Table A.2.

G. B. Rossi, *Measurement and Probability*, Springer Series in Measurement Science and Technology, DOI: 10.1007/978-94-017-8825-0,

Table A.1 Definition of some key terms

Object	The carrier of the characteristic to be measured; it may be a physical object, an event or a person
Property or characteristic	(Of an object): what we want to measure
Measurable property (or characteristic)	(Or quantity): a property that may be measured
State	(Of an object, with respect to a given characteristic): particular way in which an object manifests a characteristic
Empirical relation	A relation that may be observed between two or more objects, with reference to the characteristic of interest
Comparator	A device (or a person) that performs the comparison of objects, in respect of some characteristic of theirs
Empirical structure	(Or empirical relational system): a set of objects and a set of empirical relations on it
Nominal, difference, interval, intensive, extensive	Different types of empirical structures
Numerical structure	(Or numerical relational system): a set of numbers and a set of numerical relations on it
Scale	(General meaning): the set of formal conditions for measurement (an empirical structure, a numerical structure and a measure function constituting an homomorphism between them)
(Reference) scale	(Specific meaning): a series of standard objects with corresponding numerical values properly assigned
Resolution (of a scale)	Minimum variation in the quantity that can be properly represented by the scale
Nominal, ordinal, interval, ratio	Different types of scales
Measuring system	(Or instrument): an empirical system capable of interacting with objects carrying the property under investigation and, as a result of such interaction, of producing an observable output according to which it is possible to assign a value to the object to be measured
Measurement process	The process by which a value is assigned to a measurand, normally based on the use of a measuring system
Measurand	A property of a specific object to be measured, in a specific situation
(Measure) value, measurand value	A number that may be assigned to an object in order to express how the object compares, in respect to some property, to all the other objects that carry the same property
Measurement value	The value that is actually assigned to an object as the result of a measurement process
Calibration	The operation by which the characteristics of the measuring system are assessed
Resolution (of a measuring system)	Minimum variation in the quantity that can be detected and properly quantified by the measuring system
Model	An abstract system that, to some extent and from a certain standpoint, represents a real system (or a class of real systems); a scientific theory may be sometimes viewed as a very general model

(continued)

Table A.1 (continued)

(Natural) law	A functional relation (or a model) linking one or more properties of real objects
Measurement model	Model on which a measurement process is based
Fundamental scale	A scale that can be constructed on the basis of the internal properties (or intra-relations) of a quantity
Fundamental quantity	A quantity for which a fundamental scale exists; in the international system of metrology the term "fundamental" has a looser and more conventional meaning
Fundamental measurement	The measurement of quantity based on a comparison with a reference scale
Derived scale	A scale that is obtained on the basis of relations linking the quantity under consideration with other quantities (or inter-relations)
Derived quantity	A quantity for which a derived scale (only) exists
Derived (or indirect) measurement	The measurement of a quantity based on relations linking the quantity under consideration with other quantities

Table A.2 List of the main symbols in the book

Generic terms	
x, $\mathbf{x}$	Parameter(s) to be estimated, measurand
y, $\mathbf{y}$	Observation(s), instrument indication(s)
$f, g, \mathbf{f}, \mathbf{g}$	Scalar functions, vector functions
v, $\mathbf{v}$	Random errors affecting the indications of a measuring system
$\triangleq$	Equal by definition
$\square$	End of proof
max, min	Maximum/minimum operators
i, j, k, l	Integers, often indices
n, m, p, q, N, M	Integers, often numbers of elements in a set or of terms in a series
A, B, C	Sets
a, b, c, d	Elements of a set
$\mathbf{A}, \mathbf{B}$	Matrices
Terms concerning physical quantities	
m	Mass (not to be confused with the measure function, also denoted by "m")
T, T_0	Temperature
h, h_0	Height, but h also may denote Planck's constant
α	Thermal expansion coefficient
ρ	Density
V	Volume
x, y	Coordinates along reference Cartesian axes
v	Velocity

(continued)

Table A.2 (continued)

p	Pressure
p_{rms}, p_{rms}^2	Root mean square and mean square value of pressure
I	Intensity
$f, \Delta f$	Frequency, frequency interval
L_p, L_I	Sound pressure level, sound intensity level
$i(f)$	Sound intensity density
τ	Time
Terms concerning psychophysical quantities	
ϕ	Intensity of a stimulus
ψ	Intensity of a sensation
$\delta\phi, \delta\psi$	Stimulus variation, sensation variation
$\alpha, \beta, \gamma, \delta, \eta, \theta$	Parameters of psychophysical law
LL, L	Loudness level, loudness
L_d, L_p	Loudness estimates obtained by difference and ratio tests, respectively
Generic probability and statistics	
$\mathcal{S} = (\Omega, \mathcal{F}, P)$	Probability space
Ω	Sample space
A, B, C, D	Events; $\bar{A}$ is the complement of A
E	An experiment
$\cup, \cap$	Set theoretic "union" and intersection operators
$\vee, \wedge$	Logical "or" and "and" operators
$\mathcal{F}$	Algebra of events
P	Probability function, discrete probability distribution
p	Probability density function, also called probability distribution
$\{f_1, f_2, f_3, f_4, f_5, f_6\}$	The results in die rolling
$\{h, t\}$	The results in coin-tossing
$x, x(\omega)$	Probabilistic variable x
f, f_ω	Probabilistic function
E, var	Expectation operator, variance operator
σ, σ^2	Standard deviation, variance
$\hat{x}$	The "hat" symbol indicates an estimator or an estimated value; if applied to the measurand, it denotes the measurement value
e	Error, usually defined by $e \triangleq \hat{x} - x$
$\bar{y}, \tilde{y}$	Arithmetic mean of y, mean value of y
φ	Standard normal (Gaussian) distribution, with zero mean and unitary variance defined by $\varphi(\xi) = (2\pi)^{-1/2} exp(-\xi^2/2)$
$\psi(t, \nu)$	t-Student probability density function, with ν degrees of freedom
$\zeta(\xi; \alpha, \beta)$	Inverse-gamma distribution
n_A	Number of occurrences of event A in a series of observations
p, λ	Parameters of probabilistic distributions

(continued)

Table A.2 (continued)

The measurement scale (deterministic)	
A, B	Set of objects manifesting the quantities x, y
$a, b, c, d, \ldots a', b', c', d', \ldots$	Objects, that is, elements of sets A or B
$a^*, b^*, c^*, \ldots$	Equivalence classes, containing objects $a, b, c, \ldots$, respectively
A^*	Class of all the equivalence classes in A
m, m'	Measure function, $m : A \to \mathbb{R}$
$\mathcal{A}, \mathcal{B}, \mathcal{C}$	Relational systems or structures
$\mathcal{S}$	Measurement scale; note that $\mathcal{S}$ also may denote a measurement scale
R_i, S_i	Generic relations
$\succsim$	(empirical) weak-order relation on A; also, informally, weak-order relation on intervals
$\succsim_d, \succsim_r, \succsim_\delta$	Empirical weak-order relations amongst intervals, concerning difference, ratio or distance, respectively
$\sim$	(empirical) equivalence relation on A
d, δ	Distance as a numerical function (d) or as an empirical property (δ)
$\sim_d, \sim_r, \sim_\delta$	(Empirical) equivalence relations on intervals, concerning difference, ratio or distance, respectively; also, informally, equivalence relation on intervals
Δ_{ab}	Interval between elements a and b of A
$\circ$	(Empirical) operation of concatenation of elements of A or of intervals
$(A, \sim)$	(Empirical) nominal structure on A
$(A, \succsim)$	(Empirical) order structure on A
$(A, \succsim_\delta)$	(Empirical) distance structure on A
$(A, \succsim_d)$	(Empirical) difference structure on A
$(A, \succsim_d, \succsim_r)$	(Empirical) intensive structure on A
$(A, \succsim, \circ)$	(Empirical) extensive structure on A
$(A, B \succsim)$	(Empirical) cross-order structure on $A \cup B$
$(A, B \succsim_d)$	(Empirical) cross-difference structure on $A \cup B$
ϕ, g	Functions
S	Series of standards, $S = \{s_i \mid i = 1, \ldots n\}$, or $S = \{s_i \mid i = 0, \ldots n\}$
R	Reference scale, $R = \{(s_i, m(s_i)) \mid i = 1, \ldots n\}$ or $R = \{(s_i, m(s_i)) \mid i = 0, \ldots n\}$
Δx_r	Resolution of a reference scale for x
The measurement scale (probabilistic)	
$\mathcal{A}\ (\mathcal{B}, \mathcal{C})$	a structure on A (B, C), for example an order structure, in which case $\mathcal{A} = (A, \succsim)$
$\succsim_\omega$	Empirical weak-order relation associated to ω
$\mathcal{A}_\omega$	A structure associated to ω; for example, in the case of order, $\mathcal{A}_\omega = (A, \succsim_\omega)$
$\mathcal{E}$	A finite collection of structures: $\mathcal{E} = \{\mathcal{A}_1, \mathcal{A}_2, \ldots, \mathcal{A}_n\}$
$\mathcal{S}_\mathcal{E}$	A probability space based on $\mathcal{E}$
$x_a(\omega), x_a$	Probabilistic variable associated to the element a
$m_\omega(a), m_\omega$	Measure function associated to point ω of the sample space
φ, g	Functions

(continued)

Table A.2 (continued)

The measurement process [a]	
x	Measure value (or measurand value)
x'	Probabilistic measurement value
$\hat{x}$	(Expected) measurement value
X	Set of the possible values of the measurand, image of A in $\mathbb{R}$, through the measure function m, that is, $X = m(A)$
X'	Set of the (probabilistic) measurement values
$\hat{X}$	Set of the expected measurement values; note that $X = X' = \hat{X}$
Y	Set of the output values (indications) of the measuring system
φ, f	Calibration or observation functions: $\varphi : A \to Y$, $f : X \to Y$
γ, γ', h, h'	Functions that describe the measurement process: $\gamma : A \to \hat{X}, h : X \to \hat{X}$; $\gamma' : A \to X', h' : X \to X'$
θ	Influence quantity, typically giving rise to a systematic effect
Θ	Set of the possible values of θ
f_θ	Calibration or observation function affected by θ
Uncertainty evaluation, the GUM and risk analysis [b]	
u, U	Standard uncertainty, expanded uncertainty
g	Function governing the GUM evaluation formula
$z, \mathbf{z}$	Correction(s) of influence quantities
$\theta, \boldsymbol{\theta}$	Influence parameter(s) producing systematic effects
w, w_i	Probabilistic (or random) variable describing additive measurement noise series of realisations of w
k	Sensitivity of a measuring device
s	Term accounting for the resulting effect of influence quantities on a measuring system
$\mathbf{v}$	Vector of influence quantities
$\mathbf{a}$	Vector of sensitivities to influence quantities
$Q(\cdot)$	Quantisation operator
q	Quantisation interval
h	Term describing a hysteresis phenomenon
$A, B, \mathbf{A}, \mathbf{B}$	Safe and acceptance regions, scalar or vectorial
$\mathcal{P}, \mathcal{M}$	Production and measurement process
a, b	Thresholds
$\tilde{x}, \tilde{\sigma}$	Weighted mean and its standard deviation
σ'	Standard deviation of an inspection process: $\sigma'^2 = u^2 + \sigma^2$
$R, R(\hat{x}), R(\hat{\mathbf{x}})$	Consumer's risk, global or specific
$R^*, R^*(\hat{x}), R^*(\hat{\mathbf{x}})$	Producer's risk, global or specific
R_0, R_0^*	Global consumer's or producer's risk, when no measurement is taken
C, C_1, C_2	Various types of cost
Q, Q_1, Q_2, Q_3, Q_4	Various types of flowrate, for water meters

(continued)

Table A.2 (continued)

Dynamic measurement	
$x(\tau), x_t$	Time function (or signal), time series, τ is a continuous variable denoting time, t is a discrete time index
y_t, z_t	Time series of the indication and for the state variable
$\mathbf{z}_t, \boldsymbol{\xi}_t$	Time series of the state and of a generic driving input in a Markovian model
$x^t = (x_1, x_2, \ldots, x_t,)$	A series of dynamic values
Δt	Time-sampling interval ($\tau = t\Delta t$), frequency
f, f_s	Frequency, sampling frequency: $f_s = 1/\Delta t$
T	Observation (time) interval: $T = N\Delta t$
T_p, f_0	Period and fundamental frequency of a periodic phenomenon
θ	Time constant
$\tilde{q}(f), \alpha(f), \beta(f)$	(Complex) frequency response, modulus (amplitude) and phase of frequency response
j	The imaginary unit, that is, $j^2 = -1$
$\Delta\alpha, \Delta\varphi$	Tolerance on the frequency response components
c_i, φ_i, a_i, b_i	Terms of the Fourier series for a periodic function

[a] Note the vector version of the following variables are simply denoted by the same symbol in bold
[b] The terms used in this book are sometimes quite different from those adopted in the GUM

Index

G. B. Rossi, *Measurement and Probability*, Springer Series in Measurement Science and Technology, DOI: 10.1007/978-94-017-8825-0,

MIX
Papier aus verantwortungsvollen Quellen
Paper from responsible sources
FSC® C105338

If you have any concerns about our products,
you can contact us on
ProductSafety@springernature.com

In case Publisher is established outside the EU,
the EU authorized representative is:
Springer Nature Customer Service Center GmbH
Europaplatz 3, 69115 Heidelberg, Germany

Printed by Libri Plureos GmbH
in Hamburg, Germany